Praise for
Notes from a Dying Planet, 2004–2006:

"*Notes from a Dying Planet* is a concerned, passionate, and informed survey of the environmental problems facing our planet, with gems of succinct essays interspersed with the reality of daily news. Written from the perspective of a very well informed citizen with a solid scientific background, it conveys in clear language the many challenges we face, convinces us of the folly of continuing on our present course, and suggests useful steps that each of us as individuals can take toward making our planet a sustainable habitat for all species."

—Jeffrey A. McNeely, Chief Scientist, IUCN-The World Conservation Union, Gland, Switzerland

"Brown explains in his outstanding book how human overpopulation is diminishing the well being of all humans, including their food, water, energy, and biological resources. Clearly overpopulation is reaching crisis levels and is now the number one environmental problem worldwide."

—David Pimentel, Professor of Entomology and Ecology and Evolutionary Biology, Cornell University, Ithaca, New York

"This is a remarkably wide ranging series of short essays on almost everything. It embraces the origins of the universe and of life, the impact of human activity on the planet, and the ways in which we must change our behavior. It is a highly personal exploration by a mature scientist with strong feelings. It is not a scholarly text, but rather a collection of ideas that should awaken the curiosity of, say, promising young minds at about their transition from high school to college."

—Lindsey Grant, writer and former U.S. Deputy Assistant Secretary of State for Environment and Population.

"From the end of cheap oil to the burden of overpopulation, Paul Brown's wonderful book details each step of our headlong rush towards collapse. However, it also suggests some ways out. Terrifying one moment and inspiring the next, a more comprehensive and necessary account of our planetary emergency can scarcely be imagined."

—Mark Lynas, columnist and climate change author.

"Paul Brown has written a book that is at once unique, scary, and exciting. It is unique because of the format, scary because the author tries to be honest with the reader and exciting because if we all become as committed as Brown is we will reverse the environmental trends that are undermining our future."

—Lester Brown, Founder and President, Earth Policy Institute, and author of *Plan B 2.0: Rescuing a Planet under Stress and a Civilization in Trouble.*

Notes from a Dying Planet, 2004–2006

Notes from a Dying Planet, 2004–2006

✦

One Scientist's Search for Solutions

Paul Brown, PhD

iUniverse, Inc.
New York Lincoln Shanghai

Notes from a Dying Planet, 2004–2006
One Scientist's Search for Solutions

Copyright © 2006 by paul b. brown

All rights reserved. No part of this book may be used or reproduced by any means, graphic, electronic, or mechanical, including photocopying, recording, taping or by any information storage retrieval system without the written permission of the publisher except in the case of brief quotations embodied in critical articles and reviews.

iUniverse books may be ordered through booksellers or by contacting:

iUniverse
2021 Pine Lake Road, Suite 100
Lincoln, NE 68512
www.iuniverse.com
1-800-Authors (1-800-288-4677)

ISBN-13: 978-0-595-40094-2 (pbk)
ISBN-13: 978-0-595-84489-0 (cloth)
ISBN-13: 978-0-595-84478-4 (ebk)
ISBN-10: 0-595-40094-9 (pbk)
ISBN-10: 0-595-84489-8 (cloth)
ISBN-10: 0-595-84478-2 (ebk)

Printed in the United States of America

To my wife Sally, and to the thousands of scholars who have devoted their lives to approximating the truth.

CONTENTS

INTRODUCTION...xv

The State of the Earth
- **Chapter 1**   HOUSTON, WE HAVE A PROBLEM....................3
- **Chapter 2**   JUGGERNAUT..5
- **Chapter 3**   POOPING IN OUR NEST....................................7
- **Chapter 4**   CONFLICTS OVER LAND...................................8
- **Chapter 5**   HUMAN IMPACT ON THE WORLD'S WATER.....10
- **Chapter 6**   OUR IMPACT ON THE WORLD'S CIRCULATORY SYSTEMS..................................12
- **Chapter 7**   OUR IMPACT ON THE ATMOSPHERE.................14
- **Chapter 8**   THE ENERGY CRISIS..16
- **Chapter 9**   POLLUTION..19

Alternative Worlds
- **Chapter 10**   THE BEGINNING OF THE END..........................25
- **Chapter 11**   IF WE FAIL...27
- **Chapter 12**   IF WE SUCCEED..29

A Better Approach
- **Chapter 13**   FIXING IT BEFORE IT FIXES US........................33
- **Chapter 14**   A STRATEGY..35
- **Chapter 15**   THE SCIENTIFIC METHOD...............................37

Origins and Evolution
- **Chapter 16**   THE BIG BANG..41
- **Chapter 17**   THE ORIGIN OF LIFE.....................................44

Chapter 18	THE PRINCIPLES OF EVOLUTION	47
Chapter 19	THE EVIDENCE FOR EVOLUTION	50
Chapter 20	CREATIONISM AND THERMODYNAMICS	52
Chapter 21	THE EVOLUTION OF SPECIES	54
Chapter 22	PRE-HOMINID EVOLUTION	56
Chapter 23	HUMAN EVOLUTION	58
Chapter 24	RACE	60
Chapter 25	BIOLOGICAL EVOLUTION OF BEHAVIOR	63

The Human Mind and Cultural Evolution

Chapter 26	THE BIG BANG OF HUMAN CULTURE	69
Chapter 27	THE ORIGIN OF LANGUAGE	72
Chapter 28	THE NATURE OF LANGUAGE	74
Chapter 29	LANGUAGE AND THE BRAIN	76
Chapter 30	CULTURAL EVOLUTION	78
Chapter 31	PALEOLITHIC HUNTER GATHERERS	81
Chapter 32	BIRTH OF THE ARTS	83
Chapter 33	THE EMERGENCE OF RELIGION	85
Chapter 34	MODERN HUNTER GATHERERS	87
Chapter 35	HUMAN MIGRATION	90
Chapter 36	THE ORIGINS OF AGRICULTURE	93
Chapter 37	SETTLING DOWN	95
Chapter 38	WHY AGRICULTURE?	97
Chapter 39	CHIEFS	100
Chapter 40	CHIEFS AND MILITARY POWER	103
Chapter 41	IDEOLOGY AS A SOURCE OF CHIEFS' POWER	105
Chapter 42	EARLY NEOLITHIC CULTURE	107

CHAPTER 43	EARLY CIVILIZATIONS	109
CHAPTER 44	HUMAN NATURE 101	111

How our Planet Works

CHAPTER 45	ECOLOGY	115
CHAPTER 46	PREDATORS, PREY AND COMPETITION	119
CHAPTER 47	POPULATION GROWTH AND CROWDING	121
CHAPTER 48	GOOD LOOPS AND BAD LOOPS	125
CHAPTER 49	HUMPTY DUMPTY	127
CHAPTER 50	THE HORNS OF OUR DILEMMA	130
CHAPTER 51	DEMOGRAPHIC TRANSITION THEORY	133
CHAPTER 52	SMALLER WILL BE BETTER	136
CHAPTER 53	THE BIGGER WE ARE, THE HARDER WE FALL	140
CHAPTER 54	THE DEMOGRAPHIC TRAP	144
CHAPTER 55	THE SOCIAL COSTS OF POPULATION GROWTH	147
CHAPTER 56	ENVIRONMENTAL COSTS OF OVERPOPULATION	149
CHAPTER 57	THE PSYCHOLOGY OF FERTILITY	152
CHAPTER 58	POPULATION AND ITS MISCONCEPTIONS	155
CHAPTER 59	DEMOGRAPHIC MISCONCEPTIONS	158
CHAPTER 60	ABORTION GAG RULE, by Frank Carpenter, DMin	165

Energy

CHAPTER 61	ENERGY—WHAT IT IS	171
CHAPTER 62	OIL WARS	173
CHAPTER 63	A BRIEF HISTORY OF HUMAN ENERGY USE	176

CHAPTER 64	CARBON FUELS	179
CHAPTER 65	THE END OF OIL	181
CHAPTER 66	RENEWABLE ENERGY	183
CHAPTER 67	THE NUCLEAR OPTION	188
CHAPTER 68	HYDROGEN	191
CHAPTER 69	ENERGY POLICY AND POWER	194

Global Climate Change

CHAPTER 70	GROWING A DESERT	199
CHAPTER 71	GLOBAL WARMING	202
CHAPTER 72	AEROSOLS AND GLOBAL WARMING	204
CHAPTER 73	THE CLIMATE TIPPING POINT	206
CHAPTER 74	RUNAWAY GLOBAL WARMING	208

The End of Life

CHAPTER 75	EXTINCTION	213
CHAPTER 76	MASS EXTINCTIONS	216
CHAPTER 77	PLATE TECTONICS AND MASS EXTINCTIONS	219
CHAPTER 78	THE GREAT PLEISTOCENE EXTINCTIONS	224
CHAPTER 79	GLOBAL COLLAPSE OF ECOSYSTEMS	226
CHAPTER 80	IN MEMORIAM	228
CHAPTER 81	THE END OF HUMANITY	234

Mythology and Religion

CHAPTER 82	PALEOLITHIC MYTHOLOGY	239
CHAPTER 83	NEOLITHIC MYTHOLOGY	251
CHAPTER 84	ZARATHUSTRA AND THE MAGI	254
CHAPTER 85	MITHRA	256

CHAPTER 86	THE OLD TESTAMENT	259
CHAPTER 87	THE GREAT CHAIN OF BEING AND TELEOLOGY	261
CHAPTER 88	CAN RELIGION FIND MORE OIL?	264

What to Do?

CHAPTER 89	GETTING OUR PRIORITIES STRAIGHT	269
CHAPTER 90	REDEFINING SUSTAINABILITY	272
CHAPTER 91	HOW MUCH TIME DO WE HAVE?	275
CHAPTER 92	REMEDIES FOR OVERPOPULATION	277
CHAPTER 93	HUMANE POPULATION POLICY	279
CHAPTER 94	THE NEW LEADERS	281
CHAPTER 95	DO THESE THINGS NOW	283
CHAPTER 96	RISING ENERGY COSTS	287
CHAPTER 97	MAKING YOUR BUILDING ENERGY-EFFICIENT	289
CHAPTER 98	HOME SOLAR ELECTRICITY	291
CHAPTER 99	PASSIVE SOLAR HEATING	293
CHAPTER 100	IT'S THE CARBON, STUPID	298
CHAPTER 101	A RATIONAL ENERGY ECONOMY	301
CHAPTER 102	AN EMERGENCY NATIONAL POLICY	303
CHAPTER 103	WHAT KATRINA TAUGHT ME	305
CHAPTER 104	NATIONAL LEGISLATORS	308

CONCLUSION	313
REFERENCES	323
ABOUT THE AUTHOR	329
INDEX	331

INTRODUCTION

They say you can boil a frog by raising the temperature slowly so it doesn't notice. I doubt that anyone's tried that, but we're performing a much larger-scale experiment. We're boiling ourselves. An extraterrestrial observer would conclude we haven't noticed, because we're not doing anything significant about it.

We don't have access to the observations of extraterrestrials, but we do have access to observations by some very special terrestrials. For the last few centuries, especially in the last few decades, thousands of very dedicated and intelligent people have devoted their lives to refining our methods of observation and reasoning, which have produced, among many things, an astonishingly clear picture of how we're killing our planet. These people aren't politicians, religious leaders, entertainers, or corporate magnates. They're scientists.

Scientists, and scholars in general, have devised an approach to learning about the world that works very well. Curiously, it's based on the assumption that absolute certainty cannot be attained about anything. We can frame hypotheses about what causes what, preferably using unambiguous mathematical equations, but we can never prove them right all the time, under all circumstances. Through careful observation, preferably using numbers, we can disprove wrong hypotheses and gradually zero in on ones that stand up to all our tests, which are therefore more likely to be right. These hypotheses we call theories, and they're the highest form of scientific certainty. Disciplined scholars have also devised a mechanism for self-policing called *peer review*, which helps weed out the fraudulent and the foolish. It's not foolproof, but it works better than other methods. Surely, the meticulously obtained information and painstakingly constructed theories of these dedicated individuals merit our serious attention.

As I approach the end of my career, I'd like to apply my perspective as a brain scientist to our involuntary and destructive planetary experiment. The first result is this weekly journal over a period of two years, in which I look for useful questions and answers regarding our ongoing suicide.

This book is inevitably about sustainability: using resources only as fast as they can be replenished. The concept is as simple as balancing a checkbook. By living unsustainably, overusing our credit, we've caused global warming and habitat destruction for the living creatures that provide our life support in the form of drinkable water, edible food, breathable air, and even habitable land. Our planet is going through the worst mass extinction since the death of the dinosaurs, possibly the worst in the planet's history. To make matters worse, we've crossed the threshold of a positive feedback loop in which global warming is accelerating. It may be too late to prevent a runaway process that only microbes will survive.

A key component of our unsustainable lifestyle is overpopulation. If it weren't for wars, famine, drought, and pestilence, we would have used up our planet's resources long ago. There are now more than six billion people on this planet. In the worst case, that number could increase to twelve billion by the middle of this century. In some underdeveloped areas of the world, populations are growing completely out of control and, as far as many are concerned, the last days of humanity are upon them right now. They are harbingers of the developed world's fate if we don't do enough, soon enough.

I'm afraid our society is very sick. It's such a cliché, but it's true for reasons you may not suspect. Our entire worldviews have become so distorted it's hard for us to see the desolation we've created. Instead, we measure our security in terms of large families, material wealth, and false purposes.

We have a chance for survival, if we respond globally and immediately by reducing the size of our population, switching to clean energy sources, and protecting our remaining biodiversity. As I write this, the world's leaders and the people of many nations are awakening to the peril and beginning to take action. However, so far the response is far too little, far too late, especially in the United States.

The choice between survival and extinction isn't dependent only on our determination, although that's an essential component. It also depends on universal understanding and agreement about the processes we've set in motion, and the means to reverse them. These processes aren't limited to narrow areas of concern, which can be addressed independently. Physical, biological, psychological, and social phenomena interact in a complex fashion, which we all have to understand before we can hope to modify our behavior adaptively enough to survive. I'd like to apply my knowledge of the brain and behavior to help us reach that understanding.

There are many admirable books, written at all levels, about pollution, global warming, overpopulation, the current mass extinction and the energy crisis, and they contain many of the components of a realistic solution to our dilemma. I'll be referring to these excellent sources throughout this book because they are the sources I have used, and they go into more detail than I can in this limited space. However, I don't think any of these books takes a broad enough approach to come up with a workable plan that integrates all the relevant information. This is my personal, informal attempt to move in that direction.

My goal is to build a scientific foundation that will be helpful in healing our planet and ourselves. The choice of topics is purely my own. I'm a neuroscientist with some additional expertise in a variety of areas. I'm not an active researcher working in any of the other areas I'll be discussing. Much of this material is as new to me as it is to you, which means I may be able to relate what I've learned more clearly and concisely than an expert in, say, climatology.

For over two years, I've been writing a weekly Web column on these topics called *The Clock is Ticking* (http://clockticking.com). Each column is a short essay on a small enough topic to be discussed in about 600 words. This book is based on the columns released between March 2004 and March 2006. It will be followed by an additional volume every two years, as long as my own mental clock keeps ticking.

Each chapter is based on one column. The date of each chapter will place it in a real-time context. This book is thus a chronicle of one person's attempts to find a path to sustainability. In the original columns, I provided hyperlinks to current news stories and editorials. In this book, I replace those links with summaries of selected news stories from that week. These summaries provide you with a chronicle of events over the two-year period. For brevity, the sources are listed without full citations.

I've cheated a little. In their original sequence, the columns hopped back and forth from one topic to another. At the urging of my editors at iUniverse, I've rearranged them in a more logical sequence, grouped in sections, with better continuity. However, the news items have been kept in their original order, so the chapter dates still correspond to the news dates.

All this said, please don't misunderstand: I can't give you a full-blown plan for recovery, any more than anyone else can. However, I can alert you to the problems, the complexities, many measures that must be taken at the individual and societal levels, and the approaches that have no chances of success. We can and should be doing a number of things immediately, most of them over the long haul.

I've also pointed out a number of areas where we need more knowledge. One of them, climate science, has been dealt a severe blow just this week: George W. Bush, who refuses to take meaningful action without more "sound science," has directed NASA to cancel or delay indefinitely the launch of scientific satellites for climate research. For Dubya and his cronies, less science is preferable to science that contains inconvenient truths.

We're already embarked on an arduous and dangerous journey. Whether we survive it or not, we'll do better if we make the trip together, exploring along the way and trying to meet the challenges we encounter. Think of this book as notes from a friend who's traveling with you in spirit.

Morgantown, West Virginia, June 10, 2006

The State of the Earth

1

HOUSTON, WE HAVE A PROBLEM

(March 14, 2004)

Everyone I know is uneasy about this space ship we live on. We have good reason to be edgy. Scientists around the world are warning us of growing dangers to our survival. We only have about 150–250 years left in which to fix them. We, *Homo sapiens*, the people of this world, must make some major changes in our lives in the next 5–10 years, starting this year, or we're toast.

The threats we face stem from overpopulation and environmental degradation. The resulting climate change and mass extinctions are leading to ecological collapse, in which the once-robust tapestry of interrelationships among living creatures, climate, and our physical environment has been weakened and is starting to unravel. Clinical indicators of our planet's serious illness are illustrated in the graph. I've adjusted the vertical scales for population, carbon dioxide (CO_2), methane, temperature, and extinction of species per year so they all have a common minimum and maximum.

HUMAN GLOBAL IMPACT

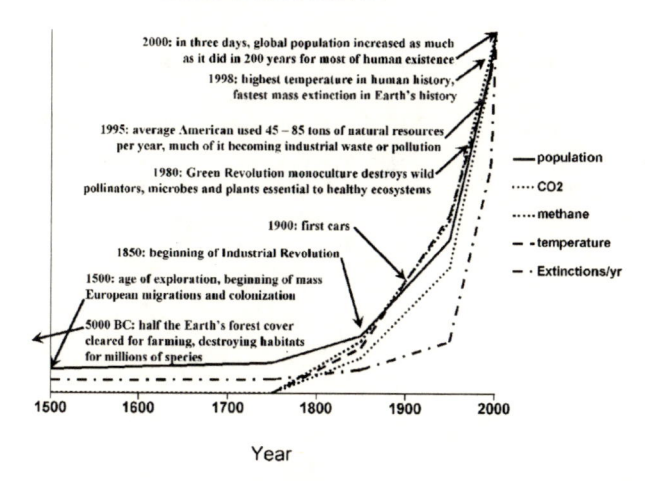

2000: in three days, global population increased as much as it did in 200 years for most of human existence
1998: highest temperature in human history, fastest mass extinction in Earth's history
1995: average American used 45 – 85 tons of natural resources per year, much of it becoming industrial waste or pollution
1980: Green Revolution monoculture destroys wild pollinators, microbes and plants essential to healthy ecosystems
1900: first cars
1850: beginning of Industrial Revolution
1500: age of exploration, beginning of mass European migrations and colonization
5000 BC: half the Earth's forest cover cleared for farming, destroying habitats for millions of species

— population
····· CO2
····· methane
— - temperature
— · Extinctions/yr

Year

All the minima occurred tens of thousands of years BC, and all the maxima are now.

Notes from a Dying Planet, 2004–2006

The state of the Earth today is unique. We're consuming the world's resources faster than they can be restored. The world's population is now doubling in less than fifty years. Around mid-century, the world's population is expected to level off at eight to twelve billion people. The lower number is far too high: population must start to decline before 2050 if we are to survive. The upper limit, to put it simply, will never be reached because we would all die first.

Because of population growth and increasing consumption, concentrations of greenhouse gases such as carbon dioxide and methane in our atmosphere are the highest in human history, as are global temperatures. This is not normal climatic fluctuation, as fossil-fuel industry shills would have you believe. The rate of species extinctions is comparable to mass extinctions that have occurred only five times before, and is likely to exceed those. The total decline of species since the Industrial Revolution will soon be worse than the mass extinction caused by the asteroid impact sixty-five million years ago off the Yucatan peninsula, which wiped out 83% of species including the dinosaurs.

Before we came along, species evolved and went extinct for billions of years, creating and filling a diversity of ecological niches. Organisms used energy from the sun to grow and reproduce, recycling the materials needed for life through an interdependent worldwide ecosystem. Mechanisms existed to maintain ecological stability, ensuring that the environment didn't change too fast for evolution to keep up. Our biosphere recovered from calamitous events like asteroid collisions, even though only a minority of species made it through some of those catastrophes. Today's ongoing catastrophe may eliminate all but the smallest and simplest of life forms.

Our species has flourished, but without realizing it, we've changed our environment too fast for other species to adapt. A system's stability can only be eroded so far, after which it becomes unstable. We're approaching a point where the world's ecosystem will change too fast even for us to adapt. We will become extinct.

It's already too late for us to return to the world as we found it or even as it was ten years ago. We've wiped out too many species. However, we can protect the remaining fragile stability. In a word, we must seek sustainability, which means consuming resources only as fast as they're replenished. All the trends on our graph have to be reversed, until they're all back to pre-industrial levels or lower. This doesn't mean returning to a pre-industrial quality of life—in fact, we should all be able to live much better *once there are fewer of us*. But we have to take effective action very soon, before it's too late.

This week's news:

NEW ENGLAND: A warming trend that appears modest is actually causing later winters and earlier springs, causing die-offs of hemlocks. Models predict 6-degree warming by 2100. A regional conference of governors and premiers has been scheduled. (*Boston Globe*)

Recommended reading:

Board on Sustainable Development, *Our Common Journey: A Transition toward Sustainability*. Washington, DC: National Academy Press, 1999.

2
JUGGERNAUT
(March 21, 2004)

In 1996, Lindsey Grant published *Juggernaut*, a book about overpopulation. He expressed a simple but sobering principle:

> *"Perpetual physical growth is impossible on a finite planet."*

That's hard to argue with. He explained the book's title:

> "In South India, at a certain festival, it was customary to roll out a giant ceremonial chariot in honor of Lord Vishnu. The chariot was wheeled, but no provision was made for steering it. Pulled and pushed by devotees in religious ecstasy, it crushed everything in its path: people, animals, buildings. It was called the Juggernaut...If we can bring those driving the population Juggernaut to consider the consequences of what they are doing, perhaps we can turn it from its present destructive course."

This story's Indian origin is grimly appropriate, because India is one of the areas with the fastest population growth, in spite of a population control program that has been in place for decades. It's obviously far from adequate, but the situation would be even worse without it. Yet you rarely hear about it.

Cities like Los Angeles have filled all their available space to overflowing. The world's largest cities now have populations over 15 million, a number unthinkable fifty years ago. By 2050, several cities' populations will top twenty-five million. Forests, farmlands, and wetlands are falling before the advance of new towns. In many areas in our own country, the water supply is being overwhelmed by the growing demands of residents, farming, and industry.

Most Americans are blissfully unaware of this problem, the signs of which are all around us. Most economists and policy makers actually assume population growth is necessary for prosperity. In most circles, you would be ridiculed for suggesting that we must go to one child per couple to survive, but that's what we have to do.

The total mass of human flesh now exceeds that of any other mammal on Earth. We have passed the "carrying capacity" (number of people that can be supported) of the planet. As a consequence, we're using up resources faster than they can be replaced. The more of us there are, the more we damage the planet, and the more we damage the planet the fewer of us it can support. Intensive farming has depleted the land, so fertilizers, herbicides, and pesticides are essential to producing the food needed by the world's teeming billions. These substances are now major pollutants. Our energy use continues to climb, as supplies of oil go down and costs go up. National policy responses to this crisis are absent or dead wrong.

As we resort to more hazardous and inefficient sources of energy such as shale oil and oil sands, rates of environmental degradation will increase until population growth levels off through a combination of decreased fertility and increased mortality rates. The number of people dying each year will equal and then exceed the number of babies born each year. At the same time, other species will continue going extinct, non-renewable resources will be exhausted and renewable resources will be consumed faster than they can be replaced. Massive famine, drought, epidemics, and pollution-related deaths will be some of the factors increasing mortality rates and lowering fertility rates for all species.

Now open your eyes: Look around you for the signs of population growth. If you're over 40, how big is your town compared to when you were a child? If you're under 40, ask some of your older friends what your town was like when they were children. Notice all the new construction and the disappearing "undeveloped" land. Listen to the rationalizations people use to justify large families and population growth, wasteful consumerism, and environmental deregulation. Some of these arguments may seem reasonable. Keep them in mind as you read this book.

This week's news:

WASHINGTON, DC: The United States has refused to participate in an international program on family planning and reproductive health due to the Bush administration's opposition to abortion. (Truthout.com)

Recommended reading:

Grant, Lindsey, *Juggernaut: Growth on a Finite Planet*. Santa Ana, CA: Seven Locks Press, 1996.

3

POOPING IN OUR NEST

(March 28, 2004)

Watch where you step. Yuck.

Limits have been imposed on fishing for many popular species because they're endangered. In spite of advice to pregnant women not to eat too much tuna or swordfish because of mercury contamination, thousands of babies are born with mercury-induced brain damage each year. The United Nations estimates that 150,000 people die each year due to global climate change caused by CO_2 pollution. Blooms of plankton caused by fertilizer runoff are depleting the oxygen in our estuaries, killing fish and shellfish. Air pollution is killing people, plants, and animals. Skin cancer rates are up in Alaska and Tierra del Fuego due to the polar ozone holes caused by man-made pollutants.

Increasingly scarce wood and metal are being replaced by plastics in manufactured goods, resulting in toxic pollution from the petrochemical industry. More and more often, commonly used plastics are found to cause health problems. As fossil fuels are exhausted, we'll be forced to obtain energy from nuclear reactors, a solution that's worse than the problem because of radiological pollution: radioactive waste. We do have an innovative solution to disposing of spent nuclear fuel, though: U.S. use of depleted-uranium munitions in warfare has spread thousands of tons of nuclear waste in places like the Balkans, Afghanistan, and Iraq.

Before we run out of fossil fuels, it's likely that the carbon dioxide we pollute the atmosphere with will have produced a runaway global climate change. The recently reported increase in the rate of atmospheric CO_2 pollution is the first sign of such a runaway process. If this happens it's likely that the biosphere will quickly shift at some point to a new stable state, probably too cold or too hot for most species to adapt. Our planet could become more similar to Venus (too hot) or Mars (too cold). Surviving organisms would probably be limited to microorganisms like bacteria, which can live in very harsh environments. It's unlikely that intelligent (sic) life could evolve again.

This week's news:

CALIFORNIA: Environmental restoration has brought in more than $65 million dollars to Humboldt County between 1995 and 2002 (*Sacramento Bee*).

Recommended reading:

Reid, Stephen J., *Ozone and Climate Change: A Beginner's Guide*. Amsterdam: Gordon and Breach Science Publishers, 2000.

4

CONFLICTS OVER LAND

(April 4, 2004)

There have been conflicts over land since before we evolved. Often different species compete for land, because most land is occupied by multiple species and organisms proliferate faster than space becomes available for them. The losers must move or die out.

Many animals are territorial. For some species, territories can be many square miles. Competition for territories within a species selects for the most robust individuals. Losers of territorial contests must move. The number of available territories largely determines the population size.

Humans compete for land all the time. Crowding due to overpopulation has been a major factor in human history, resulting in famines, plagues, wars, migration, colonization, the establishment of castes and other forms of discrimination, genocide, and the collapse of civilizations.

The world's human population is limited to about 25% of the world's surface, the rest being water-covered or uninhabitable. Average density is about 120 people per square mile, with lows in places like Mongolia of about four people per square mile and a high of about 34,000 people per square mile in Tokyo. We're aggregating in cities more than in the past, because other regions are either marginal or used too intensively for farming, industry, or lumber. We're crowded into areas susceptible to natural hazards like volcanic eruptions, earthquakes, avalanches, mudslides and landslides, hurricanes and tornados, and floods. A few areas are supposedly wilderness, but they're especially vulnerable under the Bush administration to commercial exploitation, squatters, poachers, and recreational users, marginalizing their utility for the preservation of biodiversity.

As our population grows, we need more room to put people. We also need to intensify our use of non-residential land to support that population, to grow more food, mine more minerals and fossil fuels, and cut more trees. As a result of overpopulation we're reducing the amount of usable land (e.g., by formation of dust bowls and deserts) through intensive farming, overgrazing, and practices like mountaintop removal, in which the tops of mountains are strip-mined for coal and the waste is simply dumped into valleys, destroying the streams that run through them.

The very fact that restoration of land through natural processes isn't keeping up with our destruction of it means that we've exceeded the Earth's *carrying capacity* for our species: the ability of the ecosystem to replenish natural resources faster than we deplete them.

Without population control and reduced use of renewable energy supplies to sustainable levels, we've produced a positive feedback loop that accelerates the depletion of our resources and our land. We now have serious competing demands for land use, including farming, housing, forestry, extractive industries, manufacturing, and lowest priority of all, preservation of our last remaining biodiversity. We try to apply technological fixes, with mixed results. For example, the increased use of fertilizers for the higher food yields of the Green Revolution has produced serious pollution problems. We would have been much better off just controlling the size of our population.

We're also decreasing the amount of available land due to global warming, which causes rising ocean levels. Many sources, including the Pentagon, are now predicting that large coastal areas including many major cities may be under water in a few decades. It's no coincidence that this year's monsoon flooding has left much of Bangladesh under water. At the same time, global warming will increase the rate of desert formation in equatorial regions, causing further crowding. A seemingly paradoxical effect may also come into play, namely drastic cooling of northern Europe and America due to changes in ocean currents. These changes will reduce the land useable by our still-expanding human population even further.

This week's news:

JEJU, KOREA: The U.N. Environment Program says excessive nutrients, mainly nitrogen from human activities, are causing "dead zones" in the oceans by stimulating huge growths of algae. Since the 1960s, the number of oxygen-starved areas has doubled every decade, as human nitrogen production has outstripped natural sources. Sea areas starved of oxygen will soon damage fish stocks even more than unsustainable catches. (BBC News)

Recommended reading:

Daly, Herman E., and Townsend, Kenneth N., *Valuing the Earth: Economics, Ecology, Ethics.* Cambridge, MA: MIT Press, 1993.

5

HUMAN IMPACT ON THE WORLD'S WATER

(April 11, 2004)

The geographic distributions of the world's water and of people are not well matched. Millions of people live in arid places and have too little water. Water wars have been with us for millennia, and won't go away until we learn to distribute our populations in accord with available resources.

In most of the world, surface water is much more important for humans than groundwater. Municipal use exceeds domestic use in modern urban areas by about ten to one, and farming and hydroelectric power generation both exceed municipal use several-fold. While most of the water for hydroelectric power is returned to the surface water system for other uses, water for farming isn't. In some nations, irrigation accounts for over 95% of water use. In the U.S., when we count water used to irrigate forage crops for cattle, as much as 25,000 gallons of water are needed to produce a pound of beef, compared to 250 gallons to make a pound of bread.

Runoff accounts for 10–100% of rainwater and snow melt, and it's the source of surface water. The rest of precipitation is absorbed into the ground to join the water table. A mature deciduous forest traps more groundwater than an urban area; a slow rain has less runoff than a torrential one. Extreme levels of runoff resulting from violent storms due to global warming are accelerating erosion, overwhelming storm sewer systems and dams, and causing flash floods and mudslides more frequently. Areas of destroyed rain forests are particularly hard hit because of the loss of tree root systems to stabilize the soil and take up the water.

The construction of dams for reservoirs, diversion of streams, flood control, and hydroelectric power has consequences beyond the intended ones, for locations both upstream and downstream. Silt deposition shortens the useful lifetimes of many dams. Upstream, resulting lakes submerge land, making it unavailable for other purposes. Downstream, less water may be available if some of the dammed river water is diverted.

Groundwater is particularly important for humans in regions where runoff doesn't provide an adequate supply of surface water. The High Plains of the U.S., about three quarters the size of France, depend heavily on groundwater, especially for farming. Pumping water out faster than it can be replenished has decreased the volume of the Ogallala aquifer about 50% *since 1986*. Extrapolating, we might expect the aquifer to dry up in another decade or two. However, we don't know if rainfall will remain the same, increase, or decrease in that time. It's a safe bet that, unless we change our ways, the demand for water will continue to increase.

We do know that the Aral Sea is a small fragment of its former self, due to extraction of groundwater for farming. The surrounding land has become a dust bowl, two-thirds of the region's inhabitants have developed illnesses, and infant mortality has risen to 10%, compared with less than 1% in most of the world. The sea and its watershed lie within the boundaries of five central Asian countries. All five nations must agree to protect the sea until it recovers, and then use the water only as fast as it can be replenished. This hasn't happened so far.

If the world's population increases by 2050 as expected by most experts, the destruction of aquifers and conflicts over water will become far more serious. Those conflicts can be eliminated only if we bring our population down to a sustainable size and geographic distribution.

This week's news:

BRAZIL: Almost 10,000 square miles of the world's largest continuous forest was lost last year, 40% more than in the previous year. The destruction is being driven by a growing demand for Brazilian beef in Europe because of the fear of mad cow disease and foot and mouth disease in European herds. (*The Guardian*)

Researchers from nine nations compared maps of more than 100,000 protected areas around the globe to maps of the ranges of 11,633 animal species. They found that for about 12 percent of the species, their ranges did not include parks or nature preserves that would protect them from human activities such as logging, hunting, or mining operations. Among 3,896 species deemed threatened, no protection was found for 20%, including 149 threatened mammal species, 411 types of amphibians, 232 bird species and 12 turtle and tortoise species. About 300 of those species are on the verge of extinction. (*Nature*)

Recommended reading:

Ward, Diane Raines, *Water Wars: Drought, Flood, Folly, and the Politics of Thirst.* New York: Riverhead Books, 2002.

6
OUR IMPACT ON THE WORLD'S CIRCULATORY SYSTEMS
(April 18, 2004)

Earth has countless cycles, in which materials are transformed many times and recycled for re-use. Examples include the oxygen, carbon, and nitrogen cycles, which we'll consider in later chapters. Our planet also has many circulatory systems, in which materials are moved around the planet. Life on Earth depends on the stability of these systems.

The Earth has two sources of energy: internal and external. The external source is of course the Sun. Solar radiation is most intense at the equator, least intense at the poles. This unequal heating, and the rotation of the Earth, largely establish the global wind patterns underlying our weather and climate. The sun's heat also causes evaporation of water, mostly from the oceans, which condenses when cooled and falls to earth, replenishing fresh water supplies, and causing erosion and the weathering of rock on its way back to the oceans.

The Earth's internal energy source is the heat of its core, predominantly generated by the impact and compression of the matter that coalesced by gravitational attraction during the formation of the planet, and by the decay of radioisotopes. This heat makes the Earth's *mantle* (the molten layer just beneath the surface *crust*) plastic: hot inner mantle rises toward the rigid crust, cools, and sinks. This convection current stirs the mantle, causing stress lines in the crust and movement of the more stable areas bounded by these lines, called *tectonic plates*. The movement of these plates is the basis for the migrations of landmasses over geological periods. Continents break up, landmasses crash (very slowly by our standards) against each other, and the stresses at the boundaries cause earthquakes, volcanic activity, and the birth of mountain ranges. These changes have major impacts on climate and on evolution. Humans may even be influencing the movements of tectonic plates, by changing the weight of ice masses on continental plates as global warming melts them.

Human activity has influenced Earth's water circulation for tens of thousands of years. We create dust bowls and deserts by over-intensive farming. We redirect and dam large rivers. Now, with human-induced global warming, there's evidence that we're modifying the ocean current that keeps northwestern Europe temperate. Melting polar ice caps are diluting Arctic saltwater, and increased evaporation is concentrating equatorial saltwater. This is slowing the current of equatorial water that warms northwestern Europe and northern North America, which will cause temperatures in those regions to drop sharply, while equatorial temperatures will rise. The temperate zone will shrink, especially in the Northern Hemisphere.

Global warming will cause more evaporation, raising global humidity. Water is an even more potent greenhouse gas than CO_2 or methane, so a positive feedback loop will result in which global warming and evaporation will mutually reinforce each other.

Latitudinal variations in relative humidity and temperature are responsible for global wind patterns including the *tropical trade winds* and *polar fronts*. These wind patterns are essential for the maintenance of climatic stability. The patterns of evaporation and condensation they produce largely determine the locations of deserts and rain forests. A sudden shift of these wind patterns due to global warming would produce a shift of evaporation and precipitation patterns. Suddenly many people, and the entire populations of many species, would find themselves with too much or too little water. Many people would have to relocate quickly or die and many species would suddenly go extinct.

Stay tuned.

This week's news:

SEATTLE: Researchers at Berkeley's Renewable and Appropriate Energy Laboratory announced that using renewable sources to meet new energy needs would create three times as many jobs as relying on fossil fuels. The Bush administration wants to expand oil exploration in the United States. A coalition of environmental and labor groups, calling itself the Apollo Alliance, is pushing for federal incentives to promote wind, solar, or biomass power plants. They called for a federal push similar to the Apollo space program. (*San Francisco Chronicle*)

Recommended reading:

Graedel, Thomas E., and Crutzen, Paul J., *Atmosphere, Climate and Change*. New York: W. H. Freeman, 1997.

7

OUR IMPACT ON THE ATMOSPHERE

(April 25, 2004)

Most modern organisms, including all animals, could not have lived in the original atmosphere of our planet, because it was a *reducing* atmosphere, unlike today's *oxidizing* atmosphere. Instead of having an abundant amount of oxygen, the Earth's initial atmosphere consisted primarily of ammonia, methane, water vapor, carbon dioxide, and hydrogen. By the time life began, the hydrogen had leaked into space, much of the ammonia had been converted to nitrogen, and methane concentration had decreased. The atmosphere probably consisted mainly of nitrogen, CO_2, and water vapor. Living organisms were responsible for the appearance of oxygen and for decreasing CO_2 in the air, as a consequence of photosynthesis.

The atmosphere is where our climate is, and major climate variations are accompanied by the advance and retreat of glaciers. Glaciation occurs when plate tectonics have placed a continent at one of the poles and the Earth is sufficiently cool. Changes in the Earth's orbit around the sun, the locations of landmasses and the organization of ocean currents all help determine the transitions between glacial and interglacial periods. The advance and retreat of glaciers has been a key factor in the extinction and evolution of species. It's likely that we, *Homo sapiens*, wouldn't have evolved without the ice ages.

Beside the effects of manmade greenhouse gases on climate, and atmospheric pollution in general, we're modifying the atmosphere in three other major ways. One is the acidification of rain by the oxides of sulfur and nitrogen that we produce in enormous quantities when we burn fossil fuels. When dissolved in rainwater, these become sulfuric and nitric acids. Acid rain kills fish to the point where many lakes are now dead. It leaches nutrients from soil, to the point where crops are affected and the soil is no longer protected from erosion.

Smog, another serious atmospheric problem, is produced by automobile exhaust, which contains carbon monoxide, nitrogen oxides and unburned gasoline. These produce ozone. This mix, sometimes combined with sulfur oxides from industry, makes smog. It's so damaging to health that smog alerts are now used to warn people susceptible to respiratory problems such as asthma to stay indoors.

The third effect, noted only after if was fairly advanced, was the decrease in ozone concentration in the upper atmosphere. Ozone is a natural constituent that absorbs ultraviolet wavelengths from the sun. This is an important process, because these wavelengths are harmful to many organisms including humans. Chlorofluorocarbons (CFCs), used primarily for air conditioners, wind up in the stratosphere where they persist for decades. They catalyze the breakdown of ozone into molecular oxygen, depleting our ultraviolet shield. This effect is most pronounced at the poles, where ozone holes have developed. People in the far north and south are now more subject to skin cancer as a result. By the middle of this century, stratospheric ozone concentration will be half of what it was.

The United State is the major producer of these effects, and has been a leader in studying them. Nevertheless, our county has impeded efforts to reverse these effects due to two factors: Americans are poorly informed by corporate-owned news media, and our elected officials serve polluting industries better than they serve the people.

This week's news:

GHANA: 70 percent of Ghana's people lack access to clean, piped drinking water, mainly because of privatization forced by the World Bank and the International Monetary Fund. Rudolf Amenga-Etego worked with Ghanaians to oppose privatization, and the government eventually agreed to suspend the project. He is now working to make potable water available to everyone in the country by 2010.

Recommended reading:

Eldredge, Niles, *The Miner's Canary: Unraveling the Mysteries of Extinction.* Princeton: Princeton University Press, 1991.

8

THE ENERGY CRISIS

(May 2, 2004)

As elections approach, it's time to consider one of the most important political issues of our times: the energy crisis. The cost of gas is only the tip of the iceberg.

Petroleum supplies are only good for a few more decades, but natural gas will last for a century or two, and coal for several centuries. All we need is to increase production, particularly domestic, so why worry?

The energy crisis reflects a dependence on fossil fuels that's deepening in developed countries due to increased per capita demand. As population size and quality of life go up in the developing world, especially in China and South Asia, demand there will soon outpace the rest of the world combined. So? Disseminating the American lifestyle will make the world safer for democracy, right?

The main problem is a little 3-atom molecule called carbon dioxide. It's a greenhouse gas, which means it traps heat being radiated by Earth into space, warming the atmosphere. Which causes global warming. Which is one cause (not the only one!) of the current mass extinction.

Scientists have been warning us for over three decades that climate change is here, but the corporate-controlled media have downplayed its importance. Did you know that large pieces of the Antarctic ice shelf twenty thousand years old broke off and floated away as mammoth icebergs, which then melted, in 1970? It was reported and, at least in this country, forgotten. Representatives for the fossil fuel industry (like the Competitive Enterprise Institute), including their bought and paid for scientists (like the Science and Environmental Policy Project), politicians and media cronies denied there was global warming, claiming we were undergoing normal climatic variations. Did you know the polar ice caps are still shrinking? That the glaciers in the Alps and in Glacier National Park and the snows of Kilimanjaro will soon be gone? All these events—unprecedented in human history—have been duly reported and ignored, at least in the U.S.

When undeniable evidence was presented that global warming was real, the same con men told us there was no reason to believe it was caused by human activities. When the cause was shown to be rising CO_2 they scoffed, saying more study was needed. When an international panel of 2500 scientists convened from all over the world by the UN (Intergovernmental Panel on Climate Change) reported that fossil-fuel combustion must be cut severely to avoid runaway, catastrophic climate change, the professional skeptics buried the report, said global warming was good for us, and they helped bring to power a president and vice president who were themselves oil executives. We the people were hoodwinked, lured by false promises, unaware of their agenda (which was hidden in plain sight: see the Project for the New American Century).

These Manchurian Candidates have backed out of, or violated, international environmental treaties and negotiations and have done their best to consolidate U.S. global control of oil supplies, destroying nations and international relations in their drive to power. They've replaced scientists, professional civil servants, and honest judges with corrupt ideologues. With the help of their congressional henchmen (particularly Barton, Frist, Inhofe and Pombo), this gang has dismantled existing environmental protections and opposed new ones. They've provided an immense return for the industries that invested in them, particularly oil and coal companies. Once-public land goes to extractive industries for a song.

We have only a decade in which to abandon fossil fuels and start to restore the CO_2 elimination systems of the planet—the forests. At the same time, we have to lower our rate of consumption of natural resources and stop the current mass extinction. The best war on terror would be to wage peace, providing all nations with the means for independent, sustainable energy through wind, solar and hydrogen power. We could do it now, for the same cost as our oil wars, but the fossil fuel industry and war profiteers (Halliburton, Carlyle Group) have too much to lose.

Turn off your TV, radio, and rock and roll, use your newspaper for movie times and to line your birdcage, shun spectator sports and gossip, and start reading. Read the books recommended at the end of each chapter. Contrast them with the oil industry con artists. Challenge the lies and promote the truth about the energy crisis, in conversations and in letters to the editor and to your congressional representatives. Start recruiting everyone you know to do the same. Drive the fossil-fuel collaborators out of office and hold their successors' feet to the fire.

This week's news:

DAVIS, CA: Gov. Schwarzenegger wants to make hydrogen fuel cell vehicle travel a reality in six years. However, California has to depend on federal, auto industry, and energy company funding to develop the necessary infrastructure for fueling the vehicles. Bush also endorses such a plan. However, methods to produce hydrogen directly or indirectly from fossil fuels would not reduce greenhouse gas emissions. (*San Francisco Chronicle*)

WASHINGTON, DC: Conservative politicians are accusing government agencies enforcing the Endangered Species Act (ESA) of using "junk science." They are misrepresenting a report requested by the Interior Department from the National Research Council as saying enforcement decisions aren't based on adequate data. National Research Council representatives indicate that the report credited federal scientists with using the best evidence available, as required by the ESA. Republican representative Greg Walden of Oregon has introduced legislation to add so-called "sound science" provisions to the ESA,—legislation that's backed by the Bush administration. These provisions would require that the listing of species and actions taken to protect them be based on certain specific data gathering and analysis techniques. Such restrictions would greatly hamper enforcement, to the financial advantages of corporate interests. (*Legal Affairs*)

US: researchers have analyzed the scientific literature and shown that disease has increased in turtles, corals, marine mammals, urchins, and molluscs such as oysters. Disease incidence has remained level in the shark and shrimp families and in sea grasses, and declined in other fish. One possible cause is global warming. Another is human overfishing. Others include new germs from domestic animals, bioaccumulation of toxins, and new species carried across oceans in ships' ballast tanks. (*New Scientist*)

UNITED NATIONS: 61% of women aged 15–49 years, in marital or consensual unions in 160 countries or areas worldwide use contraception, with the highest percentage found in Latin America and the Caribbean and the lowest in Africa. The percentage for the 170 million women in the richer countries is 69%, compared to 59% of 873 million women in developing countries. The rate for the 117 million women in Africa is 27%, compared to a rate of 71% of the 82 million women in Latin America and the Caribbean. (UN News Center)

GENEVA: A three-day European Population Forum is discussing policies and trends in migration, aging, fertility and mortality; and sexual and reproductive health and rights, particularly in European countries in transition. The Forum comes ten years after the Cairo International Conference for Population and Development (ICPD), where 179 governments proclaimed the importance of population and reproductive health. (United Nations)

Recommended reading:

Gelbspan, Ross, *The Heat is on: The Climate Crisis, the Cover-up, the Prescription*. Reading, MA: Perseus Books, 1998.

9
POLLUTION
(May 9, 2004)

Pollution is our way of making more things, less expensively. Well, less obviously expensively. Most goods would cost more up front if we went to the expense of safe disposal of our waste materials. So we pollute first and clean up later. This is vastly more costly in the end, which is why we don't do a good job of cleaning up. It's also much less effective, and far more damaging to living things.

We generate so much bulk waste that finding places to dump it is getting harder, and there's too much to burn without seriously polluting the air. A truly concerted effort to recycle and detoxify would not be very expensive, because it takes less energy to use recycled materials than to use raw materials for most things.

Here are some of the major pollutants and their effects:

Organic solvents, the largest single category: fire hazard, damage to the nervous and respiratory systems, liver, and kidneys; cancer, birth defects.

Waste oil, second largest category: hazards similar to solvents.

Polychlorinated biphenyls (PCBs): stable, so they accumulate in soil and water. They accumulate to dangerous concentrations in fatty tissues of fish and other animals, which are toxic when eaten. Breakdown products include dioxins, extremely toxic. Liver and joint damage, birth defects.

Paints, pigments, and preservatives: old paint can contain lead, particularly harmful to the nervous system, especially in children.

Every step of **metal production** produces hazardous wastes. Major culprits include lead, mercury, chromium, and cadmium. Many metals can form toxic organic (carbon) compounds, which damage a variety of organs.

Pesticides and herbicides: harmful to many ecologically important species and to humans: nervous system, skin, bones, liver; birth defects, cancer.

Radioactive waste: The higher the dose, the greater the chance of an individual developing cancer, and the larger the number of cancer victims. Mutations and birth defects. At higher radiation levels: widespread breakdown of cellular functions, radiation sickness, and death. Plutonium, a strictly man-made element, is so long-lived there's no guarantee we can keep it from entering the environment. Plutonium is so toxic that a microscopic amount inhaled or ingested is lethal. Radioactive iodine is concentrated in the thyroid, causing damage or cancer. Military use of depleted uranium bullets and shells by the U.S. is harmful and should be outlawed. It is probably the cause of thousands of illnesses and deaths among combatants and civilians.

Biologically active waste: Contamination of drinking water with human wastes can cause diseases such as typhoid, cholera, and dysentery. Most common in areas without adequate sewage and water treatment systems. Contamination of meat with fecal material and inadequate preservation of food cause food poisoning. Fecal contamination is most common in countries with factory farms, like the U.S. Inadequate pasteurization and preservation are most common in underdeveloped countries. Most cases of "24-hour flu" are actually food poisoning.

Excess nutrients: sewage, fertilizer, and other materials that encourage blooms of blue-green algae and aquatic plants in lakes, streams, and shorelines can choke these waterways, and decaying mats of these flora can use up the available oxygen, a process called eutrophication. This has killed many lakes, contributed to a worldwide reduction of fish and shrimp, and caused massive "dead zones" in the world's oceans.

This week's news:

US: Smog and carbon dioxide in US cities are causing an epidemic of asthma, according to Harvard University researchers. (*Toronto Globe and Mail*)

OSLO: Norsk Hydro ASA has built two 600-kilowatt wind turbines to use with a hydrogen generator and a fuel cell in providing all the electricity for 10 homes on Utsira, Norway's smallest municipality with just 240 residents. When it's windy, the turbines will produce more electricity than needed by the 10 homes. The excess power will be used to produce hydrogen fuel so a hydrogen combustion engine and a fuel cell can make electricity at windless times. (NBC)

KRASNOYARSK: In Russia, up to 30% of tree logging is illegal. The wood is illegally exported to Scandinavian countries and China. Fires and the Siberian bombyx parasite have killed large areas of forest as well. Forestry codes allow local forestry administrations to hand out licenses to the lumber companies. "After the privatization experience of the 1990s, public opinion will not stand for privatization of forests." (terradaily.com)

HONOLULU: Whale Skate Island in the Northwest Hawaiian Islands no longer exists. It has been completely submerged by the rising ocean. Species that lived there have had to find other habitats or die. With global warming, mosquitoes go higher in Hawaii's mountains, carrying avian malaria with them that attacks birds that are only found in that single habitat. With the loss of each species, biodiversity declines further. (*Honolulu Star-Bulletin*)

US: Bark beetles are thriving in forests that are stressed by drought and rising temperatures in the American West. Entire forests are dying. This combination of events is entirely consistent with the expected effects of global warming, although some natives are unconvinced. (*Los Angeles Times*)

GREEN BELT, MD: Scientists at Goddard Space Center report dramatic weakening of the North Atlantic current that moves water in a counterclockwise pattern from Ireland to Labrador, possibly due to global warming. This is likely to affect climate in the entire region. (NASA)

Recommended reading:

Leakey, Richard, and Lewin, Roger, *The Sixth Extinction: Patterns of Life and the Future of Humankind.* New York: Doubleday, 1995.

Alternative Worlds

10

THE BEGINNING OF THE END

(May 16, 2004)

Here's my worst-case scenario for humankind, part one. Like any good science fiction, it's based on science fact.

In the worst case, globalization will accelerate the growth of social and economic inequities already in progress. The World Trade Organization, World Bank and International Monetary Fund already anticipate increased unrest in nations whose wealth is being extracted by the "structural adjustments" they impose as conditions for loans. These adjustments already include privatizing utilities and transportation, selling them off to multinational corporations at fire-sale prices. The privatization of public lands is accelerating in the U.S. as well. Structural adjustments also include forced belt-tightening, cutting into recipient countries' education and public health systems so they can pay interest on their mounting national debts. This too is happening in the U.S., as we go deeper in debt from war expenditures and tax cuts for the wealthy. Third world countries are forced to lift protective tariffs on locally produced goods including food, leaving them without protection from lower-cost, subsidized imports. Development contracts are awarded to international corporations, so that much of the loan money ends up back in the countries it came from. As a result, the recipient countries' economies "race to the bottom," as economists say. Governments fail and anarchy rules. People in underdeveloped countries understand what is being done to them, and they learn to despise the developed countries, especially the U.S.

Poorer nations will be less able to control population growth, stave off starvation, dehydration, and diseases like diarrhea, AIDS, malaria, and tuberculosis, preserve their environment, or provide the basic needs of life—as is happening now. Terrorism will continue to increase as the task of education is delegated to fundamentalist religions. Suicides among the hopeless will continue to increase. Logically, suicide bombings will go up.

In the worst but not unlikely case, the U.S. will continue to oppose family planning, and world population will reach eight to ten billion by 2050. The world will run out of cheap oil before then, but let's assume that somehow the developed world manages to keep going with natural gas, nuclear power, and coal. We'll accelerate destruction of the forests and degradation of the land and water in our quest for wood, fuel, and arable land.

The wealthy will continue to become wealthier and will initially be able to shelter from danger. The non-wealthy will continue to become poorer and increasingly vulnerable. Poorer nations generally will fail more quickly than wealthier nations.

Corporations and the governments they own will continue to resist reform. As population pressure increases, the poorest people will be crowded into, and the well off will abandon, the most devastated areas—as is happening today.

Temperatures will continue to increase worldwide, except for drastic cooling in northern North America and Europe due to changes in ocean currents. The oceans will rise as the polar ice caps melt, as they already have begun to do (nearly one quarter of the Arctic ice cover is gone, and two enormous ice shelves in the Antarctic recently collapsed). Coastal areas will be flooded, forcing hundreds of millions of people to move inland or die. Some areas not inundated will be poisoned with salt from violent ocean storms, forcing farms further inland, as is already happening. Coral reefs and plankton, the microscopic ocean life at the base of our planetary food chain, will continue to die off.

As wealthy nations like the U.S. begin to feel the strain of increased unemployment, leaders will blame other countries, illegal immigrants, the poor, and minorities. They'll use the corporate-controlled media to inflame fear, racism, and xenophobia, much like today. Dissent will be brutally suppressed, as has begun to happen already, even in traditionally democratic nations like ours. Despotism will continue to rise. Civil and human rights will be suppressed more blatantly. Military, corporate, and oligarchic juntas will continue to rig or suspend elections, in growing numbers of countries—including our own.

This week's news:

WASHINGTON, DC: Bush continues going easy on polluting power plants. He has received millions of dollars in contributions from utilities that operate some of the country's dirtiest power plants. The 30 biggest utility companies owning most of the 89 dirtiest power plants have given more than $6.6 million to Bush's campaign and the Republican National Committee since 1999. (*Cincinnati Post*)

US: Citizens and companies have converted 9.2 million acres of public land—an area about 42 times the size of Mount Rainier National Park—to private use, under an 1898 Supreme Court ruling giving corporations legal treatment as "persons." (*Seattle Post-Intelligencer*)

Recommended reading:

Weiner, Jonathan, *The Next One Hundred Years: Shaping the Fate of our Living Earth*. New York: Bantam Books, 1990.

11

IF WE FAIL

(May 23, 2004)

In the second installment of our worst-case scenario, the wealthy countries will seal themselves off from the poor ones by tougher immigration restrictions and enforcement, as is already happening in our country—and probably has to happen for reasons I'll explain. Already existing ghettos for the poor will be fenced or walled off, with armed sentries guarding access points, as has already happened to the Palestinians. The gated communities of today will become heavily fortified compounds, as they already have in parts of Latin America, Africa, and Asia.

Many of the poor, in order not to starve outright, will compete for jobs as servants of the wealthy, as has been common for a long time in the U.S. and elsewhere. The number of hazardous jobs will increase as essentials become prohibitively expensive, and the poor will fill them. Prisons will become more overcrowded and become prison colonies. In the U.S., 1–2% of the population is already in prison or jail, the highest percentage in the world. This will climb to perhaps 5–10%, as it will in many other countries.

The mercenary class will grow, serving the wealthy by fighting their wars, guarding their borders, cities, prisons, and elite fortified compounds, and serving as bodyguards—continuing present trends. There will be no remaining pretense of trying to help the poor. There will be no remaining pretense of democracy. The large prison populations will be reduced to gulag slaves, providing forced labor for mining, farming, fishing, and manufacturing, under increasingly harsh conditions.

As people and animals wander desperately in search of better living conditions, previously obscure diseases will spread. Travel restrictions will fail to stop the waves of human migration and epidemics. We'll become more crowded as equatorial regions become unfit for habitation. The most crowded cities will implode from AIDS, cholera, typhus, and typhoid epidemics, starvation, and anarchy.

Terrorism will increase, but let's assume that weapons of mass destruction don't come into play in order to follow our story further.

Pollution-related diseases such as asthma, cancers, and heavy metal damage to the nervous system, kidneys, and liver will become prevalent. Pollution control and public health programs will be overwhelmed and then abandoned. We'll see the return of polio, diphtheria, and yellow fever. Antibiotic-resistant tuberculosis and malaria will sweep across the world as mosquitoes invade areas previously too cold to support them.

Atmospheric carbon dioxide concentration will climb faster, as the oceans, bogs, volcanic lakes, and rocks release carbon dioxide due to the rising temperatures and as photosynthesis fails to keep up due to destruction of forests and photosynthetic ocean life. This will cause the temperature to rise yet faster, in a positive feedback loop.

Violent weather will cause damage around the world faster than it can be repaired.

The last of the poor and the gulag slaves will die. The wealthy and mercenary classes will shrink and infrastructure will continue to collapse.

The last remaining forests of the world will begin to wither and die once extinctions have reached the point that their ecologies collapse. The oceans no longer will provide food; photosynthesis will no longer replenish oxygen. Higher temperatures will accelerate oxidative processes, and atmospheric oxygen concentration will begin to drop.

Mass extinctions will skyrocket, in a last spasm of animal and plant death.

The last human survivors will succumb to dehydration, starvation, and disease in their scattered enclaves.

The year: around 2100.

This week's news:

GLAND, SWITZERLAND: The World Wide Fund for Nature warns cod may disappear by 2020 because of overfishing, illegal catches and oil exploration. The worldwide catch shrank from 3.4 to 1 million tons between 1970 and 2000. (MSNBC)

Recommended reading:

Ehrlich, Paul R., *The Population Bomb*. New York: Ballantine Books, 1968.

12
IF WE SUCCEED
(May 30, 2004)

In the best of all possible worlds, Western consumer culture will be replaced soon by a more modest one, which will still be comfortable and fulfilling without resource-intensive junk. The U.S. will stop waging wars of empire and start making democracy safe for the world. The hundreds of millions of people around the world who are malnourished and homeless will be taking steps toward decent lives. Much of the developed world's economy will be directed to restoring the ecosystem, developing new energy sources and fuels and infrastructure to support their use, and increasing energy efficiency. Industry will be refitted for zero or negative pollution of air, land, and water. There will be more than enough meaningful, satisfying jobs in every nation as we retool our societies.

We'll have to make more room for our flora and fauna, especially for the reintegration of fragmented habitats, and provide wide corridors for the north-south movement of animals as climate changes. The destruction of habitats, especially rain forests, will be replaced with massive reforestation and (since plants can't migrate fast enough) transplantation and seed planting. The rescue of endangered species will take first priority because once a species is gone it can't be replaced.

We'll abandon areas that are marginal for human habitation such as those prone to floods, seismic and volcanic activity, and droughts, to use as wilderness areas. We'll begin the long process of rehabilitating the ruined land.

Our irresponsible growth economy will shift to a steady-state one. Corporations, whose only current responsibility is to make money for majority stockholders, will have to meet responsibilities to all stakeholders. Non-renewable resource consumption will be replaced by use of renewable resources or recycling. With a declining population, we'll begin to mine our abandoned towns and our garbage and toxic waste dumps.

These changes, once dismissed as utopian pipe dreams or socialist schemes, will now be recognized as interrelated, interdependent, and inescapable if we're to have the cooperation of all the world's population and if we're to achieve sustainability. Although the nature of our economic systems will change, it won't spell the end of capitalism.

The international cooperation required for population control, environmental cleanup, and economic transformation will provide the precedents for further social evolution. Our rogue hyperpower, driven by greed, fear, and ignorance, will become an enlightened and benign world community member (remember this is fiction). Technology that really benefits humans and the planet will advance. Most couples will feel no yearning for more than one child; many will want none. Our air, water, and food will be clean once more.

With reasonable luck and ingenuity, global temperatures should stabilize and might begin to return to pre-industrial levels. Ecosystems will return to steady states, and extinctions will return to a sustainable rate. We'll have a better understanding of our place in the world. We'll be able to travel again, and enjoy our diverse cultures and the restored semi-wilderness areas while leaving the large true wilderness areas to ecology management professionals. Eventually, in a century or two, we'll be able to return to a replacement rate of about 2–2.1 children per couple.

This is, approximately, our best scenario. Remember: we must adapt or die, and adaptation must be willing: it cannot be coerced.

This week's news:

By hunting only the biggest mammals or fish mankind is unwittingly forcing many species to evolve over decades. Researchers now believe rapid evolution, affecting species from bacteria to bighorn sheep, is common. Evidence came from the fact that, because of overfishing and sparing the smaller fish, cod were smaller, matured faster, spawned earlier, and yielded weaker offspring (*Nature*, April 29). Researchers in Alberta found that over 3 decades adult rams in the wild were smaller, with smaller horns. Trophy hunters were killing the largest rams with the largest horns and the gene pool became biased toward smaller rams. (*Christian Science Monitor*)

CA: The state Senate approved legislation by Sen. Kevin Murray to require a percentage of new single-family homes in developments of 25 or more to have solar panels for 2 kW or more of power. The law would become effective in 2006, with the percentage of affected homes to increase annually until 2010. (*San Diego Union-Tribune*)

Recommended reading:

Wilson, Edward O., *The Future of Life*. New York: Knopf, 2002.

A Better Approach

13

FIXING IT BEFORE IT FIXES US

(June 6, 2004)

I've described the main threats to our survival: overpopulation and environmental degradation. In this chapter, I'll describe the obstacles to effective action.

There are humane and equitable approaches to achieving a sustainable population size and repairing environmental degradation. Many governments have implemented measures that, although inadequate, are at least a start. However, our own government is opposing international population control efforts and abandoning environmental stabilization efforts. If this continues, in the year 2100 your descendants—if there are any—will face unspeakable hardship and degradation. Even if you're rich. Imagine what they will think of you. That's what those of you under 40 should be thinking of your parents.

The longer we delay cleaning up our act, the harder (and riskier) it will be to survive and the less we'll be able to salvage, because there will be even more people to take care of and fewer resources left to work with. Scientists have been warning us about these problems for over a century. Why have we ignored their warnings? I've rounded up the usual suspects, all of which fall under the heading of "human nature": ignorance, inertia, denial, and greed. We'll be discussing those basic human traits, and how to factor them into our solutions, in this book.

Our government, industry, and many organizations including most churches and most media are lying to us about the problems we face, and we resist hearing what little we're told through other sources. A significant portion of the American population sees no reason for concern about overpopulation or destruction of the biosphere, either because they're uninformed, in denial, or because they actually welcome the end of the world as fulfillment of Biblical prophesy. Consumerism and social Darwinism have brought about unprecedented waste and greed and an unconscionable widening of the gap between rich and poor. These are all part of our problem, which we must address to reach solutions.

Our own brains conspire against our survival. We're driven by innate drives that evolved long ago because they had survival value. They're now part of the problems that we must agree to address.

What tools can we use to survive and prosper? Only the obvious: knowledge and action. We can learn enough and do enough soon enough. Our understanding of the problems has increased dramatically in recent decades. Most of us would welcome a common cause that would heal our divisions and end our suicidal behavior. We have the capacity for intelligent self-sacrifice. In any event, we really have no other option but to act, the sooner the better. The people of the world must take the initiative, however. We must be the leaders and our governments must follow our lead, not the other way around. Remember: if we fail, we have no one except ourselves to blame.

I'll describe the approach you and I are going to use in order to define what has to be done and how to do it. Of course, we'll need to use everything we know about the damage we've done and how to fix it. But more importantly, we'll have to apply everything we know about human nature in order to transform our behavior from a destructive force to a constructive one.

For now, please do yourself a favor. Start reducing your debt and saving your money.

This week's news:

LISBON: A project for 30,000 people is being constructed, which will use renewable energy only, with little waste. WWF, the global environment campaign, and a U.K. development group, BioRegional, are implementing the project. It will occupy 13,000 acres, cost 1 billion euros, and be finished in about ten years. Similar projects, for about 5,000 people each, are planned in the U.S., U.K., China, South Africa, and Australia. The Portuguese project will include a wildlife reserve of 12,000 acres, with wildlife corridors connecting it to nearby protected areas. Fifty percent of the community's food will come from nearby. The development will be built using reclaimed and recycled materials whenever possible, and at least 90% of its organic waste will be composted. (BBC News)

Recommended reading:

Rogers, John J. W., and Feiss, P. Geoffrey, *People and the Earth: Basic Issues in the Sustainability of Resources and Environment.* Cambridge: Cambridge Univ. Press, 1998.

14

A STRATEGY

(June 13, 2004)

To arrive at workable solutions we'll need knowledge—not just of facts, but also of mechanisms: why some trends are threats and others aren't; why some solutions are feasible and others aren't; what changes in our behavior are feasible and what aren't. We need a deeper understanding of right and wrong. We need to reexamine the sources of meaning in our lives, in order to restructure our ends to fit our means.

You'll soon find you want to know more than I can tell you in this short book. By far the best sources of information are books. Consider reading the books I recommend at the ends of most chapters.

You can find other titles at http://www.amazon.com/exec/obidos/ats-query-page/. Search by subject, generally after year 1990 for up-to-date material, sorted by readers' average ratings. Read the reviews, by authorities to determine accuracy, and by lay readers to determine readability. Most important: tear yourself away from the fascination of flickering images and pulsating rhythms and start reading at least one book every other week.

Consider your sources. Academic scientists have to convince their peers with the quality of their evidence and their reasoning. They're more reliable than people paid to bend the truth. Maintain a healthy skepticism of governmental, religious, or industry representatives, because they often have serious conflicts with the truth. For two examples of organizations I consider to be pseudo-scientific fronts for industry, see the background on the Competitive Enterprise Institute, http://www.disinfopedia.org/wiki.phtml?title=Competitive_Enterprise_Institute, and their Web site at http://www.cei.org/; and background on The Science and Environmental Policy Project, http://encyclopedia.thefreedictionary.com/SEPP and their Web site, http://www.sepp.org/. In contrast, here are two reputable sources: Woods Hole Oceanographic Institution at http://www.whoi.edu/ and the National Academy of Sciences, http://www.nas.edu/. To the non-scientist, it could be difficult to tell which sites to believe. But the first two have been affiliated with radical political groups, they've been funded by environmentally destructive industries, and they have few if any practicing scientists who have recently published in peer-reviewed journals. The last two are important scientific bodies, not affiliated with industry or political groups, with internationally respected active researchers who publish in top peer-reviewed journals.

You can't always find what you need at libraries. Ask your library to order some books. You may want to buy some books. You can get best buys at sites like http://ebs.allbookstores.com/ and the used books section of Amazon.com. After you read them, donate them to your local libraries.

Switch to reliable news sources where you can follow developments relating to overpopulation and the environment. Stop watching the crap on TV (except PBS) and make time for substantive news. A good broadcast news site is http://news.bbc.co.uk/. A good wire service is http://www.reuters.com/. Remember that most American newspapers are owned by a small number of large corporations with a desire to downplay news contrary to their corporate objectives. As a consequence, most have poor coverage of many important developments, but some of the better ones are http://www.nytimes.com/, http://www.csmonitor.com/, http://www.washingtonpost.com/, http://sfgate.com, and http://www.latimes.com. Some of the better foreign newspapers are http://www.independent.co.uk/, http://www.guardian.co.uk/, http://www.lemonde.fr/, http://www.elmundo.es/, http://www.welt.de/, http://www.haaretzdaily.com/, http://www.dailytimesofnigeria.com/, http://www.atimes.com/, http://www.eldia.com.ar/, http://www.iht.com/. For international health there's http://www.msf.org/. For science coverage, I recommend http://www.newscientist.com, http://www.nrdc.org/, http://www.plosbiology.org/, http://www.gristmagazine.com/, and http://www.planetark.com/. To check on the performance of your elected officials, try http://lcv.org/.

This week's news:

BONN: More than 150 nations pledged to promote wind, solar, geothermal and other sources of energy at the Renewables 2004 conference. No specific targets were specified. Germany proposed the four-day conference after the failure of the 2002 World Summit for Sustainable Development in Johannesburg. It gained urgency because of soaring oil prices and attacks on foreign workers in Saudi Arabia. China promised to draw up a development strategy for renewable energy for rural Afghanistan. The U.S. promised to develop less expensive solar and wind technology, and said it would expand production tax credits for alternative energies. (*Boston Globe*)

Recommended reading:

Wilson, Edward O., *Consilience: The Unity of Knowledge*. New York: Knopf, 1998.

15
THE SCIENTIFIC METHOD
(June 20, 2004)

The scientific method is an approach to problem solving that all of us use to some extent. We're going to use it as much as possible in our search for solutions to the problems of overpopulation, global climate change, and mass extinctions. After familiarizing yourself with it, you'll use it more often and expect others to do the same. Instead of accepting things on authority (even from me), you'll want evidence.

People of faith often believe they've achieved certainty, yet different faiths offer contradictory certainties. They can't all be right, but they can all be wrong. Scientists are often wrong too. But a scientist is supposed to accept that certainty is impossible. Scientists seek the best approximations of answers to questions, based on careful observation and logic. We try to surmise cause and effect relationships, formulating the broadest generalizations possible. We go with the simplest explanations that work (a rule of thumb called *Occam's razor*).

We can't prove a theory no matter how carefully we test it, because a new kind of test could show it doesn't fit all circumstances, in which case we've disproved the theory, and it has to be adjusted or replaced. For example, the theories of mechanics, describing the motions and mechanical interactions of objects, have been revised many times, by Aristotle (who performed no experiments), Galileo (who performed simple experiments), Newton (who applied the calculus, which he invented for the purpose), and Einstein (who concluded nothing could travel faster than light). The predictions of modern mechanics match our most precise measurements.

This process of replacing theories when they're found wanting is what has made science so useful. *Any system of thought that doesn't revise theories when empirical evidence clearly contradicts them is unsound.*

Humans have learned there are natural laws that govern the behavior of matter and energy, and *as far as we know,* those laws apply everywhere, for all time, even to living, thinking, feeling people. The accuracy of modern theories comes from the mathematical equations used to express them. The properties of mathematics are built into the universe.

It's in our nature to create hypotheses, to explain how things work and how they came into being. The odds of being right are related to how much we already know about the subject of the hypothesis. Even when they know quite a lot, scientists are used to throwing out most of their hypotheses after a few tests. Scientific knowledge is the product of centuries of painstaking work by thousands of researchers, involving many disappointments and revisions.

Faith isn't as hard. It's natural that many people prefer faith. We shouldn't feel bad about it, but we have to learn to deal with the fact that it's often wrong. Certainty is a useful feeling, but nothing more.

"Knowledge" and "understanding" thus have different meanings in science from their casual usage, as does "theory." To say something is "just a theory" is to miss the point of what a scientific theory is. It's the best approximation to truth that we have. We won't look to faith for the solutions we need. We'll apply the scientific method as much as possible, even to questions of morality and spirituality. I'm truly sorry, but I'm afraid it's much too late to worry about religious sensibilities. I'll be stepping on some toes. I apologize in advance for the anger and unhappiness this will cause some of you, but please bear with me. I think I can convince you this is important enough for all of us to overcome our differences and work together.

This week's news:

WASHINGTON, DC: Six major environmental groups and the oil industry have agreed to support ratification of the UN's Law of the Sea treaty. The pact would establish jurisdictions, rights, and controls for coastal countries regarding the military navigation, commercial exploitation, and environmental conservation of the seas. Environmentalists support oversight laws for pollution and waste dumping, guidelines against overfishing, and protections for whales, dolphins, and other marine species. The petroleum and mining industries want rights of access to mineral-rich regions. The pact is also essential for the military. The Senate Foreign Relations Committee has approved the treaty but Frist has refused to schedule a Senate vote, apparently because conservatives who reject multilateralism on ideological grounds oppose it. Russia is trying to claim seabeds beneath melting Arctic ice, which are becoming accessible due to global warming. Treaty members are begging the US to oppose Russia's claims. (*Grist Magazine*)

LONDON: Shell's Ron Oxburgh: "No one can be comfortable at the prospect of continuing to pump out the amounts of carbon dioxide that we are at present. People are going to go on allowing this atmospheric carbon dioxide to build up, with consequences that we really can't predict, but are probably not good." Oxburgh favors sequestration, in which CO_2 is captured and stored. However, the process is very expensive and requires large amounts of energy. He admits the time scale might be impractical, "In which case I'm really very worried for the planet." (*The Guardian*)

Recommended reading:

Damasio, Antonio R., *Descartes' Error: Emotion, Reason and the Human Brain*. New York: Vintage, 1995.

Katz, Leonard D., *Evolutionary Origins of Morality: Cross-Disciplinary Perspectives*. Bowling Green, OH: Imprint Academic, 2000.

Sagan, Carl, *The Demon-Haunted World: Science as a Candle in the Dark*. New York: Random House, 1996.

Origins and Evolution

16
THE BIG BANG
(June 27, 2004)

We can learn a lot about our options for the future by examining the past, even the very distant past. Let's go back to the beginning, at the origin of the universe. Sadly, there's a substantial proportion of our population that believes in a recent, essentially simultaneous origin of everything in the universe. This is a harmful belief, in that it interferes with a realistic approach to life on Earth. We can't afford that any longer.

If the colors in the light from a single star are spread out in a rainbow pattern by an optical prism, we see discrete lines spaced out along the bar rather than a continuous sequence of colors. The colors between the lines are missing.

Light has a dual nature, as particles called *photons* or alternatively as *electromagnetic waves* akin to radio waves. Light travels at a constant speed regardless of color, so the distance between peaks (the *wavelength*) of a light wave varies inversely with its frequency of oscillation. The lowest frequency (red) human-visible light wave has the longest wavelengths and the highest visible frequency (violet) has the shortest wavelength.

The electrons in an atom can only occupy certain "permitted" sets of energy levels, different for different elements. The heat of a star is the energy of its jostling atoms. Collisions knock atoms' electrons into higher energy levels. The electrons return to lower energy levels by releasing energy as photons, so the star's heat is radiated into space as light. The decrease of energy in an electron determines the emitted photon's wavelength. Since there are only certain energy levels possible for an electron, there can only be certain colors in the emitted light, depending on the temperature and composition of the star.

Light from other galaxies has the same spectral lines as our galaxy, but shifted toward the red end of the light spectrum. The red shift indicates galaxies are moving away from us. This is the Doppler Effect, responsible for the drop in pitch of a train whistle when the train passes us. The Doppler shift of a radar echo tells police how fast vehicles are moving toward them or away from them, and astronomers use the same method to tell how fast a galaxy is receding. Astronomers can also estimate distances to galaxies by the apparent brightness of certain types of stars that are always equally bright close up. The farthest (faintest) of these stars have the largest red shifts, so they're receding the fastest. This suggests all galaxies are traveling away from a common starting point: the ones that have traveled the farthest traveled the fastest to get there. The theory that best fits these results is that the universe started as a small, incredibly dense mass that flew apart in a Big Bang roughly 15 billion years ago.

Hydrogen and helium, the elements with the smallest atoms, were formed in the Big Bang. Gravity pulled these atoms together into stars. Stars are so massive they're compressed by gravity to enormous densities. The hydrogen atoms in their cores, at high temperature and pressure, undergo nuclear fusion, forming heavier elements and further heating the stars. The outward pressure caused by the heat counters the inward pressure of gravity, stabilizing a star's diameter. However, as the star changes composition, this balance shifts. Stars eventually collapse into even denser bodies or explode. The dust and gases of exploded stars coalesce to form new stars and the planets that orbit them.

Our solar system was formed 4.5 billion years ago from the ashes of dead stars. As astronomers are fond of saying, you and I are made of stardust. Our sun and planet could sustain life for a few billion years more. The universe will last much longer than that.

Earth is the only planet that we know has life. However, astronomers have found many planets orbiting distant stars. Possibly enough planets are sufficiently like Earth for intelligent beings to exist elsewhere. But don't count on them for help. They're too far away.

If our planet becomes unlivable, we don't have the option of colonizing another planet or surviving in space ships. We don't have the technology to move large numbers of people off this planet. Anyway, we don't even know how to keep a few humans alive in a closed, artificial environment for more than a few months, or live in harmony on what was once a roomy and comfortable planet.

This week's news:

SACRAMENTO: The cars of Los Angeles, pumping greenhouse gases into the air, are producing an enormous brown cloud of pollution, dust, and chemicals visible from space. Such pollution circles the globe, causing a 10–20% reduction of sunlight reaching the earth over the last 50 years. This cools the earth's surface, but heats the middle atmosphere. Carbon particles in the cloud absorb heat, scatter sunlight, and produce the haze found in both Los Angeles and Yosemite. They also form nuclei for water condensation. Clouds become thicker and darker from increased moisture accumulation, further shading the ground. Natural clouds form droplets that fall as rain, but carbon particles are often too small to produce drops big enough for rain. Also, atmospheric CO_2 will double by mid-century, increasing temperatures, especially inland. The most warming will occur in the mountains where the snow pack contains more than a third of the drinking and irrigation water for the state. Precipitation is likely to increase in the northern third of the state and decrease in the south. Heat waves will become more frequent, causing health problems. Adapting to climate change will cost California billions of dollars and cause severe problems for agriculture. (*San Francisco Chronicle*)

WASHINGTON: The Bush administration, which earlier slashed funds for UN population control efforts, is trying to prevent the U.N. Population Fund's work in China and elsewhere, according to U.N. officials. The U.S. has withdrawn funding for a conference and threatened groups like UNICEF if they cooperate with the agency. The Bush administration also hopes to coerce the U.N.'s Latin American caucus to abandon a position on population that could be interpreted to advocate abortion. Bush has reinstated the Reagan-era policy withholding money to groups that even discuss abortion as an option, except in cases that threaten life or involve rape or incest. U.N. officials and family planning advocates have argued that advances in education and awareness on reproductive issues are being undermined by the United States, even though abortion is legal in the U.S. Supporters of UNFPA say it has made considerable progress in reducing the number of abortions in China through family planning programs in conjunction with the Beijing government. A fact-finding trip for the State Department found "no evidence [that the Population Fund] has knowingly supported or participated in the management of a program of coercive abortion or involuntary sterilization" in China. The Population Research Institute in Front Royal, Va., which calls itself a research and education group that exposes human rights abuses in population control programs, is a leading opponent of the Population Fund. They claim China's population control policy coerces women through mandatory use of contraception, forces abortions for those younger than 20 and imprisons those who do not appear for examinations. Fund supporters contend they have nothing to do with abortion policy and cuts have harmed poor women. (*New York Times*)

The blackout last August in the eastern U.S. and Canada cleared the air. One day into the power outage, the air was strikingly blue and the visibility unusually high. In Pennsylvania, atmospheric sulfur dioxide was down 90%, ozone 50%, and light scattered by particles 70%. (*Chemical and Engineering News*)

Recommended reading:

Hawking, Stephen W., *A Brief History of Time*. New York: Bantam Books, 1998.

17
THE ORIGIN OF LIFE
(July 4, 2004)

After we're extinct, what's the chance of intelligent life evolving again on Earth? Sorry, that's a trick question: if we were so intelligent, why'd we go extinct? The short answer is, it will probably take hundreds of millions of years for intelligent life to evolve again, if ever. Here's why.

There are no signs of life in the earliest terrestrial rocks, 4 billion years old, but rocks 3.5 billion years old have been found that contain fossil remains of early bacteria. Given the conditions then, we know from laboratory experiments that small *nucleic acids* called *RNA* can self-assemble from random interactions among four different subunits called *nucleotides* in the soup of the early oceans, strung end to end in chains. These persist for a while and then disintegrate into their constituents.

Judging from laboratory evidence, eventually some of these RNAs could act as templates for the formation of copies of themselves. The process is called *replication*. Eventually some RNAs probably came upon the scene that could also act as templates for the formation of small *proteins*, molecules made from 20 or so different *amino acids*, strung together end to end. The sequences of amino acids in proteins are specified by the nucleotide sequences of RNAs. These RNAs were arguably the first *genes*, units of heredity, although they were not yet parts of organisms.

Variations of nucleotide sequences in genes led to parallel variations of amino acid sequences in the proteins they specified. Eventually, some proteins with *enzymatic activity* arose: by virtue of certain amino acid sequences, they could speed up certain chemical reactions. Some enzymes could speed up the replication of genes, protect the RNAs from damage, allow RNA chains to grow longer, or facilitate translation of nucleotide sequences in RNA into the amino acid sequences of proteins. Sometimes an enzyme produced by one gene would favor the replication of other genes, or combinations of genes. Eventually there would be enormous advantages of certain combinations of genes over others, in that they were more likely to replicate as a group.

Eventually *DNA*, another form of nucleic acid, was formed. In most, but not all, modern organisms, DNA is the hereditary material and different forms of RNA are used as intermediates in the formation of proteins. The exceptions are some types of viruses that use RNA as their hereditary material, and prions, which are proteins that can direct their own self-replication in cells they infect.

It can't be proved yet that this is exactly what happened, because these molecules didn't leave any fossil evidence. However, all these steps have been demonstrated to be possible, and this means that life could have arisen spontaneously. The only viable alternative theories, such as the arrival of life forms via meteorites, would still require that life started somewhere.

There is one piece of circumstantial evidence for our hypothesized sequence of events that is pretty powerful: In every organism today, the same sequences of three nucleotides, called *triplets*, form a 64-letter "alphabet" to specify the same amino acids, with very few variations. There's no intrinsic reason why this has to be so. This strongly suggests that all modern organisms are descendants of a common set of primordial ancestors.

This scenario for the origin of life suggests that, if we had to start all over from the same starting conditions, the origin of life would follow a similar course. Unfortunately, the conditions after our demise will much less favorable to starting over.

This week's news:

WASHINGTON, DC: According to a National Wildlife Federation study, from 2001 to 2003 the government cut 42 million acres out of 82 million acres requested by government biologists to protect critical habitat of threatened and endangered species. In 2001, the administration cited economic reasons for cutting about 1% of the requests; that percentage rose to 69% in 2003. The federation maintains the administration is trying to undermine the Endangered Species Act. The Act requires the government to set aside areas that vanishing species need to survive and recover. (*New York Times*)

WASHINGTON, DC: A new US supercomputer model shows that temperatures could be rising faster than scientists had thought. Temperatures could rise by 4.7°F if carbon dioxide emissions continue at current levels, as opposed to earlier estimates of 3.6°. Atmospheric carbon dioxide has now reached 370 parts per million and continues to increase. Scientists hope to refine the models further, to predict climate change effects in specific regions such as Africa or the American Midwest. (terradaily.com)

SINGAPORE: Smoke from Sumatra's forest fires has reached Singapore, raising the pollution index from "good" to "moderate." The smoke has also hit Malaysia, and schools may have to close, to keep children from breathing the air. Burning forests in Indonesia to clear land for farming often causes problems in other Southeast Asian nations, despite its being unlawful. In 1997 and 1998, haze caused by forest fires in Indonesia caused serious health problems and disrupted airline schedules in Singapore, Malaysia, and the Philippines. It caused $9.3 billion in economic losses. (terradaily.com)

BRAZZAVILLE: Central African forestry officials will meet to formulate regional plans for sustainable management of their eco-systems. The United States, European Union, Britain, Germany, Canada, and South Africa have pledged $300 million for replanting forests and preserving endangered woodland and marine animal species in the Congo Basin. None of these funds have been released yet, pending completion of the plan. More funds will be needed in the future. Cameroon, the Central African Republic, the Republic of Congo, the Democratic Republic of Congo, Equatorial Guinea and Gabon announced a plan in August last year, to create a network of natural parks and protected areas covering 37,000 square miles and managed forests covering about 80,000 square miles—about three-quarters the size of Japan. The region's forests have dwindled by 10% or more since the 80s due to over-grazing, illegal logging, mining activities, civil wars and environmental destruction. The forestland comprises 18% of the globe's tropical rain forest and 70 percent of Africa's plant cover, including 600 tree and 10,000 animal species. (terradaily.com)

PARIS: U.K. Foreign Secretary Jack Straw and Secretary of State for the Environment Margaret Beckett, and French Foreign Minister Michel Barnier and Ecology Minister Serge Lepeltier made a joint appeal for action on climate change. "The (European) heatwave of summer 2003, repeated floods, the advance of desertification, the melting of the ice sheets and glaciers are an illustration of the first effects of climate upheaval," they said in a statement published in *le Monde*. They called for leadership by the industrial nations in reducing fossil-fuel greenhouse emissions, indicating global warming would impose an "uncalculable" cost on health, the environment, and national economies, and would have a devastating impact on future generations. The cost "will clearly be higher than the economic cost of measures to tackle the phenomenon," they argued. The U.S., the worst CO_2 polluter, has deserted the Kyoto Treaty.

Recommended reading:

Dawkins, Richard, *The Blind Watchmaker: Why the Evidence of Evolution Reveals a Universe Without Design*. New York: Norton, 1986.

18

THE PRINCIPLES OF EVOLUTION

(July 11, 2004)

Two of our most pressing problems are mass extinctions of species and the apparently suicidal behavior of our own species. In order to understand either, it's time for us to sit down and have a little talk about where species come from.

Even before Darwin, scientists realized that contemporary species didn't always exist, and that species that existed in the past weren't around anymore. Darwin concluded that species change with time, because of inheritance of characteristics with small variations from one generation to the next, and a process he called *natural selection*.

Nothing was known then of the molecular basis of genetics, in which genes (segments of DNA or RNA that specify the amino acid sequences of proteins) are replicated down the generations, with occasional inaccuracies leading to changes in the organisms. Most errors of replication (*mutations*) produce changes in proteins that are neutral, deleterious, or even lethal, but occasionally such a change benefits the organism, so it's more likely to reproduce or give rise to more offspring. This is the process Darwin called natural selection.

Because of mutations, different varieties (*alleles*) of a gene coexist in the population of a species. The alleles of organisms that reproduce more successfully become more common (frequent), because increased reproduction or survival increases their proportion of the *gene pool* (all the genes in all the individuals of a species). It turns out this genetic diversity is what protects life on earth from environmental challenges.

Environmental challenges such as climate changes can shift the selective balance for or against alleles, such as those controlling hair or feather growth. If a species has sufficiently diverse alleles to respond to such challenges, or if mutations of beneficial alleles can keep up with the challenges, the species can *adapt* to the challenges. Environmental challenges have to be slow enough for adaptive processes to keep up, or the species will die out. The shifts in allele frequencies in response to environmental factors are the basis for natural selection.

New species generally arise because of the isolation of two subpopulations of a species from each other, e.g., because of separation by a mountain range or an ocean. Their genes are kept from mixing, and over thousands of generations of gradual change, they drift apart. This can lead to the two populations becoming genetically incompatible: different species, whose members cannot crossbreed.

Every species needs an *ecological niche*, or a suitable environment in which it can reproduce. Species evolve over time to fill the available niches. The complement of species in a region partly determines the niches available (e.g., ticks that live on mammals can't survive in the absence of mammals). So niches partly determine the species, and the species help determine the niches.

Humans are different from other species in an important way. We use our intelligence, language, culture, and ability to use energy to do work and build things. This enables us to adapt to a wide variety of environments. As a result, we can out-compete many other species in the contest for resources. We also eliminate the niches of many other species. The good news is these abilities have had enormous survival value ever since our species was young. The bad news is we've broken free of the natural controls that normally prevent species from wrecking their environments and killing off the majority of other species.

This week's news:

MANILA: The International Rice Research Institute (IRRI) in the Philippines has found from 15 years of research "strong evidence of a reduction in rice yields caused by rising temperatures consistent with trends in global warming." Yields fell by 15% for every 1°C increase in daily mean temperature. Experts predict a global temperature increase of 1.5–4.5° in the coming century, 3–9 times more than in the past century. Global warming threatens to erase productivity gains that have kept pace with population growth. (spacedaily.com)

RIO DE JANEIRO: Brazil announced a plan to save twenty-six carnivorous mammals, including pumas and other wild cats, wolves and other canines, raccoons, and ferrets from extinction. Experts warned, for example, that the disappearance of the jaguar could trigger a population explosion of prey species such as deer and wild boar, which would then put pressure on plant species. (terradaily.com)

MADISON: Infectious diseases are appearing in new places and new hosts, posing increasing risks to humans and animals, as humans cut forests, drain wetlands, build roads and dams, and expand cities. Scientists warn in the journal *Environmental Health Perspectives* that deforestation and ecosystem changes provide new niches for diseases such as malaria, dengue fever, Lyme disease, yellow fever, cholera, influenza, foot and mouth, and hemorrhagic fevers. Detailed understanding of the effects of human activities on the spread of pathogens exists for only a few diseases. In the northeastern U.S. forest fragmentation, urban sprawl and the erosion of biodiversity contributed to the spread of Lyme disease. Scientists think the AIDS virus may have first infected "bush meat" hunters drawn to Africa's tropical forests by new logging roads. World travel then spread the disease. The nipah virus appeared in Malaysia and Singapore in 1999, possibly when El Nino-fueled fires drove fruit bats from forests to farms where it was transmitted to pigs and humans. The report recommends linking land use to public health policy, expanding research on deforestation and infectious disease, developing policies to reduce "pathogen pollution," and establishing centers for research and training in ecology and health research. (spacedaily.com)

LONDON: The Stockholm Environment Institute's center at the University of York reports that increasing demand for air travel is one of the most serious environmental risks facing the world. "This growth has been fuelled by generous tax breaks and state aid and is contrary to the objectives of environmental policy, especially efforts to prevent the worst consequences of climate change," they said. The study recommends that aviation fuel no longer be tax-free, trips of less than 400 miles should be made by train, businesses should use video conferencing rather than travel to meetings, and airlines should pay an environmental charge equal to the damage they cause. High-speed rail services such as Eurostar, linking London with Paris and Brussels via the Channel Tunnel, need to be improved so that every British city can be linked by train to mainland Europe. "At the moment we have cheap flights and some of the most expensive railways in the world. That is the wrong way around." (terradaily.com)

World Wildlife Fund International reports only about 700 mountain gorillas remain in the world. Any further loss of habitat in Rwanda, Uganda, and the Democratic Republic of the Congo can push them closer to extinction. In just 2 months, several thousand illegal settlers cleared 15 sq km of 425 sq km of gorilla habitat in Virunga National Park for farms. The forest was entirely cut down and turned into timber or charcoal. (*Al Jazeera*)

LONDON: The Archbishop of Canterbury, Dr Rowan Williams, warns the survival of humans is at risk because of the "addiction" of rich nations to fossil fuels. He forecast the emergence of "fortress societies" able to possess all the natural resources such as oil and water they require, with the rest of the human race excluded. Dr Williams adopted the approach of the Eastern Orthodox Church that destroying the environment is a sin, and that Christians have a duty to protect it. He said, "We should be able to see that offences against our environment are literally not sustainable. The argument about ecology has advanced from concerns about 'conservation'. What we now have to confront is that it is also our own 'conservation', our viability as a species that is finally at stake. While the survival of the human race is the long-term threat, in the shorter term, what is at stake is our continuance as a species capable of some universal justice." He criticized a society "in denial" about the destruction of the environment. He cited economic theories that don't factor in environmental factors such as soil degradation, deforestation, and a disrupted food chain as costs of activity. He remarked that since "the oil production of relatively stable and prosperous societies is fast diminishing, these countries will become more and more dependent on the production of poorer and less stable nations. How supplies are to be secured at existing levels becomes a grave political and moral question for the wealthier states, and a destabilizer of international relations. This is a situation with all the ingredients for the most vicious kinds of global conflict—conflict now ever more likely to be intensified by the tensions around religious and cultural questions." He backed a plan by the Global Commons Institute in which every person on the planet has an equal right and quota to emit carbon dioxide. He pointed out that in the first 48 hours of 2004, an average American family would have produced as much emissions as an average Tanzanian family over the whole year. (*The Guardian*)

Recommended reading:
Edey, Maitland A., and Johanson, Donald C., *Blueprints: Solving the Mystery of Evolution*. Boston: Little, Brown, 1989.

19

THE EVIDENCE FOR EVOLUTION

(July 18, 2004)

The main threat to our survival is human behavior, which is shaped by human culture and the biology of our brains. Our brains and our culture are the products of biological and cultural evolution. So that we all start on the same page, let's review why the theory of biological evolution is so well established.

Over the past 3.5 billion years, organisms sometimes became trapped in mud or volcanic ash that later turned to rock. Their hard body parts became fossils. Remains of ancient organisms are also found in amber (petrified tree rosin), tar pits, ice, and permafrost.

Layers of rock are laid down over time, so the farther a layer is from the surface the older it is. The fossils nearest the surface are most like modern species, and as we go deeper, we find less similar species, as we would expect if species evolved. Radioisotope dating confirms these relative ages. There are enough cases where continuous series of species were laid down (e.g., in the case of horses) to make it clear that evolution followed a course of gradual changes.

The fossil record shows that there were branch points, where different kinds of descendants diverged from a common ancestor. Similar features are found among descendants along both branches but the similarities decrease with time. In this way new taxonomic divisions arose, for example some reptiles evolved into birds and mammals. Many branches have died out entirely. The theory of evolution could be disproved if more modern forms were found in the same strata as more primitive ones. This has never happened.

These findings explain why the classification system developed by Linnaeus has worked so well. In the Linnaean system, all organisms are classified into a few kingdoms, kingdoms are subdivided into phyla, phyla into classes, and so forth through orders, families, genera, and species. The Linnaean scheme is a way of describing modern species according to their similarities and differences. We now know that modem species are the tips of branches in the evolutionary tree. The fossil record extends the classification scheme back in time, filling out the patterns of branching. If such a scheme had not been confirmed, the theory of evolution would have been disproved.

If animals of closely related species developed embryologically along very different lines, it would be impossible to explain from evolution how such differences could have arisen from gradual changes. However, there are broad developmental similarities not only across related species, but also even across the entire animal kingdom, and the genes responsible for embryological development can show remarkable consistency, even between vertebrates and invertebrates.

The geographic isolation that is often the basis for branching evolution also explains why large groups of organisms may be found only in one place. The timing of the separation of continents due to tectonic plate movements coincides well with the divergence of what became very different types of animals. Recently separated regions have less dramatic differences between their species than those that separated longer ago.

Finally, there is the molecular evidence. Species evolve because their molecules evolve, in particular their nucleic acids and proteins. The more closely related two species are, the more similar their molecules are. The closeness of relationships determined from fossil evidence agrees very well with the molecular evidence.

No other proposal has come close to explaining all these phenomena.

This week's news:

The Gobi desert grows 10,400 square kilometers annually. In Nigeria, around 3,500 square kilometers of land are turning to desert each year. Tuvalu, the Netherlands, and Denmark are at risk from rising ocean waters. In the next 100 years, the sea level will rise at least one meter due to global warming. Already there are 5,000 new environmental refugees every day. They aren't recognized as refugees under international law and cannot request protection or asylum. The Intergovernmental Panel on Climate Change, the UN scientific group that studies the causes and impact of climate change, anticipates 150 million environmental refugees by 2050, due to soil erosion, global warming, and water pollution. The poorest peoples don't contribute significantly to global warming, but they pay the highest price. (alternet.org)

Recommended reading:

Mayr, Ernst, *What Evolution Is*. New York: Basic Books, 2001.

20
CREATIONISM AND THERMODYNAMICS
(July 25, 2004)

"Creation science," a contradiction in terms based on a particular reading of Genesis, has been refuted in detail, as described by Wilson (1983). Nevertheless, creationists continue to argue that evolution is impossible. They continue to claim there are no intermediate forms that could explain the evolution of the eye, even though intermediate forms are well known. They persist in arguing that modern organisms couldn't arise by chance, cloaking the role of non-random, deterministic natural selection.

Most religions have no problem with evolution. But if evolution is true, creationists can find no way to uphold their literal interpretation of scripture. This threatens their beliefs that there is purpose in the universe, humans are the apex of god's creation, and moral law is handed down to man by god. They try to force their religious beliefs on us, lying to themselves and to us.

This is dogma, the enemy of reason, science, and democracy. When lies are institutionalized, we can't solve the problems created by those lies. This is well illustrated by the Soviet sanctification, based on flawed evidence, of the crackpot theories of Lysenko, which led to the collapse of Soviet agriculture. If they had not clung to their dogma for so long, they could have recognized their error and saved their wheat crops.

Consider the creationist claim that evolution is impossible thermodynamically. They argue that it's a violation of the second law of thermodynamics for complex species to evolve from simple ones. They aren't bothered, incidentally, by the alternative that life originated as a miracle, which would break just about all the laws in the book of nature.

The first organisms were single-celled, like some bacteria of today. They developed in an environment much different from today's, however. The main difference was the atmosphere, in which oxidation, necessary for respiration in modern animals, wasn't possible. These organisms apparently evolved from self-replicating nucleic acids that had arisen in aquatic habitats from chance variation in their sequences of nucleotides, and evolved through random genetic variation and non-random natural selection. They acquired cell membranes composed of lipids (fats and oils) that created an intracellular environment favorable to developing the machinery of life. The ingredients needed for growth and reproduction were absorbed through cell membranes and processed, and waste products were passed back out through the membranes. Intracellular mechanisms extracted energy from some chemical reactions and used that energy in other reactions to build complex molecules from simple ones and to transport substances across the cell membranes.

The process of evolving structures like cells from small molecules might appear at first to violate the second law of thermodynamics. This law says that systems run down and disintegrate over time. All orderly structure, such as molecular structure, eventually degrades to disorder. Lack of order is called *entropy* by physicists, and the second law of thermodynamics dictates that entropy must increase over time. Does this mean that the spontaneous origin of life and the evolution of species are impossible?

No. This reasoning must be flawed because it implies that life itself is impossible, since every organism builds complex molecules from simple ones. The error lies in the misapplication of the second law of thermodynamics, which applies to closed systems that exchange neither energy nor matter with anything outside them. The Earth is not a closed system, because it receives useful energy from the sun. Order can arise by chance, at least locally, for a while, and to a limited extent. Under the right conditions, it can bootstrap itself to a higher level of organization, as in the case of natural selection at the level of individual nucleic acid molecules and genes. Organisms use energy to build order from disorder. They decrease entropy locally by increasing the entropy in their environment. Evolution, which produces new types of organisms, is an entropy pump also.

Creationists can do themselves and the rest of us a favor by learning some real science and switching to religions less at odds with science. There are plenty from which to choose. People change religions every day, just as scientists discard theories when the evidence is against them.

This week's news:

The University of Maine's professor Hooke has compared the earth moved for residential subdivisions, roads, and mining with the natural processes. He concludes that humans move more of the planet around, about 45 billion tons a year, than rivers, glaciers, oceans, or wind. Hooke calculated that rivers move about 39 billion tons of sediment a year; rivers dump about 24 billion tons in the oceans annually. That's partly attributable to people, due to human-induced soil erosion. Hooke estimated that over the last 5,000 years of human history, people have moved enough earth to build a mountain range 13,000 feet high, 25 miles wide, and 62 miles long. At current rates, that could double in the next 100 years. Resulting environmental damage includes acid mine drainage and river sedimentation. "The Dust Bowl of the 1930s resulted in part from clearing the land," followed by severe drought, he remarked. (spacedaily.com)

Recommended reading:

Wilson, David B. (ed.), *Did the Devil Make Darwin Do It? Modern Perspectives on the Creation-Evolution Controversy.* Ames: Iowa State University Press, 1983.

21
THE EVOLUTION OF SPECIES
(August 1, 2004)

We're reviewing evolution in order to understand the normal place of extinctions in nature, to comprehend ecology, and to examine the biological evolution of human behavior. It would be foolish to think that we can arrive at a formula for survival without this basic information under our belts.

The first single-celled organisms (*prokaryotes*) had no nuclei or intracellular *organelles* (little structures inside of cells that perform specialized tasks like making energy-rich compounds or synthesizing cellular proteins). Eventually, some prokaryotes developed the ability to cleave oxygen from molecules in their aquatic environment and release it into the atmosphere. Gradually the atmospheric concentration of oxygen increased to the point where Earth had an oxidizing rather than a reducing atmosphere. This change took place slowly enough so organisms could evolve to adapt. Eventually, two types of metabolic processes came to predominate. The oxidizing type added oxygen to carbon and hydrogen in energy-rich compounds in order to use the released energy for growth and reproduction. This process, called *respiration*, took oxygen from the atmosphere and released carbon dioxide and water into the surroundings. The reducing type used the energy in sunlight to separate the oxygen and carbon in carbon dioxide via *photosynthesis*, built the carbon into energy-rich compounds, and released oxygen back into the atmosphere. These two processes complemented each other, maintaining life by recycling carbon and using sunlight to power the whole shebang. This elegant carbon cycle was stable for about 3–3.5 billion years until less than 50,000 years ago when we started screwing it up.

Around 2.7 billion years ago, *eukaryotes* branched off from the prokaryotes. They sequestered their DNA in membrane-enclosed nuclei. The prokaryotes continued to evolve and their total biomass may still exceed that of the eukaryotes today.

The eukaryotes developed chromosomes and several specialized organelles. Some formed true multicellular organisms with a variety of specialized cells and tissues, and complex reproductive and developmental cycles. They developed sexual reproduction, shuffling their genes in a large variety of combinations among their offspring. These refinements allowed the evolution of organisms with a larger variety of specialized tissues made up of many kinds of cells, leading to complex anatomies. Their biochemistries and physiologies became more complex, supporting the specialized roles and coordinating the activities of their different tissues. In the course of further evolution, they diverged into different groups of organisms that created and filled a wider variety of niches.

At first, all life lived in water. The fish evolved over 500 million years ago, and amphibians evolved from fresh-water fish around 370 million years ago, living lives partly in the water and partly on land. Reptiles evolved from amphibians 310 million years ago, from which evolved birds and mammals 225 million year ago. The first land plants evolved around 450 million years ago, and the first flowering plants over 200 million years ago.

Many life forms became extinct because they couldn't reproduce their genes as successfully as others could. New niches evolved along with the new life forms, and life became much more diverse and interdependent. Now it's common for a species to be dependent on many other species for food, habitat, oxygen, or carbon dioxide, defense from predators, and even reproduction (e.g., pollination of flowering plants). Changes in one species can threaten the existence of numerous other species, which must evolve to keep up or go extinct.

Some land organisms returned to the sea or took to the air. Land organisms spread across the continents and the continents broke up. Their fragments drifted apart and attached to other continents, taking species with them. These shifting *tectonic plates* pushed up mountains like the Himalayas when they collided, adding new barriers between members of species. These processes not only moved species around, isolating members of species from each other and producing new species, but also brought about global geological and climatic changes, including the Ice Ages.

This week's news:

WASHINGTON: The US will withhold $34 million from the UN Population Fund, about 10% of the agency's budget, for the third year in a row because the agency cooperates with activities in China that promote abortion. The agency says it does not condone abortion and advocates voluntary family planning. US conservatives and some religious groups have condemned China's policies for regulating family size and coercing women to have abortions. The UN agency responds that their work has greatly reduced coercive family planning. Their program has helped reduce female sterilizations by 16% since 1998, and increased the use of contraceptives to 90%. The ratio of abortions to live births in counties where the program operates is now below that in the United States. (*New York Times*)

Two new studies report Earth's oceans have absorbed half of the CO_2 emitted by humans since the beginning of the industrial revolution. Atmospheric CO_2 has increased from 280 parts per million (ppm) in 1800 to 380 ppm now. But CO_2 would be over 435 ppm without the oceans' absorption. Consequently, the oceans are becoming acidic, to the possible detriment of marine life. When CO_2 dissolves in seawater, it forms carbonic acid, which dissolves shells and skeletons of marine life. (NewScientist)

MAURITANIA: Locusts from North Africa are swarming to Chad, Mali, Mauritania, Niger, and Senegal. Major farming areas are being invaded. The rural sector produces 20% of the gross national product and accounts for 60% of jobs. (Al Jazeera)

BRASILIA: Scientists report that Brazil, a signatory to the Kyoto Protocol, produces around 300 million tonnes of CO_2-equivalent per year, 200 million of which comes from logging and burning the world's largest tropical forest. That makes Brazil one of the world's top 10 greenhouse gas producers. However, Brazil's are only 3% of the world's emissions. Brazil maintains that wealthy nations need to make the greatest cuts in emissions. The US is not a signatory to the Kyoto Protocol, and generates the highest amount of CO_2, 25% of the world total. (planetark.com)

Recommended reading:

Patterson, Colin, *Evolution*. Ithaca, N.Y.: Comstock Pub. Associates, 1999.

22

PRE-HOMINID EVOLUTION

(August 8, 2004)

As our prehuman ancestors might say if they could see us today, "Where did we go wrong?" Of course, being prehuman, they didn't have the cognitive equipment to ponder such questions. We humans have that equipment, which turns out to be both the cause of, and the potential solution to, our problems. At what point in our evolution did we become separated from the checks and balances of the rest of nature? Was it in the early years of our species when we threw rocks at our prey, or later when we developed tools and language, or still later with the advent of agriculture?

The first mammals appeared when dinosaurs still ruled the earth. The early primates are of particular interest because they were our ancestors. These small creatures lived an arboreal life, in the canopies of jungles. They and their descendants developed grasping fingers and an unprecedented range of movement in their shoulder joints, and like the monkeys of today, they could swing quickly from tree to tree using their powerful arm and chest muscles.

Around five million years ago, some of these primates' descendants started coming out of the trees onto the ground, until eventually, several part-time ground-dwelling species arose. Some were the prehominids, the genus *Australopithecus*, ancestors of the genus *Homo*. They were ape-like creatures who retained the prehensile fingers and universal shoulder joints of monkeys. Their brains were bigger than their ancestors' brains but smaller than ours. Their anatomies changed, adapting to their part-time upright posture.

The Earth's climate was cooling substantially, and much jungle and forest was changing to grassy savanna. The pre-hominids first lived at the margins of the shrinking forests and then moved into the savanna along with other evolving species. There was less edible vegetation, particularly fruit, in their new environment. Rather than developing new mechanisms to digest the new vegetation, they started eating herbivores. Like most predators, they preferred prey roughly their own size, because catching sufficient smaller prey is more time-consuming and they would incur too many injuries chasing prey that are too large. They hunted in packs, cooperating in running down individual animals they could separate from the herd.

Several Australopithecine species arose, spreading from Africa into Eurasia. Their main hunting method changed to throwing rocks, which allowed them to keep their distance and minimize the chance of injury to themselves. Their strong arms, dexterous hands, and mobile shoulders were essential to successful rock throwing. So was their enlarging frontal cortex, the brain area involved in motor (movement) control. Rock throwing involved more precise motor control.

These animals probably had no cognitive activity resembling human consciousness. They were just one more species gradually evolving to meet the challenges of a world whose climate was beginning to fluctuate considerably. Their use of tools was no more advanced than the tool use of our closest living relatives the chimpanzees, although their brains were slightly larger than those of chimps. They had no language, although they undoubtedly used a few calls with specific meanings, a behavior common to many birds and mammals.

A cognitive revolution has permitted many humans to develop an attitude of exceptionalism in which we hold ourselves above the rules of nature. This cognitive revolution did not begin with the genus *Australopithecus*. It began with some of their descendants, of the genus *Homo*.

This week's news:

CAMEROON: A new type of malaria-carrying mosquito may complicate the battle against the disease. Malaria kills 3000 people/day and comes at an economic cost to Africa of $12 billion/year. In Cameroon, it constitutes 35% to 40% of hospital deaths and 40% of deaths among children zero to five years of age. Public health authorities urge the use of insecticide-treated mosquito nets, but the average household cannot afford their cost: $6.50. (Al Jazeera)

Recommended reading:

Tudge, Colin, *The Time Before History: 5 Million Years of Human Impact*. New York: Scribner, 1996.

23

HUMAN EVOLUTION

(August 15, 2004)

We're the only existing species of the genus *Homo*, but there were other hominids before us. *Homo habilis*, a contemporary of *Australopithecus*, began chipping rocks to shape them into primitive tools. *Homo erectus* were hunter-gatherers, and they invented spears. These early hominid species used fire to cook food, increasing the varieties of plants they could eat. Their larynxes moved down in their throats, so their utterances sounded more like the utterances of modern humans. Spear throwing demanded more skill with their hands and arms, and the members of hunting bands probably depended more on calls to signal each other. Their frontal lobes, the part of the brain used for motor (movement) control, planning, and later on for speech, were larger than their predecessors', yet smaller than ours.

Homo erectus was thin, without much body hair, black-skinned, with a human nose—all useful for thermoregulation on the hot savanna. They were fully committed to walking and running on two legs (*bipedalism*): tall, with a narrow pelvis. The modified pelvis and enlarged brains forced them to give birth earlier, so they had to care for their young longer. Their family groups cared for the mothers and infants, and for their sick and elderly, and developed more complex social structures. Gradually hominids displaced the prehominids.

The first modern humans, *Homo neanderthalensis*, evolved in Europe during an ice age. They were short and stocky, an adaptation to the cold, and they wore clothes. They were such expert hunters and such good problem solvers they probably had at least a crude protolanguage. They may have developed religion, because they buried their dead with flowers. Their brains were actually somewhat larger than ours were, and they were far stronger and more robust.

We *Homo sapiens sapiens* came later, around 120,000 years ago, with much more advanced tool making, and the first true language and art. We evolved in Africa, spreading to Eurasia. The two human species coexisted for some time, but eventually the Neanderthals were crowded out.

We acquired abstract reasoning. This led to the capacity to generalize about what worked and didn't work, and to accumulate this knowledge over the generations through an oral and pictorial tradition.

We could hunt a prey species until it was so scarce that it was more advantageous for us to switch prey, yet still taking down members of the original species whenever they were encountered. In this way, we could bring about the extinction of a prey species. Any other plants or animals that depended on these prey species for food, as a means of spreading seeds, pollinating, or maintaining ecological balances, were endangered. We were also better able to resist becoming prey ourselves, and we started killing off species that threatened our own survival.

Because of this versatility, we proliferated beyond the limits normally imposed on other species. All this progress meant a greater margin of safety in hard times, like drought and ice ages, which forced other species to move or die. At least until we started bringing on our own hard times.

We spread across the contiguous landmasses, modifying ecologies as we went. Soon after our arrival in a new area, large animals (*megafauna*) would start going extinct, resulting in the loss of such species as sabertooths and woolly mammoths. These ecological changes were so drastic that they showed up as significant extinction events in the fossil record. This started happening well before the development of farming.

We have rarely been good stewards of Earth's life forms. We have rarely realized the extent of our impact. After all, most people living today don't understand that we pose a dire threat to life on this planet. What was different about human beings that brought about this fundamental change in the relationship between our species and the world, as compared with other species?

The simple answer is that we broke the negative feedback loops that normally stabilize the balance among species. Normally, in the absence of human interference, species go extinct relatively slowly. Slow changes of climate and the evolutionary drift of gene pools allow evolution to maintain biodiversity. But evolution couldn't keep up with the losses we inflicted. Our population expanded, accelerating the damage we were doing.

It's hard to say when some of us began to realize we were adversely affecting the balance of nature.

This week's news:

WALIKALE, CONGO: Walikale is a war zone where fighters compete for control of valuable resources. Three million people have died, mostly from hunger and disease. The region, about the size of Rwanda, is rich in coltan, a mineral used in electronics. The price of coltan has now collapsed, but Walikale is now the site of mining for cassiterite, which contains tin. A sudden price increase has ignited new competition in an area where peasants in rags dig gold and diamonds by hand using hammers. Nearly all the 15,000 residents fled during the most recent battles. (planetark.cóm)

SORRENTO: The International Whaling Commission decided not to lift the ban on whaling. (MSNBC)

LISBON: Portugal's forestry department reports that almost 25% of Portugal's forests were lost to fires over the past 10 years. The country loses 2% of its forests per year. The worst year was 2003: more than 1 million acres burned. This year, fires have consumed 250,000 acres to date. (terradaily.com)

Recommended reading:

McKie, Robin, *Dawn of Man: The story of Human Evolution*. New York: Dorling Kindersley Pub., 2000.

24
RACE
(August 22, 2004)

Everyone agrees what race is. You can tell a person's race by skin color, facial features, and body type. We call people whose ancestors were all the same race pureblooded, and people whose ancestors were more than one race are mixed blood. Different races also have other differences, including intelligence, athletic and musical ability, color and food preferences, work ethic, aggression, and sexuality. We can judge people by their appearances for these reasons.

Surprise. Every one of the preceding statements is false, although there are still many people today who subscribe to them. It turns out that there is no genetic evidence for separate, distinct races. Since most people don't realize this (I didn't until I read up on the scientific evidence), the concept of race continues to plague us all. *Racism* is the belief that discrimination based on race is justified because of genetically determined differences among races. Obviously, if there's no genetic basis for racial classification, racism is scientifically unsupportable. Racism, like nationalism and a sense of religious superiority, is harmful because it pits people against each other for no justifiable reason. Unwelcome scientific findings are countered with anecdotes and myths. To make matters worse, racism is a tenacious meme (unit of cultural inheritance), because it appears to meet important human needs—even though it doesn't.

Anthropometric measurements, based on visible features like skin color or other physical features, reveal continuous gradations that make "racial" classifications arbitrary—which is why there were so many different, hotly contested classification schemes. Even more importantly, genetic studies that compared the nucleotide sequences in the genes for human proteins have shown there is no genetic basis for classifying people into races.

Genetic homogeneity ("racial purity") in a population would require intensive inbreeding, leading to a high incidence of genetic diseases. Outbreeding, on the other hand, produces what geneticists call *hybrid vigor*, a healthier population. This difference stems from the fact that genetic diseases or defects are more pronounced in individuals with the same defective allele (variety of a gene) inherited from both parents. The odds of this happening are higher in inbred populations than in outbred ones. As a consequence, individuals who are the products of inbreeding are selected against. Classical eugenics is thus guaranteed to fail.

It turns out that the most important mate-selection factor for humans is proximity: most people breed with individuals who live nearby. The likelihood of couples pairing decreases with distance. Therefore, *genetic similarity*, the similarity of DNA segments cod-ing specific proteins, between people declines with distance. Put differently, the prevalence of an allele changes smoothly with geographical location, with the largest differences occurring because of major obstacles such as oceans or mountain ranges. Migrations and modern travel further smooth the geographical distributions of alleles.

Some genetic traits are subject to geographical variation because they're favored by environmental factors such as climate, diseases, and available foods. For example, dark skin and tall, slender physiques are selected for in very hot climates because they protect against intense sunlight and assist body cooling, respectively. Light skin and squat physiques are selected for in cold climates because they encourage formation of an essential vitamin missing in the diet by ultraviolet light penetrating the skin, and assist heat conservation. Most genetic traits, including behavioral ones, are unaffected by differences in environment: they're *selectively neutral* with regard to environmental differences.

Behavioral traits are determined both by genetics and by cultural influences. It's probably most accurate to describe genetic influences (aside from a few spectacular ones that cause brain damage) as determining behavioral *predispositions*, e.g., for various mental competencies like mathematics and music as well as abnormalities such as schizophrenia and depression. Such predispositions, combined with environmental factors such as parental and social influences (especially experience during critical periods for acquisition of cognitive skills), diet and exposure to drugs and pollutants, determine an individual's personality, character and abilities—independent of "race," but dependent to some extent on ethnicity. For example, different cultures may differ profoundly with regard to conflict resolution, mutual support, violence, morality, and religiosity.

However, if a society treats different "races" differently, the expectation of behavioral differences among "races" becomes a self-fulfilling prophesy, which then perpetuates racism. This positive feedback loop is easy to understand, but hard to remedy.

This week's news:

The Department of Homeland Security is proposing to allow its agencies to skip environmental impact studies, as required by the National Environmental Policy Act, for purposes of efficiency. The plan would also permit agencies to conceal information for purposes of "national security." Environmental groups vehemently oppose the plan. NEPA mandates evaluation and disclosure of environmental and public health risks of projects, as well as opportunity for the public to comment on proposals. The plan would permit logging of as much as 70 acres of live trees and salvage logging of 250 acres or less on DHS-controlled lands without any review. The Border Patrol could build roads in national forests without public input if DHS rules them classified. There would an exclusion from review for pesticides on all properties under DHS control. Homeland Security agencies would be exempt from review requirements for operations in waterways and wetlands in their jurisdiction. Agencies would not need reviews of waste disposal. DHS could also install classified natural-gas pipelines, without the knowledge of people nearby.

In a separate announcement, the Nuclear Regulatory Commission will no longer require the country's nuclear power plants to disclose security errors. (*Grist Magazine*)

COLORADO: If approved by a statewide referendum, the Colorado Renewable Energy Initiative will be the first state vote on renewable energy and will require Colorado's principal utilities to generate one tenth of their power from renewables—wind, solar, biomass, geothermal, or hydroelectric power—by 2015. This would be an increase from the current 2%, equivalent to removing pollution from 600,000 cars. (*Colorado Daily*)

COPENHAGEN: Denmark is considering installing 100,000 megawatts of wind power turbines in the Baltic and North Seas, four times the total wind power in Europe. Denmark obtained 21.1% of its power from wind in June 2004, compared to 6% in Germany and 0.5 percent worldwide. Denmark sells 40–50% of wind turbines. Each megawatt-hour of power produced by wind energy eliminates 0.8–0.9 tonnes/year of CO_2 emissions from burning coal or oil. There are plans in the U.K. for 7,000 MW of offshore wind farms, and in Germany for 2,000 MW or more. (terradaily.com)

NEWPORT, OR: An enormous volume of water with very low oxygen and rich with nutrients has been carried south from the sub-Arctic region by a shift in the California Current and raised to the surface on the Continental Shelf. Researchers at Oregon State surmise it reflects changes in ocean circulation. There are more than 30 human-caused dead zones around the world caused by fertilizer washing down rivers, which stimulate blooms of phytoplankton. When these microscopic flora die, they decompose, consuming oxygen and causing sea life to suffocate. Natural dead zones in open water, like ones off the coasts of Peru and South Africa, are uncommon and poorly understood. (MSNBC)

India will be the most highly populated country by 2050, going from 1.1 to more than 1.6 billion, according to Washington-based Population Reference Bureau (PRB). China will grow from 1.3 to more than 1.4 billion. Japan, the world's second richest country, will shrink from 127 to 101 million. Afghanistan will jump from 28.5 to 81.9 million, and Iraq's population will increase from 25.9 million people 57.9 million. Pakistan will go from 159 million people today to 295 million in 2050. Indonesia will remain the fourth most populous nation in the world, after the US. It will grow from 218.7 to 308.5 million. Asia will have nearly 5.4 billion people. The world population will increase from 6.4 billion to 9.3 billion people. (Al Jazeera)

Recommended reading:

Wilson, Edward O., *On Human Nature*. Cambridge, MA: Harvard University Press, 1978.

25

BIOLOGICAL EVOLUTION OF BEHAVIOR

(August 29, 2004)

One long-standing argument about human behavior is the extent to which it's determined by genetics or by experience. Both obviously contribute. All members of a species display behaviors different from other species, for example, the mating calls of different frog species. There are variations of calls within a species, including regional dialects, and differences among individuals in the same pond. Females respond differentially to the males' calls, ignoring calls of other species, preferring their local dialects, and preferring some individuals' calls over others. Analogous differences exist for courtship, nest-building, and other aspects of behavior across the animal kingdom, including humans. It's safe to say the major commonalities are inherited, and the minor idiosyncrasies are at least partly learned or original.

The course of evolution is determined by shifts in the proportions of alleles in species' gene pools, which is influenced by the contributions of those alleles to successful reproduction. That success is partly determined by behavior, e.g., the ability to attract fit mates. Those alleles that contribute to behavior that enhances propagation are selected for.

Aspects of behavior directly related to survival such as eating and procreating have strong genetic components. Parrots, commonly raised by humans from the time they hatch, don't have to be taught how to sharpen their beaks, open walnuts, or split bones to get at their marrow. They can play, exhibit dominance behavior, court, and mate.

Farm animals and pets on the other hand have been bred selectively so they depend on us for survival. Most would survive poorly in the wild, partly because we've bred them for domesticated behavior.

An important aspect of behavior is competition. Just as there's competition among species, there's competition within species for food, territory, and mates. This competition occurs between groups, such as bands of monkeys competing for the same resources, and between individuals, such as moose males competing for females. Although the groups and individuals may not see it this way, they are competing for the survival of their genetic makeups (*genotypes*). If they don't compete successfully, their genotypes don't thrive.

By studying species closely related to humans, researchers are accumulating clues about those aspects of human behavior that are shared, and therefore mostly inherited, as opposed to those that are learned, i.e., cultural. For example, chimps raised in a zoo have dominance hierarchies like those in the wild, and they can form shifting "political alliances" in order to gang up on disruptive individuals. They're capable of lethal violence in their struggles for power. This suggests our own capacities for power struggles, political behavior, violence, and war are partly innate, although there's no doubt an added cultural component in such obvious forms as formal training in these skills.

There are other ways to separate the effects of nature versus nurture, such as studies of identical twins raised in different families or cultures (their genes are the same but their environments are different), or unrelated children raised in the same family. (Their genes are different and their shared environment is similar). There are also a few studies of children raised in extreme isolation or even by animals. We won't catalog the results of those studies here, but we'll be relying on them when we consider how to shape our own behavior in order to achieve sustainability.

This week's news:

Heat waves kill mostly children and the elderly. The 2003 Paris heat wave killed 7,000–15,000. Some 700 people died in the 1995 Chicago heat wave. Computer simulations now show that future heat waves in these areas will be worse, more common, and longer. Heat waves usually are caused by domes of high pressure. Future climate changes will produce major high-pressure domes over North America and Europe. Tropical ocean temperature and current changes will produce more, and worse, stationary high-pressure systems. In the computer model, the Great Basin, Rocky Mountains, and Southern United States were hardest hit in North America, and the Mediterranean Basin will be affected most in Europe.

Populations in Europe have managed in summer temperatures ranging from 13.5 to 24.1°C, and should adjust to warming predicted for the next 50 years. If humans can cool themselves by sweating and maintain their fluids, they can tolerate much higher temperatures briefly. However, when someone is exposed to heat for a long time, the heart has to pump more blood to the skin for thermoregulation. In the elderly, the additional strain on the heart can be too much. The young heat up faster, and may be at the mercy of others for fluid replacement. (spacedaily.com)

SHANGHAI: Nearly half a million people have been evacuated and boats have returned to port in eastern China as Typhoon Aere approaches. Torrential rains, gale-force winds, and landslides are anticipated. The typhoon is expected to be weaker than Typhoon Rananim earlier this year, which killed 164 and injured 1,800. Rananim was the strongest typhoon to hit eastern China in 50 years, destroying 42,400 homes and wiping out large areas of farmland. However, meteorologists fear Aere may combine with Typhoon Chaba, which has passed the Mariana Islands and Guam in the north Pacific. Aere hit Taiwan with powerful winds and torrential rain, causing landslides and seriously impairing transportation. At least seven people were feared lost. (terradaily.com)

BRUSSELS and DUBAI: According to the sanitized version of Cheney's 2001 secret energy report the US will go from importing 55% of its oil to 66% in 2020. The US, with 5% of the world's population, consumes 26% of the world's oil. At projected consumption rates, all proven oil reserves in the world will be gone by 2054. Long before then we'll reach "peak oil": The supply curve will be headed down while the price curve will be climbing. We may have reached peak oil already. Production from the largest oilfields is declining; there's been no major oil discovery for the past 18 months. North Sea oilfields have already peaked, as has Yukos in Russia. Saudi Arabia can't increase production. Iraq's oil sector will not produce large yields in the next 10 years. Even under United Nations sanctions, Iraq exported more than now. On the demand side, China is importing 40% more oil this year than in 2003. Their production grows more slowly than their demand. Oil imports in India will increase 11% this year. Indonesia will become a net importer of oil next year. (*Asia Times*)

Recommended reading:

Dawkins, Richard, *The Selfish Gene*. Oxford: Oxford University Press, 1989.

The Human Mind and Cultural Evolution

26

THE BIG BANG OF HUMAN CULTURE

(September 5, 2004)

About 40–50,000 years ago, humans indistinguishable from both our ancestors of 120,000 years ago and humans of today suddenly embarked on a series of very rapid behavioral changes. Taken together, these changes amounted to the birth of human culture. We went "from a relatively rare large mammal to something more like a geologic force" (Klein and Edgar, 2002).

At archeological sites in Africa judged to be 30–40,000 years old, archeologists have found multiple hearths, suggestive of nuclear family cooking arrangements. They also found fine obsidian tools and ostrich-shell beads with holes in them, made with those tools. Strings of similar beads are exchanged today by bands of San (!Kung) people in Africa as a means of committing to reciprocal assistance. Making these beads is a painstaking process, and the labor involved implies they had a high value attached to them—which implies their owners had some important use for them.

At a 29,000-year-old site in Russia, 10,000 bone beads, which must have taken over 10,000 hours to make, were found in the grave of two children. This is a clear mark of social privilege, indicating a stratified society. It's likely that the beads were buried with the dead as part of a religious ritual. Sites that are 50–60,000 years old never show signs of such sophisticated use of tools, capability for symbolic thought, social complexity, and religion.

The first permanent dwellings date as far back.

The first cave paintings, accurately portraying many now-extinct animal species, were rendered by artists in France around 32,000 years ago. Working by firelight, without models, these artists had the ability to form mental images of animals and reproduce them in works of great beauty. Around the same time, cave artists carved ivory figurines, including a lion-headed man, in Germany, and a slab of rock was sculpted into a 7 cm tall statuette of a woman in Austria.

Abstract thought, mental imagery, art, religion, construction of sophisticated shelters, fabrication of refined tools and jewelry, and development of complex social interactions: All of this came about in the absence of significant anatomical changes, even of brain size.

What happened? There are two main options: genetic changes could have caused modifications not detectable in fossils, for example in the wiring of the brain; or some environmental challenge could have forced a radical change of behavior permitted by existing brain architecture.

No climatic change or other environmental challenge occurred at the time of this cultural explosion. No technological advance preceded it.

The new forms of behavior all required areas of frontal cortex involved in planning, organizing, and carrying out complex motor activity. From the shapes of their skulls, we know that the frontal cortex was larger in hominids than in prehominids. Most of the previous hominid behavioral changes were clearly consequences of changes in brain size or shape, but no additional changes are observable in the fossils of 50,000 years ago. On the other hand, changes in the *wiring* of the brain would be undetectable except as manifested in changes of behavior. It's reasonable to suspect that the cultural explosion was the result of a mutation that caused such a wiring change.

This week's news:

GREENBELT, MD: An article in the August 20 issue of *Science* reports on a study that combined the results of 12 different computer models to identify locations where soil moisture may strongly affect rainfall during summertime. These include "hot spots" in the central plains of North America, the Sahel, equatorial Africa, and India. Less intense hot spots show up in South America, central Asia, and China. In wet zones, sunlight and cloud cover have more influence on evaporation than soil moisture. In dry zones, limited water means evaporation is too limited to have much effect on the atmosphere. Hot spots occur in regions between wet and dry zones. Understanding soil moisture levels and their connection to precipitation may improve seasonal rainfall forecasting. (spacedaily.com)

DHAKA: The worst monsoon flooding in 6 years has covered 60% of Bangladesh. The death toll has risen to 628. Diarrhea, dysentery, and typhoid have struck more than 100,000 people. The country needs food for 20 million people over the next 5 months. The U.N. world food program is supplying food to nearly 2 million people and plans emergency aid for millions more. The floods have disrupted Bangladesh's $4 billion textile industry, which accounts for about 80% of the country's exports. Overall losses could total $6.7 billion. More flooding is expected because fresh rains will be unable to drain into the waterlogged ground. (*The Guardian*)

Global warming will be especially fast if aerosol "cooling" has hidden a higher climate sensitivity than is generally assumed. If true, climate change in the 21st century may exceed worst estimates. The U.K. Environment Minister said the U.K. would press for action at the highest level. (*The Scotsman*)

WASHINGTON, DC: A report sent by the administration to Congress this week admitted that human activities helped cause global warming from 1950 to 1999, but denied any earlier human contribution to warming. Bush withdrew the United States from the U.N. climate pact, claiming it would penalize the U.S. more than developing nations like China. The White House has instead promoted a voluntary program for U.S. companies to reduce their greenhouse emissions. (MSNBC)

GENEVA: More than 40% of the world's population has no sanitation, and more than 1 billion drink unsafe water, the U.N. advised. "The growing disparity between the haves and the have-nots in terms of access to basic services is killing around 4000 children every day and underlies many more of the 10 million child deaths every year. We have to act now to close this gap or the death toll will certainly rise." (terradaily.com)

STOCKHOLM: At the annual Stockholm Water Symposium Tushaar Shah, head of the International Water Management Institute's groundwater station, based in Gujarat, warned the world is on the brink of a water crisis. Native farmers are now forced to pump water from 1000 feet underground. They're rapidly depleting the aquifer, desertifying once-fertile fields. This is happening throughout Asia and will lead to major famines. (*NewScientist*)

Recommended reading:

Lewin, Roger, *The Origin of Modern Humans*. New York: W. H. Freeman, 1993.

Klein, Richard G., and Edgar, Blake, *The Dawn of Human Culture*. New York: Wiley, 2002.

27

THE ORIGIN OF LANGUAGE

(September 12, 2004)

Language co-evolved with several other cognitive elements that together form the basis for human culture. But what initiated this cultural explosion, after millennia of slow changes? Was it language itself? Here we must enter the realm of scientific speculation, because language leaves no fossils or artifacts and writing is a more recent development. The purpose of this fact-based speculation is to provide a foundation for understanding the explosive origin of culture.

Chimpanzees make about three dozen sounds, each of which has a specific meaning by itself. Humans make roughly the same number of sounds, none of which has any meaning until we string them together into words, words into sentences and sentences into more complex communications. How could our *generative grammar*, as it's called, have evolved?

The basic structure of calls for bird species is inherited. The details are learned during a critical period, like the critical period for language in humans. Is there a possible evolutionary path from a few specific calls to full language (not that we're descended from birds—they, like mammals, are descended from reptiles)?

Social bonding, cooperation, hierarchy, and division of labor are common in primates. Hunting in teams by hominids could have selected for development of command-, acknowledgment-, and refusal-sounds (words). As utterances become more rapid and connected, commonly associated strings of monosyllabic words would become polysyllabic words in their own right. From there it's not a great leap to develop different types of words, parts of speech: action-words, thing-words, and modifiers, without grammar or syntax, leading to a pidgin-like protolanguage. Gradual genetic changes in brain circuitry shaped by natural selection could accomplish all of this, just as natural selection led to parallel changes in our vocal apparatus that enabled production of modern human speech sounds.

As group tasks become more complex, neural representation and communication of information have to become more sophisticated. Eventually grammar and syntax would be needed to advance further, allowing us to generate a very large number of different sentences from a reasonable number of words. But wouldn't this require a major genetic change in a single step, with no possible sequence of small changes that could be shaped by natural selection? Yes and no, maybe. In 2001, a team of geneticists identified a single gene that is probably involved in the developmental process that culminates in speech and language. People with a defect in this gene have great difficulty recognizing speech sounds, learning grammatical rules, and understanding sentences. They're not necessarily impaired in other ways, and can have normal non-verbal intelligence. This may not be the gene that made true language possible, but it illustrates the difference a single gene can make.

Of course, I could be wrong. That's OK in science. We correct our mistakes as we go along.

This week's news:

BEIJING: At least 79 people were killed and 74 were missing in some of the worst storms southwest China has seen in many years. Thirteen inches of rain fell in some areas. More than 450,000 people have been evacuated and 127,000 homes destroyed or damaged, 400 bridges collapsed and at least 440 miles of roads and thousands of acres of farmland destroyed. Parts of Dazhou were under 3 feet of water. (terradaily.com)

Hurricane Frances hit Florida with 145-kph winds. Blinding rain whipped sand into the air and enormous waves crashed ashore on the Atlantic side, where 2.5 million people were ordered to evacuate. Two million people are without electric power. Insured losses may reach $2–10 billion. Most stores were closed and few airports continued operation. (*Al Jazeera*)

VANCOUVER BC: Mike Monea, Executive Director of the Petroleum Technology Research Centre (PTRC) reported at the international Greenhouse Gas Control Technologies Conference that geological conditions in an oil field are appropriate for long-term storage of carbon dioxide. Researchers performed a long-term risk assessment, finished geological and seismic studies, compared reservoir modeling against observations, and studied the chemistry of the reservoir. The oil field has stored an estimated five million tonnes of CO_2—equivalent to one year's emissions from 1 million cars. The CO_2 comes from a 325-kilometre pipeline from a coal-gasification plant in North Dakota. (spacedaily.com)

STANFORD CA: A group headed by Stanford University biologist Elizabeth A. Hadly has published results in *PloS Biology* showing the effect of climate change on genetic diversity. The team performed a genetic analysis of the montane vole and the northern pocket gopher. They collected DNA from living animals and from the teeth of fossilized specimens dating back to 3,000 years ago. They compared genetic variability in the Medieval Warm Period (850–1350 A.D.) and the Little Ice Age (1350–1950). Knowing the habitat requirements of the modern animals, the researchers anticipated, and observed, a drop in numbers for both species during the drier warm period and an increase during the wetter ice age. From previous research, they knew that when a population shrinks inbreeding increases and subsequent generations have more homogeneous genetics. This can jeopardize the species, because every animal inherits the same vulnerabilities. The gophers displayed a drop in genetic diversity but the voles didn't, even though their population shrank. The reason is that voles find mates over much larger areas, minimizing inbreeding. The expectation is that other species will also show genetic diversity changes in response to warming. (spacedaily.com)

Recommended reading:

Knight, Chris, et al. (eds.), *The Evolutionary Emergence of Language: Social Function and the Origins of Linguistic Form*. Cambridge: Cambridge University Press, 2000.

Lieberman, Philip, *Uniquely Human: The Evolution of Speech, Thought, and Selfless Behavior*. Cambridge, MA: Harvard University Press, 1991.

28

THE NATURE OF LANGUAGE

(September 19, 2004)

Human culture is our ability to exchange, manipulate, and store information. It allows us to pass ideas, both right and wrong, down though the generations. Language, the medium of human communication, is what makes human culture different from the cultures of other animals. Some of our most dangerous misconceptions concern language and culture. These fallacies have caused great harm, and they diminish our chances of survival.

Language is a biological phenomenon, not a cultural one. Individual languages, the forms that language takes, are cultural phenomena. They all share the same *universal grammar*, an innate faculty of our brains, using the same parts of speech. The sophistications of people's languages have nothing to do with the sophistications of their cultures. This is even true among the different classes of a culture, e.g., from the least to the most educated, the poorest to the wealthiest.

What we think is not determined by what language we speak. To quote the Pinker reference at the end of this chapter, "The way we see colors determines how we learn words for them, not vice versa." For example, contrary to popular belief, Eskimos don't have more words for snow than other people do. Many studies have shown that sophisticated thinking doesn't even require language, e.g., in babies that have no words yet, in adults who have never learned language, and in primates that have never had language. However, we do have internal representations—in our brains—of facts, things, actions, sensations, and memories. These representations are the currency of thought.

A vast number of words is possible with a small number of letters or phonemes (speech sounds), just as a number with only five digits can represent 100,000 values. Similarly, *combinatorial grammar*, in which words are combined in different ways, allows us to generate a vast number of meaningful sentences with a small number of words.

Children quickly pick up language by observation and practice. They not only acquire vocabulary but syntax as well, even though they can't describe the rules. They can even invent private languages. Learning new languages is much easier for children than adults, within a critical period. Undetected deafness is serious because of this critical period. Simple, inexpensive tests can be done at birth for language disorders or deafness but they rarely are, which is too bad because intervention at an early age is essential for successful treatment.

Pidgins are crude protolanguages invented by speakers of different languages who are thrown together and need to communicate. Remarkably, children raised with pidgin-speaking people develop full-blown languages called *creoles*. In Sandinista Nicaragua, when deaf children were first put in schools for the deaf, sign language pidgin spontaneously arose among the first children. It was replaced by a sign language creole in just a few years.

There are 4000–6000 languages today. Roughly 90% of the world's languages are now threatened with extinction. These languages provide valuable data for anthropologists and cognitive scientists, data that help us learn who we are as a species. This kind of information is important if we're to devise means of shaping our behavior into a constructive rather than a destructive mode. Scientists are working feverishly to document these languages before they're lost to us forever.

This week's news:

COPENHAGEN: An almost 2-mile long ice core has been obtained that contains annual layers of ice deposited over 123,000 years. The results will be reported in *Nature*. The new core record indicates Eemian-period temperatures over the polar regions were stable and warmer by 9°F. The transition from the Eemian to the most recent glacial period took several thousand years. The core also provided evidence of air temperatures jumping 9° in just five decades roughly 115,000 years ago, just prior to the Eemian–glacial transition. Air temperature also jumped about 18° in roughly 50 years about 10,500 years ago at the end of the last glacial period. These two jumps indicate dramatic temperature changes within a single human lifetime. In late June, the research team also recovered what appear to be plant remnants nearly two miles below the surface between the bottom of the glacial ice and the bedrock. The material, several million years old, looks like pine needles, bark, or blades of grass. This would indicate a temperate or even tropical climate. (spacedaily.com)

The actual number of endangered species could be 50% higher than indicated on the International Conservation Union's Red List, because estimates have failed to take into account many species, such as parasites and beetles, that depend for their survival on animals and plants that are recognized as threatened. For example, at least three louse species depend on the threatened red colobus monkey. Unfortunately, it won't be easy to add dependent species because in most cases species' affiliates aren't known. (*NewScientist*)

Recommended reading:

Pinker, Steven, *The Language Instinct*. New York: Harper Perennial, 1995.

29
LANGUAGE AND THE BRAIN
(September 26, 2004)

Although there's nothing supernatural about language, it's a miraculous feat of evolution.

We learn specific languages, but not the essence of language: how to generate new sentences. That's innate. More specifically, the capacity to learn a language develops while the child brain develops. There's a *critical period* for acquiring language, because the brain's wiring hasn't developed sufficiently before that period and because after that period the language circuits become less flexible. Without exposure to language during this period, the necessary circuits don't develop properly. Acquisition of language in such people is far more difficult, and not nearly as complete.

Some children are born with impaired ability to learn language. This runs in families, so it's genetic, probably a single dominant gene. Conversely, fluent language is common in people with many different kinds of mental impairments. Lesions of specific brain areas affect specific aspects of language without affecting other mental performance, which you wouldn't expect if language were just variations on, or combinations of, other mental capacities.

Speech circuits are normally in the *cortex* of the left *hemisphere*. The two hemispheres are the two halves of our *cerebrum*, the top part of our brain. The cortex is a layer of nervous tissue a few millimeters thick and one square meter in area on the surface of the cerebrum, intricately folded to fit in the skull. Sensory stimuli on the right side of the body are perceived in specific sensory regions of the left cortex: electrical stimulation of these regions produces sensation on the right side of the body. Movement on the right side of the body originates in other, motor, regions of the left cortex: electrical stimulation there produces muscle movement on the right side of the body. The left side of the body is represented in the right cortex. The same with the left and right halves of our visual fields: things on the left are seen by the right cortex and vice versa.

Broca discovered in the nineteenth century that *aphasia* (the inability to speak, usually resulting from a stroke or head injury) was always associated with damage to an area in the left cortex, subsequently named after him. Wernicke's area is essential for the understanding of speech. Electrical stimulation of these areas disrupts their normal function. Sign language and the interpretation of pictographs are similarly affected by the same stimuli, so they must use the same neural circuits.

Sometimes people have the communication pathways between the two hemispheres severed neurosurgically to control a type of intractable epilepsy. Because speech centers are only in the left hemisphere, these people can't report what image is flashed in the left part of their visual fields, and can't pick out what they saw using only their sense of touch with their right hands, but they can with their left hands. They can report what objects they're shown on the right side, and pick it out with their right hand, but not with their left. It's like there are two people in the two hemispheres, and only one can talk.

In bilingual people, cortical sites used to name objects in the two languages only partially overlap. Perhaps this is why the best time to learn a second language is while one is learning one's primary language. Extra real estate in the cortex can be staked out for the second language while the circuits are still maturing.

We have a cortical area adjacent to Broca's area in the frontal cortex, essential for sequencing motor actions. It's adjacent to another area involved in analyzing sequences of sensory events. This region expanded 2.5 million years ago, when hominid brains were changing substantially. Many uniquely human skills depend on complex sequencing: tool use, art, music, dance, games—and language. Could Broca's area be modified sequencing cortex?

This week's news:

The United Nations Population Fund reports that a quickly expanding global consumer class of 1.7 billion is the principal reason for increasing environmental crises. While consumers in wealthy nations still have the most environmental impact, exploding populations in less-developed nations are increasing environmental pressures as well. A decrease in population growth reduces environmental pressures less if consumption is increasing. The report notes, "Traditional practices that may have been ecologically viable when the population was small are becoming increasingly less viable for species and ecosystems as population grows and demands rise," For example, bush meat consumption in central Africa has reached the point where forest-dwelling animals are endangered. (*Financial Times*)

According to two articles in *Nature Medicine,* multiple-drug-resistant TB could easily reach pandemic proportions. The World Health Organization's attempts at control are not working. TB kills 2 million a year. Resistant strains are also increasingly common in Russia, Eastern Europe, South Africa, China, and Israel. (Al Jazeera)

Recommended reading:

Calvin, William H., and Ojemann, George A., *Conversations with Neil's Brain: The Neural Nature of Thought and Language.* Reading, MA: Addison-Wesley Pub. Co., 1994.

30

CULTURAL EVOLUTION

(October 3, 2004)

Once human culture appeared, largely as the result of language, it developed explosively. Was this the result of additional biological evolution? It's certainly possible, but not necessary. Culture can evolve on its own, constrained only by the limits of human cognition. How is this possible?

The phenomenon of human cognition has long been regarded as a bit spooky, which may help account for the many myths about non-material minds and spirits and where they come from, how they interact with the material world, and where they go when we die. Many of these myths were credible when they were formulated, but none of them has passed the test of experimental scrutiny. Those myths have impeded our understanding of what culture is, because culture's a product of cognition.

Culture is our repository of information, which can be shared and built upon among contemporaries and their descendants. This information includes anything that can be represented in our brains and communicated to others, including questions, facts, theories, artistic expression, myths, values, errors, lies, and languages. Different groups have different cultures and to the extent that those groups compete, their relative success will largely depend on how well their cultures serve them.

We can draw an analogy between human cultures and their informational components, and genes. It's not hard to see that cultural components will spread and persist to the extent that they help the cultures that depend on them to compete. That's a lot like natural selection. The analogy to genes is sufficiently useful to our understanding cultural evolution that cultural components have been given a name: *memes*.

As groups of people have spread around the world through expansion or migration of populations and in wars of conquest, their genes and memes have spread with them, mixing with the genes and memes of other populations. Just as variation and natural selection shape biological evolution, analogous processes shape cultural evolution. Human cultures come and go, as do human varieties, and in the process, memes and genes come into existence, redistribute in human populations, and are lost or conserved.

However, there are important differences between the two processes. For example, the origins, variations, and interactions of memes depend on human cognitive processes such as perception, logic, creativity, experimental inquiry, social rules, and even dogma and altered states, as well as prostheses of the mind such as microscopes, musical instruments, and farm implements. Cultural evolution is faster than biological evolution because if unimpeded it is continually accelerating due to a positive feedback loop: the development of new memes is facilitated by the interactions of existing memes. For example, we can design and synthesize new chemical compounds more quickly every year based on our experience with compounds we've already made. The more memes we have, the more possible interactions between them there are. Hello, information explosion.

Human culture is another entropy pump, making possible more information over time.

It's very likely that the original cultural explosion of 30–50,000 year ago was the beginning of this positive feedback process that continues today.

The important implication is that we can study the parallel biological and cultural evolutions of ethnic groups, ranging from individual tribes and clans all the way up to the entire human population. Modern researchers are applying methods of archeology, anthropology, ethology (comparative studies of behavior), molecular biology, and mathematical modeling to these studies.

Our objective, of course, is to answer questions humans have realized are crucial for as long as we've been able to ponder them: How different or similar are individuals or ethnic groups? Are some superior to others? What does superiority mean? How do nature and nurture shape us? What is normal behavior? What is justice? What is a healthy versus a sick society? What is our place in the world?

As we sink deeper into species extinctions, global climate changes, resource depletion, pollution, and overpopulation, these questions gain new importance. Our scientific knowledge is expanding explosively, but unfortunately so are our problems. We'll know which will win out within the lifetimes of many people alive today.

This week's news:

The collapse of the Larsen B ice shelf in 2002 has accelerated glacial flow in the Antarctic, which has warmed 2.5°F in just 50 years. Researchers have used satellite tracking to monitor several glaciers moving into the peninsula. They are flowing 2–8 times as fast as before the collapse. (BBC News)

SNOWMASS, CO: Rocky Mountain Institute (RMI) Monday released a plan to displace oil less expensively than buying it. The US can save half its oil usage through efficiency, then substitute competitive biomass fuels and natural gas for the rest. "Unlike previous proposals to force oil savings through government policy, our proposed transition beyond oil is led by business for profit," said RMI CEO Amory Lovins. "Our recommendations are market-based, innovation-driven without mandates, and designed to support, not distort, business logic. They're self-financing and would cause the federal deficit to go down, not up." According to their analysis, the US can save more oil by 2015 than it gets from the Persian Gulf; by 2025, use less oil than in 1970; by 2040, import no oil; and by 2050, use no oil at all. "Because saving and substituting oil costs less than buying it, our study finds a net savings of $70 billion a year," Lovins said. The report shows that ultra-light, ultra-strong materials like carbon fiber can halve vehicles' weight, increase safety, and boost efficiency to about 85 mpg for a midsize car or 66 mpg for a midsize SUV. To fight better and save money, the Pentagon—the world's largest oil buyer—would accelerate the market emergence of super-efficient land, sea, and air platforms. A more efficient and effective military can protect American citizens instead of foreign oil, while moving to eliminate oil as a source of conflict. (spacedaily.com)

CA regulators enacted the first limits on automobile CO_2 emissions in the US, mandating a 30% reduction for new cars and light trucks by 2016. Manufacturers have threatened to take the new regulation to court. The only way to lower CO_2 emissions is to improve fuel efficiency. By law, only the federal government is allowed to set gas mileage requirements, but CA officials say the greenhouse gas regulation is justified under the state's authority to regulate air quality. (*Washington Post*)

GONAIVES, HAITI: In the wake of hurricane Jeanne, at least 1,316 inhabitants have been confirmed dead, with 1,097 missing and 3,000 injured in floods. 200–250 thousand people live in the city. A UN peacekeeping force has been mobilized to secure food distribution sites, to keep them from being mobbed by starving Haitians. Haiti has little protection from flooding due to the most extensive deforestation in the Western Hemisphere. (terradaily.com)

Recommended reading:

Pfeiffer, John E., *The Creative Explosion: An Inquiry into the Origins of Art and Religion.* Ithaca, NY: Cornell University Press, 1985.

31

PALEOLITHIC HUNTER GATHERERS

(October 10, 2004)

The explosive developments 30–50,000 years ago of tool technology, art, language, and religion could only have happened if they had a strong survival value. This is generally the case in biological evolution. Without a significant survival value, there would be no selection pressure for the genetic changes responsible for evolution. In fact, gradual evolutionary change usually is the result of changing the frequencies of already present alleles in a population's gene pool as they become more or less valuable for survival. I'm convinced that the same was true for the dawn of cultural evolution, when survival of our species was by no means assured.

What challenges did our hominid ancestors face, and what existing behavioral repertoire did they have that could be shaped by selection, resulting in the birth of human culture? The answer surely lies in our pre-cultural lifestyle and ecological setting.

Ever since they left the forest for the savanna, our pre-hominid and hominid ancestors led a nomadic existence, traveling in bands. They rarely encountered other groups of the same or similar species, because of low population size and dispersal over large ranges. Life was precarious: they were prey as well as predators, and their weapons were crude, often requiring that they fight at dangerously close quarters. Our ancestors had to coordinate their behavior in group hunting and defense. When they succeeded in thinning out or eliminating larger predators, they soon began to deplete the supply of smaller prey. They had to keep on the move and learn the seasonal movements of other species.

In the European ice ages, hominids took shelter in caves, after evicting the giant cave bears that often occupied them. They needed better tools and weapons, including barbed spears for hunting and sharp needles with eyes for sewing hides to make warm clothes and shoes. Every member of the band was important, so they learned to care for injured or diseased members. They needed to cope with unanticipated disasters such as the loss of important members of the band or major fluctuations in food supply. They learned to plan ahead, and the remains of large food stores have been found from this era.

These pioneers had to deal with losses that gravely imperiled their survival, and the bonding and trust that were already strong in pre-hominids strengthened. As they tried everything they could to deal with challenges, they learned from experience and inevitably developed superstitions along with valid formulas for success. The individuals who were most successful at this associative process were most likely to reproduce and raise their young successfully. Bands that learned to test innovation cautiously were most likely to survive novel challenges. They acquired the cognitive tools for mental imagery, abstract thought, and language.

Although pre-hominids had some sense of self and may have had a notion of mortality, there's no sign they had concepts of a spirit separate from the body or of an afterlife. However, *Homo neanderthalensis* buried their dead with flowers and body ornaments. *Homo sapiens* refined this process about 30,000 years ago, burying their dead with valuable jewelry and tools needed in the afterlife, sometimes including materials obtained by commerce from distant regions. There's clear evidence that they performed rites and ceremonies, and had religion. Their new art, visible in abundance for the first time on cave walls and in portable forms, had multiple purposes including ornamentation, status indication, rites of initiation, and magic.

There's no question that all these cultural developments had survival value or were natural accompaniments to memes (units of cultural information) that did. They couldn't have developed without a need for them and a preexisting skill base.

This week's news:

PRINCETON, NJ: Increased tree farming and other land use changes have increased the release of volatile organic compounds from plants, contributing more to smog formation than in the past. (spacedaily.com)

WASHINGTON: Scientists have mapped the genome of a microscopic alga that absorbs CO_2. The plant is thought to convert as much CO_2 to oxygen as all the world's jungles. Algae produce up to 40% of marine organic carbon. (spacedaily.com)

PARIS: Insurance companies estimate at least $50 billion in economic losses and $20–35 billion in insured losses from this year's hurricanes. Local Florida insurers, such as Allstate and State Farm, are expected to be hardest hit. (terradaily.com)

ATLANTA: GE Energy announced that its 2,500th 1.5-megawatt machine has been installed in Sardinia. The first 1.5-megawatt wind turbine was installed in Germany in 1996, and has become the mainstay of the GE Energy wind turbine line. (spacedaily.com)

BERKELEY: University of California paleobiologists warn that the combination of climate change and human pressure could have resulted in extinction of two-thirds of all large mammals. They drew this conclusion from studies of late Pleistocene extinctions, in which humans had a major role in these extinctions, but climate change also contributed. Climate change is occurring more rapidly today than in the late Pleistocene, meaning the impact could be greater. These would be the last of the large animal populations. (spacedaily.com)

Recommended reading:

Rudgley, Richard, *The Lost Civilizations of the Stone Age*. New York: Free Press, 1999.

32

BIRTH OF THE ARTS

(October 17, 2004)

The first human visual art, at least in a form that could endure for tens of thousands of years, appeared suddenly and in great quantity during the cultural explosion that also ushered in sophisticated tools, clothing, music, and religion, probably the same time we developed language and abstract thought. It appeared in the form of drawings, etchings, and paintings on cave walls, and as sculptures. It probably developed from more primitive, transient art, such as body adornment and drawing in the sand, because unlike the primitive art of children that we might intuitively expect, much of this work was breathtakingly beautiful, evoking the same awe and wonder as the greatest art of recent centuries.

These artists were exceptionally skilled by today's standards. Several styles coexisted, including realism, surrealism, and abstract art. If these served different functions, we can only speculate on what they were. There were also geometric forms, possible symbols, and tally marks on the cave walls. Flutes and percussion instruments have been found in the same strata at the same sites, signaling the contemporaneous origin of music.

Although there is plenty of wall art in the living areas of 30,000 years ago, there are also magnificent paintings deep within the caves, in tunnels and chambers that are difficult or dangerous even for accomplished spelunkers to reach. It's not surprising that new finds are still being reported a century after the first Upper Paleolithic (late Old Stone Age) cave art discoveries. To reach these sites, one has to crawl through mud, wade through underground streams, wiggle through tunnels barely wide enough to allow passage, skirt precipitous drops, and ascend or descend steep, slippery pitches. Why would these people go to so much trouble? They had to carry with them the pigments they painted with and the tools they carved the rock with, navigating and working by torch or lamp light in small chambers where the oxygen would soon be depleted. These sites were far from any help if their lights went out or if they broke a leg or were pinned by a rockslide. Why did they choose sites that many of their companions might never visit, perhaps never know about? How could they produce such accurate and naturalistic images of animals from memory?

What purposes did these works of art serve? These were not casual decorations. They were often placed so the passing cave explorer would miss them. If unexpectedly encountered they can be startling, especially the images of saber-toothed tigers painted on bulging or carved surfaces to give an illusion of three dimensions. Viewed by flickering lamp light, they appear to move, grimacing and menacing. Was their importance private or social? Do their specific locations have significance? If they were instructional, why make the classroom virtually inaccessible? It's conceivable that they played a role in initiation rites or other ceremonies no longer practiced or imagined. They may have been placed there as acts of sympathetic magic, to improve the odds in hunting. They may have had a religious purpose, because religion appeared at the same time and the world's religions have inspired some of our greatest art. Under the influence of consciousness-altering substances, physiological stress such as sleep, food, or sensory deprivation, or religious fervor, the images may have been used to accomplish psychological objectives, similar to techniques used to effect rapid religious conversions, brainwashing, or other dramatic changes of heart, or to evoke feelings of exaltation, awe, or reverence. We just don't know. It does seem very likely, given the powerful emotional effect of Paleolithic art on modern people, that the human mind responded to art in much the same way as today, and that the drive to create art descended through millennia to us largely unchanged. If the need for, and appreciation of, art back then was similar to today's, that may provide important clues to the survival value and social functions of art itself.

This week's news:

LAGOS: A declaration of war against Nigeria, the main oil-producing country in Africa, by a rebel group in the oil-producing Niger delta has driven oil prices up. Although 98.5% of Nigeria's external income is from crude sales, the Niger delta receives none of the oil money. Unregulated oil development has destroyed the environment. The air and water are badly contaminated and area fishermen have lost their means of subsistence. Local politicians armed pirate gangs to win the most recent elections and then didn't keep their promises, losing control of the gangs, which have overrun security. If nothing is done, there may be civil war during the 2007 elections. The major oil producers don't intend to stop production. (*Liberation*, translated from the French by truthout.com)

Over the last 50 years the annual atmospheric CO_2 increase has averaged around 1.5 parts per million (ppm) per year. In 2002 and 2003, the annual increase was more than 2 ppm. Previous unusual CO_2 increments have been explained by natural events, but there haven't been any natural events to explain the 35% increase in the past 2 years. It's too early to draw conclusions, but some scientists are concerned that we may have only a few years to reverse global warming rather than decades. The higher increases could be the first signs of a "runaway greenhouse effect," where the warming can't be stopped. (*The Guardian*)

Recommended reading:

Gardner, Howard, *Art, Mind and Brain: a Cognitive Approach to Creativity*. New York: Basic Books, 1982.

Dissanayake, Ellen, *Homo Aestheticus: Where Art Comes from and Why*. New York: Maxwell Macmillan International, 1992.

Lewis-Williams, J. David, *The Mind in the Cave: Consciousness and the Origins of Art*. New York: Thames and Hudson, 2002.

33
THE EMERGENCE OF RELIGION
(October 24, 2004)

Religion, indicated archaeologically by ceremonial burials with items to be used in the afterlife, appears to have blossomed for the first time about 30–50,000 years ago. What was its survival value?

In Europe, where religion began, climate change was a major challenge. Also, our over-hunting of food species made food scarcer, and killing off our predators resulted in a human population explosion. Competition increased among the bands of hunter-gatherers.

These factors favored increased sizes of bands, with increased complexity of social structure within and between them. Hunting became more sophisticated. Bands may have fought, but there's evidence of Paleolithic trading, sometimes over hundreds of miles. These practices all favored the development of language and abstract thought, once the necessary neural circuits were available.

We know social stratification arose, because a select few were buried with very labor-intensive jewelry. In modern Neolithic tribes, shamans are some of the most privileged members. They hoard secret knowledge and deceive other tribe members (and often themselves) into believing they have supernatural powers. They're the tribes' valued teachers, moral and spiritual guides, healers, and interlocutors with the gods. Perhaps the privileged burials included shamans or seers (literally, see-ers).

Our new mental status included an awareness of personal mortality and longing for an ideal life. Innovation was very risky in a world filled with dangers, and there must have been strong pressure to conform and cooperate. We needed new ways to teach a rapidly growing body of knowledge and social customs. In the absence of writing, these were communicated not only by language but also by art. Just as we use play activity, rhymes, pictures, and music to rehearse, store, remember, and convey information, our ancestors did the same.

Judging from Paleolithic artifacts and the customs of modern hunter-gatherers, the cave dwellers undoubtedly had an oral tradition requiring memorization and endless recitation of myths packed with history and practical information. These stories were probably filled with mythical creatures and supernatural events. They provided a mnemonic device for a large store of knowledge, including guidance on proper behavior: morality. Just as modern shamans use stress to transform people's worldviews (e.g., in brainwashing and religious conversions), Paleolithic shamans probably used the same methods to force-feed their audiences.

Our enhanced ability to generalize from experience often led to false generalizations, just as it does today. These superstitions, as long as they have a socially useful function and aren't outright harmful to survival, are passed down through the generations and absorbed at an age when we haven't yet developed critical skills. We incorporate magical concepts and we see our universe through a worldview that contains strong elements of irrationality. Beliefs in the supernatural go unchallenged, even though they're unsupported by scientific evidence, and different people's beliefs are often mutually exclusive. Religious dogmas are retained even if they're dysfunctional, like the injunction to multiply even though we're overpopulated.

Given the circumstances, it's hard to imagine that the Paleolithic cultural explosion could have occurred without the emergence of religion.

This week's news:

LONDON: Sir David King, the U.K. chief scientific advisor, told BBC News Online last year's rise in CO_2—the highest in history—is probably a normal fluctuation unless proved otherwise. Another U.K. scientist, Peter Cox, who heads the carbon cycle group at the Hadley Center for Climate Prediction and Research, noted the CO_2 increase was not uniform across the globe. He said the shift might have been caused by something unusual in the northern hemisphere. For example, Europe's very hot summer in 2003 and unusually large number of forest fires could have killed off vegetation and increased carbon releases from the soil. (spacedaily.com)

GENEVA: Michelin and a Swiss technical institute have built a fuel cell car. "It's not the first time that a prototype has been made using a fuel cell, but this one has particular characteristics: It's light, it uses oxygen and its consumption is very low," Daniel Laurent, head of Michelin's Swiss research centre, said. The Paul Scherrer Institute near Zurich said the car refuels with a mix of oxygen and hydrogen that powers electric motors located inside the front wheels. A local power company provided a refueling station that produces the hydrogen-oxygen mix by electrolysis powered by solar energy. (spacedaly.com)

Recommended reading:

Sargant, William Walters, *Battle for the Mind: A Physiology of Conversion and Brain-Washing*. Cambridge, MA: Malor Books, 1997.

34
MODERN HUNTER GATHERERS
(October 31, 2004)

One source of information about sustainable human cultures is the study of past cultures and their fates. Unfortunately, the archeological record is fragmentary and situations were different, limiting generalizations. Fortunately, another information source exists: current cultures whose lifestyles provide valuable insights about sustainability and counteract our ethnic myopia. I'm particularly indebted to Pfeiffer's *The Creative Explosion* (1982) for his descriptions of today's Kalahari Bushmen (San, !Kung) and Australian aborigines. These two cultures have lasted for millennia, longer than any nations. The mythologies of these and similar cultures accurately record events as long ago as 10,000 years.

Both the Aborigines and the San are desert-dwellers who live in nomadic bands. Having no fixed abode and traveling on foot, they carry few possessions. In fact, they teach their young not to cherish possessions and encourage them to share. The San and the aborigines are also extreme egalitarians. While they make use of available expertise, leadership is evanescent, shifting among individuals according to the task at hand. Attempts by individuals to achieve higher status are suppressed through gentle ridicule. This minimizes competitive behaviors that would be dangerous in nomadic bands living at the edge of survival.

The size of these bands—roughly five five-member families—is an optimal tradeoff between the need for cooperation and the increased aggression that results from larger groups. Bands interact in order to pass along gifts (the antidote to individual ownership of treasured objects), share information, engage in large hunts, and find not-too-closely-related mates. These bands have loose boundaries: individuals are free to move from one band to another within a tribe sharing a common dialect, numbering in the hundreds. Research has found the sizes of tribes appear to be optimal for their purposes.

A major problem for preliterate nomadic bands is the preservation of valuable information. The San and the Aborigines both use frequent retelling of stories to embed needed information in a narrative, providing a valuable mnemonic context much more effective than memorization of isolated facts. Frequently repeated stories provide the rehearsal necessary for secure memory consolidation.

The Aborigines have gone to extreme lengths to preserve information about locations where water can be found. This vast amount of information, covering hundreds of thousands of square miles of bleak desert, is passed on in the form of art, song, chanting, dance, initiation rites, and myths. Their rock art, periodically redrawn to preserve it against the elements, is ubiquitous and utilitarian, providing geographic signposts and links to their mythology.

From infancy, Aborigines sear into their minds a detailed map of thousands of landmarks in the desert. Their mythology, including their creation myth, relates the travels and adventures of totemic animal ancestors who left tracks, which are the landmarks they must memorize. The myths are drummed into their heads with tools used worldwide for religious indoctrination. Their music, dances, and rituals are charged with fear, ecstasy, and pain, to make an indelible impression in the memory. These are rehearsed over and over to the point that they are performed with incredible skill and precision. The walkabout, or tour of the desert, is another drill, associating geographic landmarks with the myths and with the locations of food and water. Even Aboriginal spear throwers are covered with art symbolizing routes to water.

These practices have their analogies with similar practices in other cultures. Few contemporary peoples recognize the original mnemonic purposes of their myths. Their logic has been lost, replaced by irrational and obsolete dogmas that hinder our ability to adapt to change. It's almost as though we're survivors of a history-obliterating cataclysm, wrongly interpreting relics from the past, as in the science fiction classic *A Canticle for Liebowitz* (Miller, 1959). This is a phenomenon referred to by some as "The Great Forgetting," subject of a future column of *The Clock is Ticking*.

The San and the Aborigines, like all long-lived cultures, have family planning, including sexual taboos and marital traditions, birth control, and abortion. Some of their methods strike us as horrific, but they work for these nomadic people in their struggle to live sustainable lives in environments with very limited resources. We, with our non-sustainable lifestyles, have no right to judge them, with their sustainable practices. Someday soon, we could find ourselves in even more trying circumstances. We all will have to fit solutions to our cultures and our depleted environments, however we can.

This week's news:

WASHINGTON: Congress has adjourned without acting on important environmental problems, including the Clear Skies proposal, an expired industry tax to pay for cleanup of Superfund sites, caps on greenhouse gases, and federal protections of endangered species. Federal programs have inadequate funding and public health threats are rising. The Bush administration is using its regulatory powers to handle environmental questions, leaving Congress out of the loop. The utility industry has donated $34 million, mostly to GOP federal candidates since Bush took office, and lobbied to minimize regulation. The odds are low for passage of a bill that would curb emissions linked to global warming. "It's really irksome," said Natural Resources Defense Council advocacy director Greg Wetstone. "What's broken is a political system that's evading science, fact, and public opinion." (*Washington Post*)

The world's population is consuming 20% more natural resources than the planet can produce, conservationist group World Wide Fund (WWF) has said. Marine and terrestrial species fell 30% and freshwater species declined 50%, between 1970 and

2000. The average amount of productive land needed to sustain one person (that person's "ecological footprint") is currently about 2.2 hectares, but the earth has only 1.8 hectares per head. The fastest growing component of the footprint is energy use, up 700% between 1961 and 2001. North Americans are consuming resources at a particularly fast rate, with ecological footprints twice as big as those of Europeans, 7 times those of Asians or Africans. (Al Jazeera)

A survey by Transparency International (TI), an anti-graft organization, rates Bangladesh and Haiti the most corrupt countries in the world, followed by Nigeria, Chad, Myanmar, Azerbaijan, and Paraguay among the world's most corrupt nations. Finland is the least corrupt country in the world, with very low levels of corruption in New Zealand, Denmark, Iceland, Singapore, Sweden, and Switzerland. There is a direct correlation between poverty and corruption. The most corrupt countries—mostly developing nations—are also unstable politically. Non-governmental organizations say the world's richer nations are partly responsible for the plight of the world's poor because of colonialism, conquest and the effects of free trade policies and globalization. They also contend the developed world's insatiable thirst for oil is fueling political instability in poorer nations. (Al Jazeera)

Typhoon Tokage produced an 80-foot wave, the largest ever recorded in Japan, off the port of Muroto on the southern island of Shikoku. The previous record was 65 feet in a wave monitored off Miyazaki on the southern island of Kyushu last August. (terradaily.com)

Recommended reading:
Ehrlich, Paul R., *Human Natures: Genes, Cultures, and the Human Prospect*. Washington, DC: Island Press, 2002.

35
HUMAN MIGRATION
(November 7, 2004)

As a population grows, it has to spread out, either to maintain a geographical density that is sustainable or to abandon a region that has been depleted to the point where it's no longer habitable. This has been the rule throughout our brief stay on Earth, with only three exceptions. A population might not grow, because it responds to crowding with either an increased death rate or a decreased birth rate. A technological innovation, like the use of agriculture, may buy some limited time for further population growth. Alternatively, a change in ecology, usually due to a shift of climate, may buy some time or cause a population to crash. Only the first case, where birth or death rates adapt to available resources, can be sustained for a substantial length of time. Often this is achieved through warfare, famine, or drought, e.g., primarily by varying the death rate. Alternatively, a population may manage to regulate its birth rate in order to live within its means.

The first expansion of territory by our species involved going where we had never gone before. The sequence of migrations was originally worked out by studying the movement of humans through their remains and artifacts such as tools; by using radioisotope methods like carbon dating; by stratigraphic methods, based on the sediment layers in which human remains and artifacts were found; and by analysis of fossil plant and animal remains in the same layers, often leftovers from human meals. More recently, we've used genetic analysis of human remains and of contemporary humans, and linguistic analysis of ancient documents and contemporary languages.

Our first migrations, from the east African region where our species originated, took place about 100,000 years ago, when we began to spread throughout Africa and into the Middle East. At that time, there were about 50,000 of us altogether. We were exclusively hunter-gatherers. We spread throughout Asia around 50–60,000 years ago, reaching Japan, Indonesia, and Australia with the help of primitive boats. We began to spread across Europe around 40,000 years ago. Last of all, we migrated into the Americas in successive waves starting about 35,000 years ago, island hopping or crossing a land bridge between northeast Asia and northwest America, where the Bering Strait is today.

Living conditions were highly varied. Under the influence of natural selection, gene frequencies in these isolated gene pools shifted, so the characteristics of the different populations diverged, producing physical differences that provided some compensation for different diets, diseases, climates, and terrains. There's no evidence of selection for differences in cognitive capabilities.

Migrations and invasions have mixed our genes and our cultures ever since, in successive population movements that have been mapped remarkably precisely using a combination of archeological, genetic, and linguistic methods. The resulting precision could not have been achieved with any one of these methods alone, attesting to the advantage of multidisciplinary approaches.

By the time we adopted agriculture 10,000 years ago, at the transition between Paleo- and Neolithic eras, our population had reached about 5 million, an average growth rate of about *50 people per year*. In the subsequent 10,000 years to the present, we've swelled to more than 6 billion, an average growth rate of about *600,000 people per year*. Hello, population explosion.

This week's news:

LONDON: James Lovelock, atmospheric chemist and author of *Gaia*, warns that global warming is the worst weapon of mass destruction. If humans continue degrading the environment, they will lose their place in the world. He argues that we must use technology to lessen our impact on the atmosphere. He recommends implementing a variety of energy sources, including nuclear; synthesizing food products to reduce the food burden on the earth; and accepting the risks of cancer from chemicals and radiation. He notes that not stopping global warming will pose a much greater chance of extinction than cancer, as shown by the 20,000 Europeans who died during the 2003 heat wave. (spacedaily.com)

NEW DELHI: The Chittaranjan National Cancer Institute reports that two out of five New Delhi residents suffer from lung, liver, or genetic disorders due to polluted air. "We have found that polluted air has also altered immunity and caused blood-related abnormalities among many of the victims tested by us," an institute official said. The institute surveyed 2,379 people through questionnaires and clinically examined 1,270 people between 20 and 75 years old. It also examined 4,671 children aged 8–16 and concluded they were as much as three times more prone to respiratory disorders compared to children in the same age group tested in the marshy forests of the Sundarbans near Calcutta. Most of the worst afflicted victims lived in the heart of New Delhi. (terradaily.com)

IOWA CITY: The Bush administration is trying to suppress scientific evidence of the dangers of global warming, a NASA scientist announced. "In my more than three decades in government, I have never seen anything approaching the degree to which information flow from scientists to the public has been screened and controlled as it is now," James Hansen told a University of Iowa audience. Hansen is director of the NASA Goddard Institute for Space Studies in New York. He has briefed Cheney's task force on global warming, and was one of the first government scientists to brief congressional committees on the dangers of climate change, back in the 80s. Hansen said the administration wants to hear only scientific results that "fit predetermined, inflexible positions." Data that would increase concerns about climate change are frequently dismissed as insufficiently interesting to the public. "This, I believe, is a recipe for environmental disaster." Hansen said scientists generally accept that temperatures are rising because of increased greenhouse emissions from burning fossil fuels. The rising temperatures will cause sea levels to rise and trigger severe environmental consequences, he said. Hansen said such warnings are consistently suppressed, while those who cast doubt on the scientific results receive favorable treatment from the administration. He also said reports that outline potential dangers of global warming are edited to make the problem appear less serious. "This process is in direct opposition to the most fundamental precepts of science," he said. (MSNBC)

PARIS: The International Energy Agency (IEA) says that unless governments switch to clean sources, fossil fuels will continue to dominate the world's energy market, increasing atmospheric CO_2 pollution and global warming. World energy demand will increase by almost 60% by 2030. Two-thirds of the demand increase will come from developing countries. About 85% of the energy increment will come from oil, gas, and coal. "CO_2 emissions will be more than 60% higher." Scientists are recommending a 60-percent cut in emissions. The IEA recommends greater energy efficiency and incentives to switch to clean energy. (spacedaily.com)

Recommended reading:

Cavalli-Sforza, Luigi Luca, *Genes, Peoples, and Languages*. New York: North Point Press, 2000.

36

THE ORIGINS OF AGRICULTURE

(November 14, 2004)

Although hunter-gatherers were quite capable of overpopulation, habitat destruction, and driving other species to extinction, they sometimes reached an equilibrium of sorts with their environment, just as they sometimes do today. Even when they didn't, they couldn't do as much harm as we do today because they didn't possess the technology (including the sophisticated uses of energy) or reach the population densities of today. A major step toward modern technology and population density was the birth of agriculture.

We now know that agriculture developed independently on several occasions, separated in space and time. Probably the first was in the Fertile Crescent, where Babylon, and more recently Iraq, used be. Archeologists date the first agricultural settlements, whose remains were found overlying older hunter-gatherer sites, back to 10,000 years ago. In other words, we *Homo sapiens sapiens* have grown our own food for only about one twelfth of our species' lifetime. Agriculture developed in China more than 8,000 years ago, in eastern North America and in South America over 5,000 years ago, and in sub-Saharan Africa more than 4,000 years ago. From those sites agricultural practices spread to Europe, the rest of Asia, Australia and the rest of the Americas by cultural diffusion across existing populations, and also by migration. In some cases, domesticated animals and plants were transported from one region to another, and in other cases, local species were domesticated. All domesticated varieties were developed from wild plants and animals, first inadvertently and later deliberately, and in the process humans acted as an agent of natural selection.

The transition from nomadic hunter-gatherers to settled farmers occurred in parallel with domestication of plant and animals. There were nomadic herders of sheep and goats, and settlements of hunter-gatherers. The transition was driven by a desire for greater security, i.e., for protection against fluctuations in food supply or for a greater yield as wild animal or plant species dwindled or human population densities rose: basic supply and demand problems. This was a continuation of survival strategies such as storing food, begun around the time of the cultural explosion 50.000 years ago, when we developed the capacity to plan ahead.

Hunter-gatherers discovered that they could expand their food supplies by such simple practices as burning off unwanted vegetation, planting seeds and the tops of tubers, transplanting food plants, and improving the browsing conditions for deer. As Smith (1995) points out, this in itself was not domestication.

Domestication is genetic modification through selective breeding. To some extent, this occurs automatically as a result of farming. Further domestication requires separating the plants and animals from wild ones, selecting those with the most desirable characteristics for planting and breeding. In the process, we shape the evolution of organisms, to the point where they can no longer compete in the wild. Most of these, the first genetically modified organisms, depend upon us for their very existence.

Ultimately, the impact of agriculture was as great as major forces of global change such as ice ages and the movement of tectonic plates. We destroy habitats of many species and drive them to extinction in order to grow our domesticated ones. Our population grows far beyond that which could be supported by hunting and gathering. Eventually it grows beyond that which can be supported by current agricultural technology, and new technology has to be developed. Rather than learning to control our numbers, we resort to technological fixes, each of which further damages the ecosphere that supports all life by replenishing the earth, water, and air.

As we shall see in later columns, there have been other impacts of agriculture as monumental as the environmental ones, including cultural and genetic changes in humans. The agri-culture that developed is tainted by harmful memes, which could prove fatal to our species without some serious *euculture* (making cultures more normal, i.e., sustainable).

This week's news:

PARIS: Gendarmes and wildlife officials sealed off part of the Aspe Valley near the Spanish border in the western Pyrenees, searching for a bear cub after hunters killed its mother, the last native female bear in the Pyrenees. The ten-month-old cub ran for the undergrowth when its mother was shot. Philippe Grégoire, the Prefect of the Pyrénées-Atlantiques département, ordered a ban on hunting and dog walking in the Aspe area. "The big problem is the destiny of the cub. It is our duty to guarantee his survival," M Gregoire said. "The cub is not tiny. If he is not disturbed, he has a serious chance of survival." (*The Times*)

CAMBRIDGE: The British Antarctic Survey reports that Antarctic whales, seals, and penguins are threatened by food shortages in the Southern Ocean. Antarctic krill, shrimp-like crustaceans central to the marine food chain, have declined by 80% since the 70s—probably due to a sharp drop in the amount of sea ice in the winter. Sea ice is the feeding ground for krill. This may explain declines seen in several species of penguins. Lead author Dr Angus Atkinson says: "This is the first time that we have understood the full scale of this decline. Krill feed on the algae found under the surface of the sea-ice. The Antarctic Peninsula, a key breeding ground for the krill, has warmed by 2.5°C in the last 50 years, with a striking decrease in sea-ice. We don't fully understand how the loss of sea-ice here is connected to the warming, but we believe that it could be behind the decline in krill." Knowing how the environment affects this food web will help scientists model responses to future changes. (spacedaily.com)

Recommended reading:

Smith, Bruce D., *The Emergence of Agriculture*. New York: W. H. Freeman, 1995.

37
SETTLING DOWN
(November 21, 2004)

By Darwin's time, most naturalists had come to accept the notion that life evolved from extinct ancestors into contemporary organisms. Most assumed this had happened for *teleological* reasons, that is, with a predestined goal. In fact, many naturalists were members of the clergy, and believed that studying god's creation would provide insights into his plan. This reflected a Christian belief system called the *Great Chain of Being*, in which all God's creations had their places in a hierarchy ranging from the inanimate through plants, lower animals, humans, angels, and god himself. We were the top of the earthly chain, and it was natural to assume that we were the goal of evolution. Victorians further assumed that the cultural evolution of humans had a goal, namely Victorians. This may seem silly, but remember that even today some people consider some other people to be lower life forms.

With reasoning like this, it was natural to assume that the emergence and eventual dominance of agriculture over foraging, and *sedentism* (being settled rather than nomadic) over nomadism, were improvements. It was assumed that this way of life led to better nutrition, better health, and modern civilization. And that's what we've been taught. However, the archeological evidence suggests something quite different.

Over the span of our evolutionary history, Earth has undergone a series of glaciations, or ice ages, alternating with interglacial periods. The Cro-Magnons of Europe, who underwent the creative explosion marked by cave paintings, more advanced tool use, and the origin of language, lived during the most recent major glaciation. At that time, a layer of ice up to three miles thick extended from the poles through what is now the northern United States, northern Eurasia, and a similar portion of the southern hemisphere. These ice layers were as thick as many of our seas and oceans of today are deep. Temperatures were 10–20° F lower than today. Since the amount of water on Earth remains roughly constant and much of the planet's water was locked up in ice, the oceans and seas shrank and the waters withdrew to the edges of the continental shelves. The weight of the ice was so great that the tectonic plates that they covered actually sank hundreds of feet. Even today, approaching what may be the end of the current interglacial, much of Canada is still rebounding, rising after the release 10,000 years ago from the immense weight of the glaciers. Much of the land bordering the ice sheets was steppe and tundra, with unrelenting cold, dry winds. In Europe, only the southern part was inhabited by humans. This zone of habitation extended through the Middle East, into Africa and central Asia.

As the ice age ended around 10–12,000 years ago and atmospheric temperatures rose, humidity rose as well from evaporating water and ice. Rains began to fall more frequently, the growing season lengthened, and forests began to spread. The lower valleys, some of which today are covered with water, were the most fertile. Hunter-gatherers who lived in these fertile valleys and forests didn't need to roam as much because food was abundant and little time or effort was needed to hunt or gather it. When they came upon particularly bountiful spots, like shorelines where aquatic food was abundant or near the migration paths of meat on the hoof, people settled and let the food come to them. *Sedentism* (being settled rather than nomadic) became more common, although this certainly wasn't the end of the hunter-gatherer lifestyle.

These settlements began to transform lifestyles. Life was less precarious, so sharing was less necessary. People could acquire and keep more personal property such as tools, clothing, and art. They became acquisitive, possessive. They developed individual, more independent, goals. People built shelters and lived together in stable family units rather than in bands whose members came and went. These were prototypes of more modern households. Some took up horticulture, some started herding livestock. These means of providing food took little effort, and their food supply became more secure.

Planting, herding and sedentism, necessary for full-time farming, had been established. But agriculture as a way of life wasn't necessary in these times of abundance, and it didn't develop. One more change was necessary to provide the incentive for agriculture.

This week's news:

The Pew Center on Global Climate Change has released its report, *Observed Impacts of Global Climate Change in the U.S.*, documenting ecological changes already under way and predicting mass extinctions, a sea level rise of 3 feet and disruption of Earth's life-support mechanisms. About 50% of the 150 plants, animals, and insects examined have already been affected by global warming. Many species are disappearing in the southern parts of their ranges, but doing better in the northern parts. Edith's checkerspot butterfly, for example, is disappearing around the Mexico-California border where it's now too warm and dry, but expanding in British Columbia, which used to be too cold. Red foxes are moving north and can now be found in Arctic regions. Unfortunately, the arctic fox may be endangered because it can't compete with the red fox. Meanwhile, tropical species are moving into Florida and the Gulf Coast. A third of all species will be extinct by 2050 according to some scientists. Entire ecosystems are being affected, disrupting soil creation, plant pollination, and natural recycling of water and air. (wired.com)

Recommended reading:

Bogucki, Peter, *The Origins of Human Society.* Malden, MA: Blackwell Publishers, 1999.

38

WHY AGRICULTURE?

(November 28, 2004)

This question is the title of the second chapter of Richard Manning's excellent book, *Against the Grain* (2004). Thinking on this issue is undergoing a revolutionary change.

The development of large-scale farming didn't happen until it was necessary. That came to pass once hunting and gathering weren't sufficient to meet people's needs. This was due to a population size that couldn't be supported by the depleted or climate-altered environment, and a lack of anywhere to move because other folks already lived next door. Then people *had* to take up gardening, herding, and major crop growing. This is when they began to domesticate cows, sheep, goats, boars, grains, beans, squash, etc. in earnest. Agriculture developed independently at different times and places around the world as the need for major-scale agriculture developed. Agriculture also spread by diffusion, migration, and conquest. In Eurasia, people also domesticated horses as beasts of burden and transport. This completed the triad of horses, wheat, and cattle that Eurasians later used so successfully in conquering other civilizations (Diamond, 1997).

The transition from hunter-gatherer to agricultural life has been likened by some to the expulsion from Eden—in fact, the biblical account can be read in that light. Cain, by the way, was a farmer and Abel was a pastoralist (herder). Plowing, sowing, cultivating, reaping, processing, and storing food were hard work that caused severe work-related disabilities. People were less well nourished by a much less varied diet. They were shorter in stature and shorter-lived. Only in recent generations, have stature and longevity returned to pre-agricultural levels.

The sedentism and changes in division of labor brought on by the agricultural revolution led to a revolution in human social structure, which we'll consider in due course. For now, it's worth noting changes in living conditions that resulted from the transition to agriculture. Weeds evolved from the pioneer plants that repopulate devastated areas such as land cleared by forest fires. People had to devote as much labor to fighting these plants as they did to sowing and reaping. The human settlements expanded as population grew, and people had to devote more land to farming: they were crowded into the first cities just as rural people are today.

Agriculture brought people and their animals into closer contact. Consequently, new disease organisms evolved and were transmitted among humans and their animals, as happens today. Lack of sanitation and the ignorance of germs led to a new phenomenon: epidemics like smallpox, cholera, typhoid, and bubonic plague. These diseases could decimate populations, and were often followed by famine, since there was no one to tend the farms. The plagues spread through trade and shipping routes, some of which went back to the pre-agricultural period. New pests infested the settlements, ruining crops and the stored harvest and spreading disease among humans, plants, and animals. Human existence had changed from the relatively healthy and carefree lifestyle of the early hunter-gatherers in lands of plenty to the crowded, unhealthy, brutish, and short lives of agricultural communities. The remaining hunter-gatherers found themselves crowded into more desolate landscapes by the expanding, aggressive farmers. There was no turning back: this pattern has continued to the present day.

This week's news:

THE HAGUE: Dutch researchers have found that even though most plankton is in the northern hemisphere, it influences climate more in the southern hemisphere. Plankton manufactures the gas dimethyl sulphide (DMS), which is a source of small atmospheric sulphur particles. The particles serve as condensation nuclei for water vapor. The resulting haze reflects sunlight, causing the Earth to cool. In the southern hemisphere plankton species produce more DMS than their northern cousins and there's a larger sea surface area and a higher DMS flux from sea to air. There's also a lower oxidation capacity in the southern hemisphere, which means that DMS is broken down more slowly. In the more industrialized northern hemisphere, more charged molecules are dumped into the air, which increase the oxidation capacity. (spacedaily.com)

BANGKOK—Over 15,000 animal and plant species face extinction, reveals the World Conservation Union (IUCN) in its *2004 Red List of Threatened Species*. Among threatened species are 1/3 of gymnosperms, 1/3 of amphibians, 1/2 of freshwater turtles, 1/8 of birds, and 1/4 of mammals. Countries with the most threatened species are mostly in the tropics. People constitute the main threat to species, through habitat destruction and degradation, over-exploitation for food, pets, and medicine, introduced species, pollution, disease and climate change. Major conflicts between the needs of threatened species and rapidly increasing human populations are predicted in Cameroon, Colombia, Ecuador, India, Madagascar, Malaysia, Peru, Philippines, and Tanzania. The report also named Brazil, Cameroon, China, Colombia, Ecuador, India, Indonesia, Madagascar, Peru, and the Philippines as countries that have a large number of threatened species and are financially unable to invest in conservation. (*Inter Press Service*)

RIO CUEIRAS, BRAZIL: Brazil has destroyed 97% of an Atlantic rain forest originally 1/3 the size of the Amazon jungle—and could destroy the Amazon as well. Brazil has a disproportionate number of threatened animal species. The government is attempting to slow destruction and theft of plants, animals, and natural medicines that lose it billions of dollars a year. Farmers use slave labor to clear tracts as large as New Jersey in a year. Environmentalists are concerned about destruction of the forest, as large as Western Europe, called "the lungs of the world" for its capacity to recycle greenhouse gases. It has 10% of the world's fresh water and 30% of the world's plant and animal species, and is an essential source for medicines. (*San Diego Union-Tribune*)

WASHINGTON—House and Senate negotiators have inserted an anti-abortion provision into a $388 billion spending bill. The abortion section would prohibit withholding money from health care providers that refuse to provide or pay for abortions or refuse to offer abortion counseling or referrals. Current law, designed for Roman Catholic doctors, provides "conscience protection" to doctors who don't want abortion training. The new provision would offer similar protection to all health care providers, including hospitals, doctors, clinics, and insurers. The provision could affect millions of women, according to Senator Boxer, Democrat of California. House Republican leaders insisted the provision be included. Some insiders considered their insistence a sign of the political strength of Christian conservatives, who played an important role in re-electing Bush this month. Congress returned to Washington after the election in order to pass the omnibus bill. The alternative is to let government funding for a wide array of agencies expire, leading to a partial government shutdown. Hawaii, Maryland, New York, and Washington pay for some abortions for low-income women through their Medicaid programs. (*New York Times*)

ROME: 300,000 square kilometers in Mediterranean Europe are currently affected by desertification, threatening jobs of 16.5 million Europeans. It's mostly the result of overfarming and land clearance, drought, and possibly climate change. Poor land management removes soil nutrients, reducing vegetation. Without plants to anchor the soil, centuries' worth of topsoil is eroded until the land becomes barren. The Worldwatch Institute estimates our planet loses as much as 24 billion tonnes of topsoil annually. Mediterranean Europe is densely populated dry land, and would be vulnerable to desertification anyway, although regular forest fires make it worse. Between 600,000 and 800,000 hectares of forest burns annually—destroying vegetation in an area almost as large as Corsica. The World Wildlife Fund figures 95% arise from negligence or arson. (spacedaily.com)

REYKJAVIK: The melting Arctic ice cap could provide a new northern waterway between Europe and Asia. A route along the Siberian coast would be nearly 40% shorter than the current one through the Suez Canal. This would come at a price: climate change and the extinction of species. Ships presently can sail through the Arctic 20–30 days per year, which could be extended to 150 days for some ships. Shipping companies are already considering the possibilities. While two weeks could be saved for some routes, travel through the Arctic Ocean would probably be approved only for ships with ice-strengthened hulls. (spacedaily.com)

Recommended reading:

Manning, Richard, *Against the Grain: How Agriculture has Hijacked Civilization*. New York: North Point Press, 2004.

39
CHIEFS
(December 5, 2004)

The transition from social groups without leaders to those with leaders marked an important turning point in human society. The relatively egalitarian structure of hunter-gatherer societies indicates humans have the capacity for living with or without leaders. What determines which way a group will go?

One factor is group size. A band of five to twenty people may be small enough to reach decisions quickly without a designated leader, and may benefit from not having one. Larger groups (clans, tribes, chiefdoms, nations) would be too unwieldy for quick responses to challenges and everyday management of activities if everyone had to be involved in every decision. Also, people in larger groups inevitably manifest more pronounced divisions of labor and develop unequal status and wealth. The increased complexity of social links allows for the formation of power groups. Managers are needed to keep track of large-scale endeavors like farms, in which many individuals work together and are rewarded according to the effort they invest. Community policy is required for matters of defense, resource allocation, dispute management, and trade, and such policy is most conveniently arrived at by a manageable subset of the whole community.

Timothy Earle has dedicated much of his career as an anthropologist to studying chiefdoms (societies of a few thousand to tens of thousands of people). His book, *How Chiefs Come to Power* (1997), examines three very different chiefdoms in order to look for common themes: Denmark during the Neolithic and early Bronze Ages (2300–1300 BC), the Incas (500–1534 AD), and Hawaii from early in its settlement until its incorporation into the world economy (800–1824 AD).

Earle starts with a few important definitions, which I paraphrase. *Authority* is the right and responsibility to lead, sanctioned by a group, in recognition of certain capabilities or social position. *Power*, on the other hand, is measured by the mastery that a leader exercises over others, the ability to force compliance with the leader's commands. It reflects unequal relationships among people. Sources of power include social relationships, economy, military might, ideology, and information.

Social relationships are rooted in our biology: bonds of nurturing, cooperation, and domination. Hierarchies of clan and lineage develop as societies become more complex. Power accrues from manipulation of social relationships, for example by strategic marriages between powerful families. *Economic power* arises from the ability to buy compliance. *Military might* provides the ability to coerce compliance. It's problematic, because the military may decide to shift its support from one leader to another. *Ideology* presents a rationale for the authority structure of a society. *Information* can be manipulated by leaders to make it appear that the ruling elite have both the right and the might to retain power.

In the Hawaiian case, economic power didn't come from the need for centralized management of their complex of irrigation canals the way it did in societies like Egypt and Mesopotamia, because the complex consisted of independent systems, each of which was simple to manage. Instead, the chiefs came to own the land and canals through wars and inheritance. The chiefs allocated plots to farmers in return for a fraction of the food they produced. They rewarded their military with plots of land, and had managers that supervised the construction of the irrigation system. The foods they produced were the currency of the chiefdoms. Island chieftains later fought to absorb other island chiefdoms, and rule was consolidated into a small number of larger chiefdoms.

In the Andean highlands, economic power was more limited because the agriculture was marginal. Canals were used to water the farmland. Population expansion led to land wars, forcing populations into large fortified settlements on hilltops and ridges. The chiefs had less economic power because there was less surplus food. The Inca came to power through military strength, and maintained their power by forcing farmers to produce more foodstuffs and clothing for them. The Incas achieved greater productivity with more highly planned irrigation, farming, and storage complexes. They imposed an ideological base for rule as time went by.

In Denmark, development didn't proceed from egalitarian to ranked to stratified societies, or from equality to political authority to coercive state. In the early Neolithic period, there was no economic or other source of power or authority that could support chiefdoms. Only in the Bronze Age did prestige wealth begin to accrue from raids on the former Roman Empire. Wealth in the form of metal became a measure of personal valor and connection. Chiefdoms were based on military might and ideological legitimacy. Late in the Viking period states grew up around trade centers, the development of currency, and a new urban growth focused on trade.

This week's news:

WASHINGTON: Twenty-two national feminist, health, and population organizations denounced the decision by a National Public Radio (NPR) affiliate not to identify a sponsor as a reproductive rights group, charging infringement of free speech. The organization, Ipas, provides family-planning and reproductive-health training, research, advocacy, and supplies in 40 countries on five continents. In October, the station told Ipas the word "rights" would not be allowed. After unsuccessful attempts to negotiate, Ipas withdrew its support. The WUNC general manager said the decision was a precaution to avoid possible action by the Federal Communications Commission (FCC). The FCC has never defined reproductive rights as "political," but can sanction stations retroactively if it decides the words should not have been aired. "Reproductive rights" has become politically charged, since the Bush administration has often tried to have the phrase removed from declarations by international conferences. Administration officials have argued the phrase implies a woman has the right to have an abortion, which is true in the U.S. However, the administration and its Christian Right base detest abortion, and the right-wing Congress and the courts have not countered their administrative sanctions. Their efforts to erode reproductive rights include a last-minute withdrawal of funds for a major international health conference in Washington because one of the featured speakers had publicly attacked the administration's abstinence-only agenda and the re-imposition of the Global Gag Rule. The gag rule prohibits foreign non-governmental agencies that receive U.S. foreign aid from engaging in any abortion-related activities—even providing information about abortion to their medical clients or lobbying their own governments to ease anti-abortion laws. Ipas, which refused to tell its overseas partners to stop abortion-related activities, lost $2 million. (OneWorld.net)

REYKJAVIK: The Arctic is being invaded by species never seen there before. Locals have no words for these new animals, insects, and plants. More species will be retreating from the warming south, into a shrinking area bounded to the north by the Arctic Ocean. Salmon are swimming farther north, hornets are buzzing north, and barn owls are arriving in zones that have never had a barn. In Arctic Europe, birch trees are advancing north and Saami reindeer herders are encountering roe deer and elk on former lichen pastures. Thrushes have been spotted in Saami areas of the Arctic even in winter. Foreign ministers from eight Arctic countries will convene in Reykjavik but there's no consensus on actions to be taken. The U.S., the only one of the eight not signed on to the Kyoto treaty, opposes any new action. The Arctic Climate Impact Assessment report says that the region will warm by 7–13°F by 2100. (Reuters)

Recommended reading:

Chance, Michael R. A., and Jolly, Clifford J., *Social Groups of Monkeys, Apes and Men.* New York: Dutton, 1970.

40

CHIEFS AND MILITARY POWER

(December 12, 2004)

Our closest primate relatives live in bands, and display patterns of aggression that are similar to those in other mammals, including many aspects of human aggression. These include territorial defense, dominance behavior, protection of family members, and inter-band fighting. With the acquisition of language, humans devised more sophisticated strategies for hunting and for combat. The motives for inter-band and intertribal conflict became more complex and more pressing. During periods of over-hunting, competition undoubtedly arose among groups for game and for hunting and fishing grounds. Population pressure induces warfare among such groups today, in which ritual combat, cannibalism, raids for food and women, and contests for territory are common. In these types of warfare, all combat-capable men typically are involved.

Social organization ratcheted up as the result of sedentism, agriculture, and larger aggregations of people: chiefdoms arose. The new social specializations that came into being included a warrior class. Depending on the means by which chiefdoms arose, these fighters were engaged in different types of combat, including defense, resource wars, raiding and plundering, and annexation of other chiefdoms and the associated capture of women and slaves. Captives were forced to labor in conquerors' fields, pave their roads, and even fight in their wars.

The military was often needed to hold a chief's power over his own people. This was particularly true with regard to populations annexed to chiefdoms by geographic expansion and slavery. Chiefs often demanded tribute and forced hard labor on the conquered peoples. As castes arose within chiefdoms, lower castes often had to pay rent or tribute to higher castes.

A military is a two-edged sword. If the military is powerful enough to subjugate a population, it can also depose the chief. For this reason, in many chiefdoms the chief has been the warrior with the greatest prowess, and he has had to defend his primacy though combat against challengers. Although alliances and factions probably existed in tribes, in chiefdoms such factions could seize power by gaining the allegiance of the military.

There appears to be a direct co-causal link between the formations of chiefdoms and militaries. In many cases, chiefdoms arose through the assembly of a military, and for other chiefdoms to survive they had to form militaries as well. This was another development that, like agriculture, was irreversible. It meant the loss of personal autonomy and the enforced power of a ruling class over a subject class. Once such a pattern has been established, the only way for the ruler/ruled structure to be abolished has been through the total collapse of the society. The only way to avoid the formation of chiefdoms has been to remain a hunter-gatherer band or tribe, eschewing the trappings of sedentism and agriculture. Humans have tried to devise societies in which the people can select their own leaders, such as various forms of democracy, but the burdens imposed by chiefs and their militaries remain.

This week's news:

CHICAGO: 8,200 years ago as the last ice age was ending, a volume of glacial melt water over twice that of today's Caspian Sea entered the North Atlantic from a glacial reservoir in just a few months, the largest flood in 100,000 years. As the fresh and salt water combined, the Gulf Stream flow, which warms the North Atlantic climate, changed abruptly. Temperatures dropped in a decade or less for 100 years or more before fresh and salty water were back in balance and ocean currents and temperatures returned to normal. These conclusions are based on a new set of core samples from the swamps and marshes in the Mississippi River delta in Louisiana indicating a sudden rise in sea level of less than 1.2 meters. (spacedaily.com)

REAL, PHILIPPINES: Typhoon Nanmadol is approaching the east coast of the main island of Luzon, where an earlier storm caused floods and landslides that washed away entire villages, killed 400 and left nearly 200 missing. Nanmadol has 108 mph winds. Residents blame the flooding on illegal logging. (terradaily.com)

45 million children will die by 2015 because rich nations are reneging on their aid promises, according to Oxfam. The U.S., Germany, and Japan had pledged in 1970 to give 0.7% of their gross national incomes (GNI) in aid. The aid budgets of rich nations are half what they were in 1960 and poor countries are trying to pay $100 million a day in debt repayments. The U.S. is giving just 0.14% of GNI in aid, one-tenth of what it spent to invade Iraq, and much aid from the European Union arrived a year late. UNICEF reports that the lives of more than 1 billion children are at risk due to poverty, war and disease. (Al Jazeera)

Recommended reading:

Earle, Timothy K., *How Chiefs Come to Power.* Stanford, CA: Stanford University Press, 1997.

41

IDEOLOGY AS A SOURCE OF CHIEFS' POWER

(December 19, 2004)

Chiefs, whose specialized roles took form to meet the requirements of settlements (sedentism), have four primary sources of power: economic, military, ideological, and informational. The role of ideology is to shape subjects' conception of their place in society, the society's place in the universe, and a basis for moral behavior. These are usually based on myths and "revealed" cosmic laws. They're inculcated through government, churches, schools, and legal institutions. To the extent that a chief can control the ideology of his subjects, he can convince them that he has the rightful authority to rule them, and their proper behavior is to obey him. As long as there have been chiefs, religious or secular, they've used ideology to control their subjects.

To the extent that chiefs' opponents can use alternative ideologies to challenge their authority, chiefs are vulnerable to overthrow through ideological means.

Current anthropological thinking holds that ideology is not just inculcated at the knee of a teacher or in textbooks. Rather, for a populace to share a common ideology, it is most effective if they share in common the experience of public expressions and affirmations of the ideology. Mass meetings led by the chief or his representative provide one such opportunity, especially if the leader is a great orator, and the spectators are encouraged to respond verbally, in unison (responsive reading, cheers). Mass singing, chanting, and prayers are powerful means of gaining shared adherence to an ideology. Each participant takes note of the apparently willful participation of acquaintances, people whose opinions he or she respects, lending further credibility to the ideology. Failure to comply leads to ridicule or other negative reinforcement. Restrictions on who may participate in public rituals create a desire to belong. Initiation rites serve the same purpose.

Shared "volunteer" work that reinforces the ideology is another effective mechanism. Working together in missionary work, in the construction of a monument, in the maintenance of a communal farm, park, or nature area, in combat or in humanitarian effort, all are effective means of imprinting and reinforcing an ideology. Awe-inspiring ceremonies or the public bestowing of awards for good behavior are very effective. Mass rallies to welcome home soldiers or the hometown football team accomplish similar ends. Many "socialization" experiences fall in this category.

Symbolic objects such as monuments, public parks, gardens, churches, and imposing government buildings all provide daily reminders of a population's values. Quarrels over flag desecration or the public display of the Ten Commandments in this country are a sign of the importance of visual symbols of ideology. Lesser symbols include uniforms, badges, medals, and logos. Art, music, dance, drama, and literature can reinforce religious and patriotic ideology with their emotional power.

It should be noted that one ideology doesn't fit all in a society. It makes sense to have different variations on the common ideology, or even quite different ideologies, for the rulers and the ruled, for the rich and the poor, for the soldier and the doctor, as long as they don't sow discord.

Finally, ideologies have to be plausible and comforting, e.g. by promising rewards (national security, eternal bliss, material comfort) to their adherents.

Ideology is a means of codifying what is right and wrong. To the extent that this can be used to establish the authority of a person or an institution, this is a means of control: power. Given the special importance of this kind of mind control in our lives, we should recognize attempts to impose ideologies upon us (including this book) for what they are. In fact, it's our responsibility to recognize ideological influences and to evaluate critically the worldviews that they reflect. None of us is impervious to such influences, but if we hope to use reason to resolve the pressing problems of our age, we must reshape our ideologies to ensure survival.

This week's news:

An organization called the Lavoisier Group has published a book by Melbourne climate change skeptic William Kininmonth, who maintains that global warming is natural and not caused by humans burning fossil fuels. *Climate Change: A Natural Hazard*, attacks models used by climate scientists. The Lavoisier Group was formed in the 90s to challenge greenhouse science and the Kyoto Protocol. The 90-odd members are mostly retired engineers and scientists from the mining, manufacturing, and construction industries. Many others are retired captains of industry. The skeptic posture has also been adopted by right-wing think tanks. The skeptic network has managed to foster the illusion that climate change science is mired in uncertainty. Climate experts worry that skeptics will delay action on climate change using the same methods as the tobacco industry did to deny cancer links. Meanwhile, we are seeing the most rapid warming of the planet in 10,000 years. CO_2 concentration is nearly 380 ppm, the highest in 400,000 years. Still the skeptics persist. Some can't believe humans are capable of modifying the atmosphere. Many, like the Lavoisier Group, are more concerned about the cost of remediation. (*The Age*)

KINSHASA: Over 1,000 Congolese civilians a day are dying from disease and malnutrition. Almost half of those who die are children under five. "If the effects of insecurity and violence in Congo's eastern provinces were removed entirely, mortality would reduce to almost normal levels," an aid agency representative said, pointing to the case of Kisangani, a town where fighting has ended, basic services have been restored and mortality has dropped by four fifths. (Reuters)

Recommended reading:

Harman, Gilbert, *Change in View: Principles of Reasoning*. Cambridge, MA: MIT Press, 1986.

42

EARLY NEOLITHIC CULTURE

(December 26, 2004)

Neolithic civilizations developed at different times around the world; the advent of farming cultures, cities, pottery, and writing in the Tigris-Euphrates and Nile valleys occurred around 7,500 BC. Originally, archeologists thought the local hunter-gatherers had been forced to settle down and take up farming due to overpopulation and climate change. But some findings indicate much earlier instances of Neolithic lifestyle, including sheep domestication in North Africa (17,000 BC), grain domestication in Palestine (14,000 BC), domesticated horses in southwest Europe (12,000 BC), and sophisticated architecture in Jericho (8,000 BC). Indeed, the symbols in Paleolithic cave paintings resemble too closely for coincidence the later Indus valley, Greek and runic alphabets. They were probably a form of proto-writing that evolved later into true writing.

Remarkably, Plato (600 BC) refers to a tenth millennium BC war between an ancient Greek society and an Atlantic maritime empire, both of which were destroyed by earthquakes and floods. There is some archeological evidence to support this legend, including the first signs of human violence in the Ukraine, the earliest paintings of battle scenes in Spain, and unprecedented numbers of arrowheads from northern Europe to the Negev, all of which came from this period. If the centers of these civilizations were originally on the coasts of the Mediterranean, they may have been submerged by the subsequent rise of sea levels by three hundred feet that inundated the old coastlines.

Neolithic settlers arrived in what are now Palestine/Israel, Turkey, Syria, and Iran late in the eighth millennium. They were not simple nomadic hunter-gatherers. They brought with them many of the skills of later civilizations: advanced architectural design, fully domesticated animals, grains, pottery, and even metalwork. It's not known where they came from. Perhaps they were the remnants of Plato's destroyed cultures. These thriving new cultures subsequently collapsed in the early sixth millennium BC, for a variety of reasons including drought.

A second wave of Neolithic settlements occurred around 6,300 BC, almost precisely when Aristotle, Plato's student, dates the life of Zarathustra, the Iranian prophet who sent missionaries abroad to spread agriculture.

It appears from anthropological research that by this time there was an Indo-European culture, which, although manifesting many regional variations, shared a common linguistic and mythological heritage. It extended throughout Europe, the Middle East, and India and was the precursor to modern civilizations in these regions. It produced the many religions of these areas, as well as most of their languages. Modern Mosaic religions, Hinduism, Buddhism, Jainism, Zoroastrianism, and their offshoots all derive from religions of this period.

The culmination of the early Neolithic period was apparently the city of Çatal Hüyük (6200–5300 BC). It occupied about 30 acres in central Anatolia (in Turkey), and 5000–6000 people lived there. The art and ornamentation found there reveal a standard of living more advanced than other sites of the same period. These people had mastered copper and lead work, and had extensive trade with other societies. The city was apparently a center of power and religion. It contained art styles that mirrored the art of many contemporary societies, from Iran to the Aegean Sea. Indeed, some of the population may have migrated from those areas when their societies collapsed. Their ancestors were probably Paleolithic, yet they anticipated the Bronze Age in many ways. Their culture came to an end when the city was destroyed by an intense fire.

For us, the most interesting developments by the various early Neolithic societies include further spiritual, artistic, technological, and agricultural refinements, an early form of writing, and the earliest evidence of war. Societies rose and fell in a cycle of birth, death, and rebirth that continues to this day.

This week's news:

The Wildlife Society has released a study showing that climate change is already having visible effects on ecosystems. The golden orange prothonotary warbler has been returning a day earlier each spring for almost 20 years as temperatures have climbed. Warming has changed migration routes, blooming cycles, and breeding habits of animals and plants across the continent. Several types of warblers have been shifting north, leaving the budworms they used to eat to attack local firs. Movements of the mountain pine beetle and the oak processionary caterpillar have depleted some forests. The decoupling of communities could cause serious problems. Species could be at risk if they're trapped in places they can't escape. (*Washington Post*)

WASHINGTON: More than $4.5 million from the corporate world enriched Bush's inauguration fund. Northrop Grumman, world's largest shipbuilder and second-largest U.S. defense contractor, donated $100,000. Occidental Petroleum, which will benefit from the president's actions concerning Libya, donated $250,000, as did Exxon Mobil, which reported record profits. Texas oilman T. Boone Pickens gave $250,000. Former Enron President Richard Kinder also gave $250,000. Energy provider Southern Co. gave $250,000. The Nuclear Energy Institute, policy organization of the nuclear industry, gave $100,000. (Associated Press)

The Altamont Pass has the world's largest wind farm, producing enough electricity for 200,000 households. It's also the worst in the country for bird kills. Altamont Pass is a prime hunting ground for golden eagles and other raptors, and scientists estimate conservatively that the turbines kill some 4,700 birds every year. Unable to see the whirling blades, the birds fly into the turbines—nearly a third golden eagles, red-tailed hawks, American kestrels, burrowing owls, and other raptors. Environmental groups favor clean wind energy, but they join with state energy commission officials in calling for taller turbine models, moved to the leeward side of the hills. (*San Francisco Chronicle*)

Recommended reading:

Settegast, Mary, *Plato Prehistorian: 10,000 to 5000 B.C. in Myth and Archaeology*. Cambridge, MA: Rotenberg Press, 1987.

43

EARLY CIVILIZATIONS

(January 2, 2005)

Human social organization has evolved considerably from the first bands of hominids roaming the hot African savanna. With each biological and cultural advance, starting with tool use and the acquisition of language, social organization became more sophisticated. The transition from tribes to chiefdoms occurred with the development of sedentism and agriculture. Chiefdoms are often referred to as *transegalitarian* societies, because they were the less egalitarian transition between egalitarian hunter-gatherers and inegalitarian civilizations.

We all think we know what a civilization is, but anthropologists admit there's no good (scientific) definition. One proposed by Childe in the early1950s is as good as any, as long as we recognize its imperfections. His criteria are surprising in some ways, but they help define the transition from pre-civilized to civilized societies in a cultural evolutionary sense. He proposed that civilizations are characterized for the most part by permanent settlement in dense aggregations (urbanization), nonagricultural specialists, taxation and wealth accumulation, monumental public buildings, a ruling class, writing, predictive science, artistic expression, trade for vital materials, and a decline in the importance of kinship. Many characteristics such as permanent settlement, non-agricultural specialization, taxation and wealth, monumental buildings, a ruling class, art, and trade were already common. But in the context of the new elements of writing, predictive science, and the decline of the importance of kinship, the older elements took on new forms and significance. One universal phenomenon was a growing inequality gap.

Caution: the word "civilization" and its cognates are loaded terms. We often assume civilized is better than uncivilized. But that's a value judgment not shared by some individuals, bands, tribes, and chiefdoms. They can know what civilization is, they can interact routinely with civilized people, and they can appreciate its comforts and conveniences but they may still prefer their own forms of social organization. In our search for ways to live sustainably we need to consider what advantages and disadvantages are inherent in different forms of social organization.

Most civilizations have cities, which differ from pre-civilized settlements in a number of ways. They're typically larger, more socially complex, and are population centers, depending on and controlling large territories for their support. As a nation's population density increases, there's pressure to move to cities for work. Today, most of the human species lives in cities. As cities grow, there is often a disproportionate growth of their slums.

The first civilizations were the first states, defined by Brumfiel (1994) as powerful, complex, institutionalized hierarchies of public decision-making and control. They provide the structure for intricate economies, enact laws, and enforce them. The necessity for these functions provides the rationale for great concentration of power in a small ruling class and further loss of autonomy by the citizens. Most early states depended on tribute from the subjects, a continuation of chiefdoms' practices.

States collapse. In many cases, collapse involves growth of the population to an unsustainable size. In others, collapse is held off by extending dominion over larger territories until the state is overextended. Invasions and internal conflicts often contribute.

A powerful factor separating civilizations from chiefdoms was the development of writing. The earliest forms of proto-writing were little more than tally marks used for inventories. Numbers arose from these. Symbols for concrete objects arose to meet similar needs. Writing systems emerged independently around the world, some based on ideograms, others on alphabets. The degree with which they could represent increasingly abstract concepts probably lagged well behind the developing capacities of humans for abstract thinking, judging from advanced capacities for abstraction in illiterate societies, but there is no question that the increasing sophistication of written material has contributed to cultural advancement, for example in all areas of modern scholarship. The volume of information to be stored and retrieved soon exceeded that which could be handed down in an exclusively oral tradition. Modern electronic information processing has greatly expanded this capacity. When an elite group has exclusive access to literacy, they can control much of the information flow in their society, a source of great power indeed.

This week's news:

The rainforest of the Amazon and the Brazilian savanna straddle the state of Mato Grosso. One of the most biodiverse in the world, the area is home to parrots, jaguars, pumas, the maned wolf, anteater, and giant otter. An explosion of soybean fields and cattle ranches is now crowding them out. The governor of Mato Grosso is Blairo Maggi, the owner of the Maggi group—the Soybean King. The Maggi Group is the largest private soy producer in the world. Maggi has also established transportation infrastructure that further opens the Amazon to development and deforestation. In his first year as governor, the deforestation rate in Mato Grosso more than doubled. However, under pressure from non-government organizations, World Bank president James Wolfensohn has called for an audit, stating in a letter to Brazilian NGOs that, "the audit will provide an independent review of the issue and the results will be made public." (Alternet)

Recommended reading:

Tainter, Joseph A., *The Collapse of Complex Societies*. Cambridge: Cambridge University Press, 1988.

44

HUMAN NATURE 101

(January 9, 2005)

One of our objectives has been to learn about our nature from our history so we can understand why we have so heedlessly destroyed our environment, and how we can modify our behavior on the chance it's not too late. What have we learned so far?

First, we share a family tree with other animals, and we've inherited their anatomy, biochemistry, physiology, and behavior, with modifications due to evolution. Our predecessors were ape-like prehominids, with bigger brains and more versatile arms and hands than their apelike ancestors had. Prehominids developed from vegetarians with an opportunistic appetite for meat, to true omnivores with meat as a major component of their diets. Their behavior progressed a little beyond their simian forebears. They lived in nomadic bands, and had few if any possessions beyond simple tools.

The hominids were brighter, more dexterous, and acquired new skills such as building better shelters and the use of fire and improved hunting strategies and tactics. They learned to adapt to greater environmental challenges brought on by climate change and the varied environments they penetrated. They became generalists, able to switch diets as needed. They began to extinguish other species, not because they constituted a new, killer species or because they were evil or cursed by gods, but because they had better survival skills, eliminating predators and competitors and capturing prey. They had no understanding of human-caused environmental destruction.

The cultural explosion of 50,000 years ago ushered in language, art, religion, and more sophisticated tool use and social interactions. Cultural evolution became a major factor in human development. Although this was a relatively abrupt behavioral change, the biological change probably involved only small modifications of our genome. All the new behaviors were recognizable as new forms of old behavior. Language differed in a number of ways from previous communication, but the differences probably arose from minor changes in old cognitive skills. Religion, art, and technology were also recognizable variants of older human traits.

Migrations distributed people across the globe. They adapted to highly varied conditions from arctic ice to tropical jungles, to arid deserts to oceanic archipelagos. Their genetics changed slightly to adapt to new challenges, but the constant intermixing of gene pools prevented the emergence of truly distinct races.

The agricultural revolution brought about further major changes in lifestyle, including a greater degree of property ownership, possessiveness and acquisitiveness, more pronounced division of labor, and social stratification and inequality. Hunter-gatherers, herders, gardeners, large-scale farmers, tradesmen, shamans, leaders, and warriors were distinct groups with different lifestyles, family structures, diets, and religions. Again, these diverse groups and their diverse behaviors all resulted from easily comprehensible natural processes, not the capricious schemes of fickle gods.

Although we humans occupied a special place because of our remarkable versatility, we were a part of the evolved ecosystem. There was nothing in our natures that set us apart with any unique authority over life on earth. We were not "higher" beings than our companions on Earth were, because all life forms had evolved from the same beginnings to fit their ecological niches with varying degrees of success. However, our greater ability to affect the ecosystem for good or bad exceeded our wisdom to an increasingly perilous degree.

This week's news:

DOVER, PA: Was the landscape around Dover created in just six days? Yes, according to many of the townspeople. Next month their high school will start to preach creationism as science. The school board voted 6-3 that the biology course for 15-year-olds should include "intelligent design" or ID. A bitter debate is raging across the US: will Christianity be incorporated into government or will a meaningful separation of church and state be sustained? Two members of the school board have resigned in protest. A lawsuit filed by the ACLU accuses the school board of violating the First Amendment. Born-again Bush was helped into office largely by creationists. (*The Independent*)

Recommended reading:

Durham, William H., *Coevolution: Genes, Culture, and Human Diversity.* Stanford, CA: Stanford University Press, 1991.

How our Planet Works

45

ECOLOGY

(January 16, 2005)

Why do some species become endangered and not others? As you can imagine, answers to this question are vital to stopping the current mass extinction and ensuring our own survival.

For species with short doubling times like insects, adaptation to an environmental challenge may be quick. For those with long doubling times such as eagles, adaptation is slower and they're more likely to go extinct. The reason is it takes a number of generations for a shift in frequencies of different alleles in a gene pool, the principal mechanism of natural selection. Generally, larger species have longer generations (months, even decades) than smaller ones (weeks, even minutes). This is one reason ecologists are particularly concerned about the high rate of extinction of large animals (say, larger than a rabbit): since they have long doubling times, they're most vulnerable.

Some species have strategies that improve their chances of survival. Ants, other social insects, and mole rats have *reproductive castes* that provide all of the colony's progeny, and sterile castes that provide food, care for young, and defend the colony. Such a system depends on some individuals sacrificing reproductive fitness in favor of relatives. It improves the likelihood of propagating the species' genes.

Another type of individual sacrifice that has survival value is *altruism*. Sharing food (dogs and humans) and defending one's country (humans) are forms of altruism. Again, individuals sacrifice for the good of the genes and the behavior is selected for.

Organisms live in *communities* consisting of multiple species: Community + habitat = ecosystem. Environment acts on species and species affect the environment. Humans have long had more power than other species to make changes in the environment due to mechanization and the use of energy from combustion of fossil fuels to do work. One of our most calamitous effects on the environment is our impact on nature's carbon cycle (redrawn from Brewer, 1979):

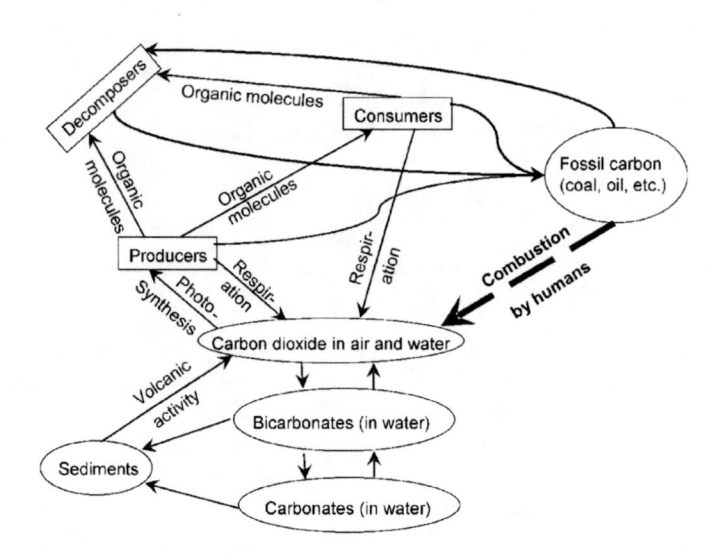

[Figure: Carbon cycle diagram showing Producers, Consumers, Decomposers, Organic molecules, Fossil carbon (coal, oil, etc.), Carbon dioxide in air and water, Bicarbonates (in water), Carbonates (in water), and Sediments, connected by processes including Photosynthesis, Respiration, Combustion by humans, and Volcanic activity.]

With less fossil fuel combustion and a greater abundance of plants, CO_2 concentration would not be building up in the atmosphere. But we're both increasing fossil fuel combustion and depleting Earth's flora. Dumb. CO_2 is a major greenhouse gas that absorbs heat being radiated by Earth into space. The trapped heat warms the atmosphere and the planet. Hello, global warming.

There are many other cycles in which materials are recycled among organisms and the environment, for example the nitrogen cycle in which nitrogen is captured from the atmosphere by soil microbes, and the roots of plants absorb the resulting compounds. The nitrogen is mostly incorporated into proteins, which are ingested by herbivores and plant decay organisms. Decay organisms and carnivores eat the herbivores. Eventually the nitrogen is released back into the atmosphere, in the process of decomposition of dead organisms.

An obvious form of interaction among species is the food chain, which we all learned about in grade school. It's more aptly called a food web because a species typically eats several species and is eaten by several other species:

A food web (incomplete!)

The diagram (redrawn from Brewer, 1979) shows some (not all!) of the species upon which the barred owl depends for food, directly or indirectly. The initial loss of a small fraction of the species in a community may cause a chain reaction in which many more species are lost. In the diagram, if enough insects were lost due to leakage from a toxic chemical dump, all the animals in the second row from the downy woodpecker through the Acadian flycatcher might decline, leaving less food for the barred owl.

A decline in the barred owl population may be a warning of more widespread ecological havoc brewing. Be concerned about insignificant-sounding endangered species. We don't want to find out the hard way, which ones are necessary for our survival, or are indicators of the overall health of an ecosystem.

This week's news:

WASHINGTON: A report from the National Academy of Sciences raises by 20 times the amount of rocket fuel pollution in drinking water considered safe. Perchlorate, the pollutant from rocket fuel, can affect thyroid function. The Academy's National Research Council was asked by the Department of Defense, NASA, and other agencies to determine if perchlorate in drinking water or food is harmful and if so, how much is safe. The Natural Resources Defense Council (NRDC) says federal agencies tried to influence the report's conclusions and has published documents as evidence. White House and EPA officials were not available for comment. (MSNBC)

Bat carcasses are turning up at a wind farm on Backbone Mountain and a nearby wind farm in greater numbers per turbine than at any other wind facility in the world. The bat problem could worsen, conservationists fear, as more wind turbines go up in the area.

Bats serve important roles in nature, and their populations are believed to be in decline. Researchers, who are trying to develop recommendations to avoid the problem, don't know whether bats are attracted to the spinning blades or if their sonar fails to detect the turbines. The wind industry confronted its biggest environmental challenge when early model turbines in Northern California killed large numbers of birds. Newer turbines and more attention to site selection have dramatically cut the number of bird deaths in subsequent projects around the country, though some environmentalists say too many birds are still dying. In West Virginia and Pennsylvania, the turbines are positioned on wide paths cleared amid maple, oak and other hardwood trees, which may have something to do with the bat deaths. Bats might be attracted to the open areas cleared by the wind developers because they can more easily find insects there. Most of the deaths occur between July and September, which includes the months of peak migration. (*Washington Post*)

Recommended reading:

Goodland, Robert (ed.), *Race to Save the Tropics: Ecology and Economics for a Sustainable Future*. Washington, DC: Island Press, 1990.

46

PREDATORS, PREY AND COMPETITION

(January 30, 2005)

Loss of biodiversity (variety of species and of the alleles within their gene pools) can cause the collapse of ecosystems. One factor that affects biodiversity is the patterns of predator-prey and competitive interactions in an ecosystem. The study of predator-prey relationships is another way of looking at the food web. If we loosely define predators as eaters and prey as eatees, we can include all organisms. There's an arms race in this jungle. Prey develop defenses such as camouflage, flight, thorns, and poison. Predators develop acute senses of vision, hearing, and smell and the ability to outrun prey, consume thorns, or resist poisons. Inadequately defended prey are eaten. Predators that can't consume enough prey starve.

Poisonous butterflies often develop distinctive markings that birds soon associate with illness due to poisons, to the benefit of both. Palatable butterflies may evolve markings similar to poisonous ones, to fool the same birds. Turnabout is fair play. Predators can fool prey by looking harmless or even by luring them into traps.

There have been some spectacular successes and disastrous failures when humans have attempted to use natural biological responses for pest control. Garden and farm pests can be repelled using plantings such as garlic and marigolds. Playing alarm cries of birds sometimes works to keep them away from airports. Pheromones can be used to trap insects. However, there are risks to this approach. For example, many pests are in fact species once introduced to control other pests.

Diseases can be considered from an ecological perspective. Infectious diseases are an ecological interaction between pathogen and host. An epidemic is the spread of a communicable pathogen through a population that depends on a threshold number of individuals (*susceptibles*) lacking immunity, and the end of an epidemic depends on the decline of susceptibles due to acquired immunity, death, or quarantine, to below the threshold.

Deficiency diseases generally arise from a cultural dislocation relative to the environment. For example, peoples living on long-term traditional diets rarely show deficiency diseases. But the introduction of corn (called maize outside North America) into southern Europe and Africa caused the eventual appearance of pellagra because corn is deficient in niacin.

Cancers are largely dependent on environment. Perhaps as many as 90% of cancers are environmentally produced. Stress-related diseases are environmentally produced. One source of stress is crowding. It's possible, though still not certain, that increases in crime, alcoholism, divorce, and mental illness in crowded inner cities are similar responses on the human level as are increased aggression, unusual sexual behavior, and breakdown of normal social function in other crowded animals.

In an evolving ecosystem, negative feedback loops tend to favor coexistence of interacting species. Positive feedback can cause extinction of one or both species. If the ecosystem has a complex enough, and therefore redundant enough, web of interdependencies, it can fill the ecological functions of a lost species. The ecosystem is said to have sufficient biodiversity. Conversely, an ecosystem without sufficient biodiversity is at serious risk of degradation or collapse. Current mass extinctions are causing extensive loss of biodiversity all over the world. In fact, they may reflect a collapse already in progress. We'll know soon.

This week's news:

KOBE: The US objects to references to climate change in UN talks aimed at setting up a disaster early warning system. Forty-five hundred experts and officials from 150 countries are meeting in Kobe and are expected to make a list of targets to be met by 2015 on ways to reduce the risks of disasters. UN relief chief Jan Egeland in his opening address to the conference said that besides natural disasters, "We now face threats of our own collective making: global warming, environmental degradation, and uncontrolled urbanization." (spacedaily.com)

WASHINGTON: The Sierra Club responded to Bush's call to build more nuclear power plants in a press release: "In his interview published in today's *Wall Street Journal*, President Bush calls for building more nuclear power plants as a solution to the environmental dangers he now admits are posed by coal burning power plants. Nuclear power is a dangerous energy source that creates more problems than it solves. It is irresponsible for the President to advocate building new nuclear power plants that will endanger communities and serve as potential terrorist targets. Switching from dirty coal plants to dangerous nuclear power is like giving up smoking cigarettes and taking up crack. President Bush's energy policy makes America less safe by increasing our dependence on polluting fossil fuels, nuclear power, and foreign sources of energy. It puts our energy future in the hands of foreign countries, while funneling billions of taxpayer dollars to polluting energy companies. The Bush energy policy does not aggressively pursue energy efficiency and renewable energy, even though these solutions are our quickest and cleanest alternatives. President Bush should scrap his energy bill and replace it with a bill that serves the American people, not just giant energy companies. It's time for America to harness its own innovation to move towards energy freedom. The technology exists today to make our cars and trucks go farther on a gallon of gas, produce more of our energy with clean, renewable sources like wind and solar power, and become more energy efficient." (Common Dreams)

Owens Lake held water for at least 800,000 years. Beginning in 1913 the Owens River flow was diverted to Los Angeles; in the mid-1920s, Owens Lake dried up. It became the single largest source of particulate-matter pollution in the United States, emitting almost a million tons annually. Air pollution around the lakebed reached 25 times the limit allowed by clean-air standards. In 1998, the city was forced to implement dust-control measures including shallow flooding with some Owens River water and planting irrigated fields of salt-tolerant grass. (*Grist Magazine*)

Recommended reading:

Wilson, Edward O., *The Diversity of Life*. New York: W. W. Norton, 1999.

47

POPULATION GROWTH AND CROWDING

(February 6, 2005)

Before we can begin to understand the effects of humans on our planet, we need to learn a little about the ecology of population growth, important for understanding human overpopulation and the evolution and extinction of species.

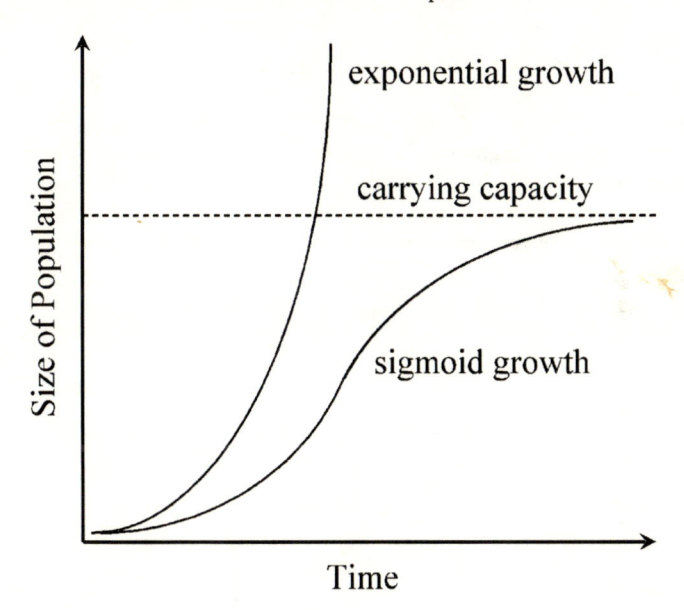

*The graph above shows population size over time, for two different situations. The "exponential growth" curve shows the pattern of increase for an uncrowded population, where each individual has enough resources to reproduce as fast as it can. The population size doubles at uniform intervals (the *doubling time*). The "sigmoid growth" curve shows what happens to exponential growth as the population becomes crowded. Growth slows and population size approaches a limit (the *carrying capacity* of the environment). A favorite argument of those opposed to population control is that human population growth will slow once we're sufficiently crowded. This hasn't happened yet, but we're sufficiently crowded to endanger our survival.

The diagram below shows how the age distributions of populations are influenced by their patterns of increase. If a population is increasing in size, the young outnumber the old and the average age is low. A stable population has a higher average age.

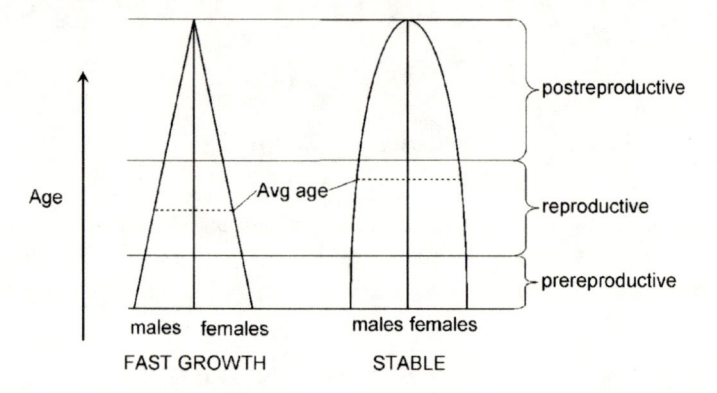

[Figure: Age distribution diagrams showing FAST GROWTH and STABLE populations, with labels for males/females, and brackets indicating prereproductive, reproductive, and postreproductive age groups, with Avg age marked.]

The prevalence of a characteristic (e.g., insecticide resistance in an insect population) can change over time due to growth of a subpopulation possessing or lacking a particular gene or allele:

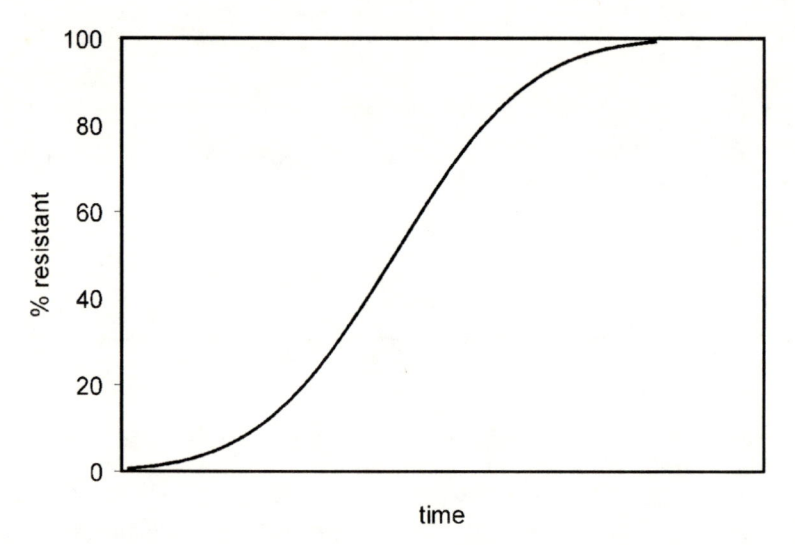

[Figure: S-shaped curve graph with % resistant (0–100) on y-axis and time on x-axis.]

A few insects may be resistant to start with, as illustrated at the left end of the graph. For example, they may have an enzyme that detoxifies a poison produced by the leaves they eat, and one particular allele of the enzyme gene makes a form that happens to detoxify the insecticide. The individuals with that allele will be more likely to survive and reproduce in an area being treated with the insecticide. Their numbers will increase and the number of insects without the allele will decrease.

Some species, like bacteria, can evolve rapidly because they have short generations, sometimes just a few minutes. A generation is the average length of time from birth to reproduction. Others, like large mammals, evolve much more slowly because their generations take years. When there's a sudden environmental change that affects all animals, like a serious drought, the rapidly reproducing organisms are more likely to survive. That's one reason why a global ecological disaster is more likely to wipe out plants and animals, including us, than bacteria.

Uncrowded species evolve differently from those living near carrying capacity. In short-lived communities, e.g., river sand bars and forest edges, species with large numbers of young, eggs, or seeds and early or long reproductive periods are selected for. In a community where species are living near carrying capacity, e.g. mature forests, the species selected for are generally larger and longer-lived.

From these simple principles you can see that the response of a community of species to an environmental challenge will depend on multiple factors, including the genetic diversity (without the right pre-existing alleles, the insects won't develop insecticide resistance) and whether the community is transitory or stable (which determines the types of animals selected for). There are many such factors, not all of them well understood. Because there are so many factors involved, expert ecologists can be wrong in their predictions of the effects of environmental changes. For this reason, they advise a conservative approach, in which species that seem unimportant are protected.

This week's news:

PORTO ALEGRE, BRAZIL: 50,000 people marched through Porto Alegre on the opening of the annual World Social Forum on developing country issues. A report from ActionAid indicates that thirty companies produce 33% of the world's processed food, five companies control 75% of grain trade, six control 75% of the pesticide market, two control 50% of bananas, three control 85% of tea, and one, Wal-Mart, controls 40% of Mexico's retail foods—and Monsanto controls 91% of GM seed. Nestlé, Monsanto, Unilever, Tesco, Bayer, and Cargill have expanded size, power, and influence in the past decade because of trade liberalization advanced by the U.S., Britain, and other G8 countries whose leaders are meeting this week in Davos. "A wave of mergers and business alliances has concentrated market power in very few hands," the report says. It says the companies have excluded local companies from the market, forced prices down, set trade rules to suit themselves, imposed standards that poor farmers can't meet, and are now charging consumers more. Agrifood companies paid eighty-five percent of recent fines on global cartels, with three of them forced to pay $500 million to settle price-fixing lawsuits. "It is a dangerous situation when so few companies control so many lives," said ActionAid's John Samuel. Many food giants are richer than the countries in which they do their business. A coalition including Greenpeace, Friends of the Earth, Amnesty, Via Campesina, and Focus on the Global South are collaborating to press for corporate accountability. (*The Guardian*)

A U.N.-backed conference on biodiversity in Paris has called for a top scientific panel focusing on species loss. A petition by some of leading authorities in the field said, "Biodiversity is being destroyed irreversibly by human activities." The millions of species on Earth are the product of more than three billion years of evolution, "a natural heritage and a vital resource upon which humankind depends in so many different ways." Organisms everywhere are threatened by habitat loss, pollution, over-exploitation, "and, more recently, signs of long-term climate damage." Edward Wilson, a leading U.S. biologist, said it would cost $3 billion and take twenty-five years to compile an inventory of the world's species. Saving the twenty-five most threatened, most biodiverse areas would cost $25 billion. Experts are convinced human interference is driving the planet toward the worst mass extinction in 65 million years, when most of the world's species went extinct. (Al Jazeera)

Recommended reading:

Brewer, Richard, *Principles of Ecology*. Philadelphia: Saunders, 1979.

48

GOOD LOOPS AND BAD LOOPS

(February 13, 2005)

To understand how conditions can remain stable and recover from fluctuations for billions of years, and then suddenly shift rapidly to a new (and unfavorable for us) steady state, we need to understand the concept of feedback. We've all heard the word "feedback" to describe the howl in a public address system that results from putting the microphone too close to the loudspeaker or turning up the volume on the PA amplifier too high. That's positive feedback.

Negative feedback generally supports stability. In a predator-prey relationship, if the number of predators goes up for some reason, they consume more prey. This decreases the prey population, so fewer predators can survive, and the number of predators comes back down. In other words, the increase in predators causes a negative effect: a decrease in prey, which *feeds back* to decrease the number of predators. If the number of prey goes up for some reason, the number of predators increases because they have a larger food supply. The increased predator population consumes more prey, returning the prey population to previous levels. Notice that the effect of increased or decreased prey is decreased or increased predators, respectively. The sign of the change of predator numbers is the reverse (negative) sign of the change of the prey numbers. For that reason, it's called negative feedback.

Positive feedback generally disrupts stability. The PA system was one example. Another example from nature is the capture and release of CO_2 by rocks and water. CO_2 dissolves in water to form bicarbonate and reacts with minerals to form carbonates. Warming decreases the ability of rocks and water to store CO_2, and causes CO_2 to be released back into the air. CO_2 is a greenhouse gas: it decreases the radiation of heat into space from Earth. Increased atmospheric CO_2 therefore increases the greenhouse effect, causing further warming. Warming causes more warming in this positive feedback loop. Since the air temperature change is the same sign as the air CO_2 change (both increase), this is called a positive feedback loop.

Many positive and negative feedback loops affect atmospheric temperature. There are many feedback loops affecting the survival of species as well. As long as the negative feedback loops dominate, stability is maintained. However, if there is such a large perturbation that the negative feedback loops are overwhelmed, then the positive feedback loops may dominate, rapidly driving the system to a new steady state.

Species can adjust to large environmental changes by evolutionary adaptation. But evolution is a slow process, and if the environmental change is too rapid, species fail to adapt quickly enough and go extinct. In a later chapter, I'll explain why even small changes in the environment can lead to massive extinctions, and how humans have caused major extinctions past and present. If extinctions are sufficiently massive, most of the world's ecosystem will collapse, leaving the Earth uninhabitable for all but some simple organisms. Complex life might once again evolve over hundreds of millions or billions of years, but chances are, this planet would never again be home to an intelligent species.

One property of a system in which positive feedback has overwhelmed negative feedback is an increasing rate of change (acceleration). The danger of such acceleration lies in the possibility of a runaway system that we can't stabilize. One indication of such acceleration is an increased rate of change of a global phenomenon like atmospheric temperature. Watch news reports for signs that global warming is accelerating.

This week's news:

WASHINGTON: NARAL Pro-Choice America, the nation's leading advocate for personal privacy and a woman's right to choose, said Americans are outraged that the Bush Administration will allow Medicare to cover prescriptions for sexual enhancement drugs such as Viagra while blocking efforts to give women access to emergency contraceptives that would reduce the need for abortion. Nancy Keenan, president of NARAL Pro-Choice America, issued the following statement: "Americans opened their newspapers today to read yet another example of the Bush Administration not giving women's reproductive health needs equal treatment. The president is willing to allow coverage of Viagra under Medicare, but his FDA continues to delay a decision on whether women can have access to Plan B, a morning-after pill, from their pharmacies. Despite overwhelming support for the pill's safety and effectiveness from medical experts and the Food and Drug Administration's own staff, the Bush Administration blocked such access last year. Because the FDA is now considering another petition, it has an excellent opportunity to prevent unintended pregnancy and reduce the need for abortion. Today's news is especially disturbing to women in this country who either do not have their prescription birth control covered by their health insurance or face the prospect of their pharmacists refusing to fill these prescriptions. We call on the Bush White House to drop its opposition to the morning-after pill to ensure that women have equitable access to prescriptions for their reproductive health. This issue isn't about Viagra; it is about fairness and equality." (Common Dreams)

Recommended reading:

Mayr, Ernst, and Provine, William B. (eds.), *The Evolutionary Synthesis: Perspectives on the Unification of Biology.* Cambridge, MA: Harvard University Press, 1980.

49

HUMPTY DUMPTY

(February 20, 2005)

It's 2050, we're twelve billion eggs, and there's no more room on the wall. Let's use a little parable to illustrate how we got here.

Imagine American newlyweds on their honeymoon in the South Pacific in the year 1915. Their ship goes down in a typhoon, and they're marooned on an island with a carrying capacity (the maximum number of people it can support) of about 1000 people. They nostalgically name the island America. They start making babies, and their babies grow up to make more little Americans. There's not much else to do, and the only contraception is abstinence, but why abstain? Through the generations, population size on the island doubles every fifteen years. After 120 years, in the year 2035, the island population reaches 512. It would seem they have another 120 years before they reach America's carrying capacity. But the population reaches 1024 people after only 15 years, in 2050. What happened?

The population *doubling time* was only 15 years. The population doubled, from 512 to 1024, in just 15 years, not the 120 years it took to grow from 2 to 512 people. All of a sudden, population size has begun to exceed the island's carrying capacity.

It's quite likely that they'll pass this point of no return before they know it.

Imagine that the island has the usual interdependent ecological system. America's native crops depend on native insects for pollination. Other local organisms, unnoticed, replenish the soil and provide the food chain upon which the Americans' food animals depend.

Once the carrying capacity of the island is exceeded, the system doesn't immediately collapse. In fact, the islanders notice no change, and continue making babies. But to maintain "the American way of life," they have to keep clearing land for living and farming space, and some species are crowded out of existence as their habitats are destroyed. The Americans don't even notice when the trees in which the pollinating insects live are lost to deforestation. At first, they don't notice when pests begin to increase because their other food sources have declined, and the animals they hunt and trap are no longer replaced as quickly. By the time they do notice, too many species essential to their survival have gone extinct. They don't really understand what went wrong. All they can do is clear more land, work harder hunting and farming, and bury their children and elderly as they die of malnutrition, deficiency diseases, and soon, starvation. In a few generations, America is just another uninhabited desert island.

If we do go extinct this century, the human race will have lived about 120,000 years and will have driven to extinction all of the larger organisms on Earth. Ironically, those who fought population control will have thwarted their own goal of making as many babies as possible. By shortening the span of human existence to a fraction of its potential length, they will have guaranteed that fewer humans, not more, will have been born during our stay on Earth.

It's time for you to consider seriously having no children, or limiting yourself to just one child. If you want more children, adopt. Start changing our society's attitudes about making babies. Explain to your relatives, friends, and acquaintances why we must cut back on population size.

This week's news:

WASHINGTON: Rep. Camp (R-MI) has introduced a bill that would offer tax credits of $600–$4,000 for hybrid vehicles. Bush supports the incentive, as do Democrats and environmentalists. The existing federal deduction of $1,000 for a hybrid vehicle drops to $500 next year and will be phased out in 2007. Some states also offer tax breaks for hybrids. Hybrids usually sell for $3,000 more than conventional vehicles. Other technologies covered by Camp's bill, endorsed by auto manufacturers, include leaner burning diesel fuel engines and fuel cells, which are still in the testing phase. Bush proposes extending the credit through 2008. The president would also like to make a credit of up to $8,000 available for the purchase of fuel cell vehicles between 2005 and 2012. (MSNBC)

A group of scientists advises increasing the albedo of clouds, reflecting more sunlight back into space, giving us more time to stop global warming. Stephen Salter, an Edinburgh University engineer, proposes atomizing seawater, which would evaporate, leaving tiny particles of sea salt. These would be drawn into clouds by air currents. The particles would act as nuclei for extra droplets to form, making the clouds more reflective. Filling the clouds with smaller droplets also makes clouds last longer before they disintegrate. Salter envisions thousands of unmanned boats carrying equipment to disperse atomized seawater into the air. The boats and their equipment would be driven by wind. Satellites would direct their movements, placing the boats where clouds could be modified most effectively. Latham's computer simulations predict that modifying an area covering around 3% of the Earth's surface will balance the warming from doubled carbon dioxide levels. (*The Guardian*)

More than 200 scientists employed by the U.S. Fish and Wildlife Service say they've been told to change their findings to lessen protections for plants and animals. More than 50% of researchers responding to a survey said they knew of cases where timber, grazing, development, and energy companies used political pressure to change scientific conclusions harmful to their businesses. Bush administration officials have been hostile to the Endangered Species Act, maintaining that it's hurt business interests and hasn't helped wildlife. Along with GOP cronies in Congress, Bush is trying to emasculate the act. One scientist in the Pacific region wrote, "Science was ignored and worse, manipulated, to build a bogus rationale for reversal of...listing decisions." More than 20% of respondents said they had been "directed to inappropriately exclude or alter technical information," and more than half reported they'd been ordered to alter findings to lessen protection of species. (*Los Angeles Times*)

PARIS—The Intergovernmental Panel on Climate Change (IPCC), consisting of 2500 climate experts from around the world, has concluded unanimously that man-made CO_2 is causing global warming. The 1990s was the hottest decade on record; 2004 was the fourth warmest year, with the highest temperature increases in Alaska, the Caspian Sea, and the Antarctic Peninsula. Even if Kyoto is enacted in full, industrialized signatories will at most reduce their emissions by a couple of percent over 1990. The International Energy Agency (IEA) calculates CO_2 levels have risen 15% since 1990 and will increase another 60% by 2030. A scientific conference at Exeter has concluded that in order to cap the temperature increase over pre-industrial times to 3.6°F by 2100, CO_2 will have to peak no later than 2020 and then fall swiftly. Today's annual emissions of six gigatons would have to be roughly halved by 2095. (Agence France Presse)

Recommended reading:

Hollingsworth, William G., *Ending the Explosion: Population Policies and Ethics for a Humane Future.* Santa Ana, CA: Seven Locks Press, 1996.

50

THE HORNS OF OUR DILEMMA

(February 27, 2005)

One reason most underdeveloped countries (UCs, for short) stay underdeveloped is that there are not enough resources on our planet to support everyone at the level of developed countries (DCs for short). As large developing countries like China seek our standard of living, there will be serious conflicts over resources.

See http://www.newamericancentury.org/ (Project for the New American Century) for the traditional US approach: seize and maintain global control of strategic resources (think "oil"). That immoral approach cannot work much longer. Our overextended military empire will bankrupt us. People of the world will react with revulsion to the slaughter and misery. The PNAC also ignores overpopulation, environmental degradation, and the ongoing mass extinction, for ideological reasons.

None of these problems would have reached their current levels if we had halted population growth around 1850, at the beginning of the Industrial Revolution. Malthus (http://www.ucmp.berkeley.edu/history/malthus.html) described the overpopulation problem back in 1789.

At this juncture, further population growth is obscene. If it angers you to think that our ancestors were responsible for the fix we're in, think how angry our descendants—if we have any—will be if we don't turn things around.

But how can we persuade people not to multiply? Many people expect their children to support them in their old age. Others consider having children to be a religious obligation. Many think their lives would be empty without children. None of these arguments is sensible. They're all rationalizations for a very basic instinct. Every species has to have enough reproductive drive not to become extinct. It's in our genes.

The irony is that this instinct, which once had survival value for our species, is now endangering our survival.

Population control requires global cooperation, and there's the rub:

- Most people in UCs have inadequate contraception (the U.S. fights U.N. family planning efforts). Many mistakenly view their bloated populations as assets for a variety of reasons, including safety in numbers: reserves against famines, epidemics, and colonial oppression.

- People in DCs depend for their way of life on the transfer of wealth from UCs through mercantilism (DCs buy raw materials and labor cheap from UCs, and force UCs to buy expensive DC finished products and food. See http://www.landandfreedom.org/ushistory/us3.htm). Adam Smith exposed and opposed this corrupt and ultimately self-destructive practice in 1776. The American Revolution was triggered by English mercantilism.

Today mercantilism has a new, cruel twist: as a condition for loans, DCs force UCs to cut back on education and health care and sell utilities to DC companies cheap. The new mercantilism is part of a scheme called globalization.

There is a civilized alternative. It's called sustainability, in which we adjust our population size and limit our use of resources so they're naturally replenished as fast as we consume them. We have to commit to such a program, or face annihilation.

This week's news:

JAKARTA: Environmentalists have uncovered the world's biggest smuggling of a single type of wood, from Papua New Guinea to China. The illegal billion-dollar racket is threatening the last intact tropical forests in the Asia-Pacific region. Around 300,000 cubic meters of logs/month of merbau, a hardwood used mainly for flooring, is being stolen. A small number of international crime syndicates are controlling this theft. The syndicates pay $200,000 per shipment in bribes to clear Indonesian waters. Collusion with Indonesia's powerful military is rampant. An area of forest the size of Switzerland is lost each year. More than 70% of Indonesia's original forests are gone. Communities in Papua receive around $10 for each cubic meter of merbau on their land; the same logs are worth $270/cubic meter in China, $2,700/cubic meter in North America. Smugglers have come to Papua because forests on Borneo and Sumatra have been depleted. (MSNBC)

WASHINGTON: Sen. James Inhofe (R-OK), who heads the Environment and Public Works Committee, has ordered two organizations that oppose Bush's ""Clear Skies" proposal to hand over their financial and tax records, ten days after a representative of the groups criticized the proposal before a Senate subcommittee. Inhofe is the leading sponsor of the administration bill. The executive director of the organizations charged intimidation. "Clear Skies" would establish new standards for pollutants and provide a market-based approach preferred by industry. Proponents claim it would cut sulfur dioxide, nitrogen oxide and mercury emissions by 70%. Opponents argue reductions could be achieved better through existing Clean Air Act regulations. (*Los Angeles Times*)

After a costly war with neighbouring Eritrea from 1998 through 2000, Ethiopia is struggling to feed its 67 million people. Almost 70,000 were killed in the war. About 4000 U.N. peacekeepers still police the border. Sporadic fighting between ethnic groups has left a growing number of people homeless and in need of food. The average income is $100/year, 1/5 the average for sub-Saharan Africa. Seventy-five percent live on less than $2 a day. The U.N. Office of Humanitarian Affairs (OCHA) warns that grain supplies could run out in a month in Ethiopia. Donors provided only 11% of requested aid for March. For the full year, less than a quarter of the food requested has been provided. The number of people needing food is expected to increase from the current 2.4 million to 2.9 million in March and to 3.1 million in April. (Al Jazeera)

Increased sorghum prices reflect grain shortages in many areas in Sudan, already suffering widespread hunger, according to the U.N. World Food Programme (WFP). Donor funding has fallen far short of what is needed to feed 5.5 million people, and the number of people at risk may be far more. Malnutrition is severe in the drought-affected regions. War has aggravated the food shortages. (Al Jazeera)

Recommended reading:

de Rivero, B. Oswaldo, *The Myth of Development: The Non-Viable Economies of the 21st Century.* London: Zed Books, 2001.

Greider, William, *The Soul of Capitalism: Opening Paths to a Moral Economy.* New York: Simon and Schuster, 2003.

Hartmann, Thom, et al., *The Last Hours of Ancient Sunlight: The Fate of the World and What We Can Do Before It's Too Late.* New York: Three Rivers Press, 2004.

51

DEMOGRAPHIC TRANSITION THEORY

(March 6, 2005)

All of our worst problems—environmental degradation, resource depletion, even climate change and mass extinctions—can be eliminated or alleviated, but only if we reduce our population size enough, fast enough. Population pressure has caused wars, destroyed great civilizations, and driven brutal colonization, genocide, and exploitation. Overpopulation has been recognized for many centuries, yet progress in curbing it has been negligible. Why? Let's consider the present state of affairs.

The average age in the world is 27 years. In sub-Saharan Africa, 45% of the population is under 15. In Asia, 35% is under 15, and in Latin America, 33%. Ninety-seven percent of current population growth is taking place in only six counties: India, China, Pakistan, Nigeria, Bangladesh, and Indonesia. The highest average age and lowest (even negative) growth rates are found in the economically most developed nations. Nations with the highest energy use per capita have the lowest population growth.

The *demographic transition theory* originated from the observation that in undeveloped countries, particularly in Europe, population trends went through three phases. (1) In the *pre-modern* period, both birth rates and death rates were high and population growth remained low. (2) In the *transition* period, death rates declined and birth rates increased due to an increased quality of life, so population size grew. (3) In the *modern* period, with continued high quality of life, birth rates declined and population growth slowed, stopped, or reversed. Although there was no reason to assume that this was universally applicable, the assumption that improving the quality of life would stabilize population levels strongly influenced population policies. Unfortunately, four facts ruined that prospect. First, many "pre-modern" societies were already devastated by overpopulation. Second, the increased quality of life in Europe was maintained partly by exporting its surplus population to colonies around the world, and exploiting the natives and their resources. Third, left to itself an overpopulated, underdeveloped country can neither support its existing population nor slow its growth, so its quality of life can only decline. Fourth, if outside assistance takes the form currently called "free trade" or "globalization" the country loses the last of its self-sufficiency, its economy collapses and its population continues to explode.

To this day, many opponents of population control (especially the Catholic Church and the United States) claim that populations stabilize at supportable levels automatically.

They assure us that fertility rates decline as populations begin to deplete their resources. This is a cruel theory at best, which supposes that population stabilizes only once the misery level is high enough. But the reality is even worse: mortality rates rise to match or exceed birth rates. Consequently overpopulated societies either collapse once their resources become too depleted, or if they can, they delay the collapse by resorting to expansion (destroying societies that are living within their means)—eventually collapsing anyway when they become overextended.

Social and economic obstacles impede population control and development. Poor countries have more diseases, lower life expectancies, higher illiteracy, and less infrastructure to meet basic human needs. The people work harder if they can get work at all, and they still can't make ends meet. These people aren't lazy: they work harder than you or I do given the opportunity. Immigrants from poorer countries are willing to take the most menial jobs, at the lowest pay.

Please urge everyone you know not to have more than one child.

This week's news:

House Resources Committee Chair Richard Pombo (R-CA), has produced a "report"—written not by scientists but by aides to Pombo and Rep. Jim Gibbons (R-NV)—that flies in the face of evidence that mercury from coal-burning power plants is dangerous to health. Two of their claims are especially startling, as science journalist Chris C. Mooney observes: 1) "There has been no credible evidence of harm to pregnant women or their unborn children from regular consumption of fish." 2) "Current, peer-reviewed scientific literature does not show any link between U.S. power-plant emissions and mercury in fish." The report covers up mountains of data showing that mercury causes abnormal fetal development and learning and memory problems in children, and increased cardiac problems in adults. The report also attempts to cover up the fact that coal-burning power plants are the major industrial source of mercury pollution in the U.S. A Clinton-era proposal would have required reduction of mercury emissions using maximum achievable control technology (MACT), which would cut mercury as much as 90% within a year, in 2008. Bush proposes a cap-and-trade program requiring smaller and slower reductions of 70% from 2005 levels by 2018, which critics argue wouldn't have full effects before 2025. The Pombo-Gibbons report references "data" from the industry-funded, quasi-scientific Edison Electric Institute and Electric Power Research Institute. The report also attacks greens: "As a result of the well-funded effort to push their political agenda, environmentalists have caused American citizens to become unnecessarily concerned about possible adverse health effects from exposure to trace amounts of mercury." The National Academy of Sciences and the EPA have warned of the risks of mercury exposure and determined that individuals should restrict their intake of fish and other seafood in order to limit exposure. The Food and Drug Administration and 45 state agencies have released health advisories about dangerous mercury levels in fish. A tri-partisan coalition of 45 senators last April wrote urging the EPA to strengthen their mercury plan, saying current proposals "fall far short of what the law requires and they fail to protect the health of our children and our environment....We do not believe [they] are sufficient or defensible." Gibbons revealed his true agenda, saying, "With a more restrictive, unnecessary regulation we could see a large portion of this country's coal supplies become useless." Olivia Campbell, of National Wildlife Federation, says, "If you spread the cost of applying pollution controls out to average households it would be the equivalent of about a cup of coffee a month—ranging from less than a dollar to three dollars, depending on the utility." These costs are nothing compared to the human health costs posed by the neurotoxin. (*Grist Magazine*)

In his State of the Union address, Bush stated that the U.S. needs "reliable supplies of affordable, environmentally responsible energy," and urged Congress to "pass legislation that makes America more secure and less dependent on foreign energy." However, his Fiscal Year 2006 budget includes major cuts in renewable energy, energy efficiency, clean air, and climate change related-programs at the Departments of Energy, Agriculture, and Transportation, the Environmental Protection Agency, and other agencies. The Bush budget would phase out DoE's hydropower program and the Advanced Hydropower Turbine, examining fish-friendlier turbines, just at the time when full scale testing is about to begin. The cuts come on top of earlier reductions. There would be significant increases for wind energy, the hydrogen program, and the fuel cells program. The hydrogen program as currently envisioned is not a truly renewable energy program because it supports hydrogen production from fossil fuel and nuclear sources. (*Energy Bulletin*)

WASHINGTON: Paul Hoffman, a Bush political appointee who has no biological training, has ordered the National Park Service to allow the installation of artificial water systems in California's Mojave National Preserve. Maintaining that the artificial water sources ("guzzlers") are illegal and will harm native wildlife, Public Employees for Environmental Responsibility (PEER) and the Center for Biological Diversity today filed suit to stop the plan. One of Hoffman's emails contends guzzlers enhance "coyote and varmint hunting" on the Preserve. The Mojave National Preserve covers more than 3 million acres of desert and provides habitat for over 2,500 species, 100 of which are considered threatened. Wildlife experts contend expansion of artificial watering in the Preserve hurts native species by drying up natural springs and wetlands upon which they depend; drawing concentrations of ravens and other animals that prey especially on young tortoises (dead or dying desert tortoises have been found at 30% of the current guzzlers); and potentially spreading Africanized honeybees. Fifty-seven scientists specializing in desert ecology have signed a letter opposing the guzzlers. (Common Dreams)

Recommended reading:

Leisinger, Klaus M., et al., *Six Billion and Counting: Population Growth and Food Security in the 21st Century*. Washington, DC: International Food Policy Research Institute, 2002.

52

SMALLER WILL BE BETTER

(March 13, 2005)

In order to sketch a quick picture of sustainability, I'm going to present my view, based on what we've learned in previous chapters. Remember, in science you're always welcome to disagree but in the end it's the evidence that will decide who's right, not holy scripture or an opinion poll. As the evidence comes in, you and I may change our minds more than once.

To live long and prosper, the 6.5 billion world population must come down to 4 billion by 2050 instead of the currently projected 8–12 billion, and then further decline to 1 billion (1850 world population) in 2100 and 600 million (present U.S. population) in 2150. I believe six hundred million people could sustain a very satisfying lifestyle at less than the 1850 worldwide level of renewable resource consumption using modern technology. I believe that one billion people could sustain only a bare subsistence.

If, starting today, all over the world, the birth rate were set at one child per couple, population would begin to decline, but at crucially different rates for countries with different age distributions. With a starting global average age of 35, similar to some developed countries, we could reach our objectives (dashed line in chart). I based my calculations on simple assumptions: uniform age distribution from 1 to 70 today, every couple reproduces at age 25, and everyone dies at age 70. The average age would go from 36 today to 46 in 2051 and stay there until we return to 2–2.1 children/couple. Assuming everyone works from age 20 to 65, the percent of the population of employable age would actually go up, from 66% today to 76% in 2019, and would fluctuate between 69 and 77% thereafter. The much-feared decrease in the ratio of working to dependent people doesn't materialize. However, a larger fraction of dependent people would be elderly, so average cost of care would rise. These changes are already happening in many developed nations without any upheavals, except that the Germans and Italians fear their culture will be eroded.

World Population

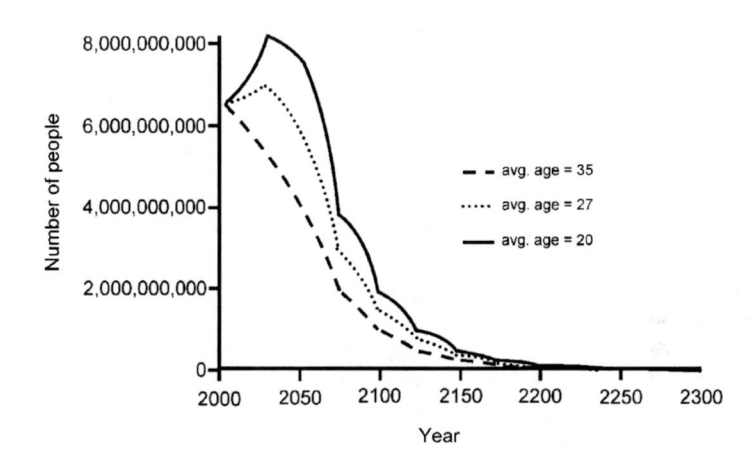

Figure: World population graph showing three projections from 2000 to 2300, with curves for avg. age = 35 (dashed), avg. age = 27 (dotted), and avg. age = 20 (solid). Y-axis shows Number of people from 0 to 8,000,000,000.

The bad news is that the current average age worldwide is actually 27. The dotted line indicates the projected population trend. We can't meet our objectives: population won't reach 3 billion until 2074, 1 billion in 2117, and 300 million in 2161. However, these are probably close enough for countries with current average ages of 27 or less.

What about countries with the worst average ages, around 20? We can get some idea by assuming the whole world has such an age distribution, illustrated by the solid line. We wouldn't reach 3 billion until 2089, 1 billion in 2123, and 300 million in 2164. We probably don't have that much time. Probably the least unjust solution would be for countries with low average ages to have fewer than one child per couple, other countries accept their refugees, and less overpopulated countries provide technical and financial assistance to more overpopulated ones so they can achieve sustainability at the same standard of living. This would help ensure universal compliance with the treaties that would be necessary.

For worldwide social and economic stability, developed nations must cede to developing nations the means independently to produce goods for which they have the necessary resources. Excess production in a nation can be traded for other goods with the nearest countries that can provide them, to minimize fuel use for transportation. The final population level of any region of the world will have to be determined by what it can sustainably produce while maintaining a zero balance of trade.

We'll need to convert to sustainable, non-polluting energy, requiring substantial capital investment. Those costs will be more than offset by shifting government subsidies from fossil fuel to renewable energy sources, distributing work, wealth, and taxes more equitably, lowering per capita consumption, and eliminating destructive and expensive burdens such as corporate welfare, prisons, the military and the associated industrial complex, and the hidden costs associated with pollution. Energy rationing need not be worse than the Second World War if we start soon. Fossil fuel and nuclear power production would be replaced by solar, wind, tidal, and geothermal, at costs made affordable by mass production and lower consumption.

Commuting distances will have to be shortened by bringing home and work closer together. We'll switch from cars to public transit for travel. Rural residence will be limited to rural workers. Farming and manufacturing will be shifted from national and international corporations to the local level. We'll use railroads, river vessels, and coastal shipping for transcontinental transport. Ships and ocean-going barges will be most economical for the greatly reduced intercontinental transport. The cost of changeover will be offset by elimination of airline and automobile industry subsidies and higher fuel costs, lowered road maintenance costs, and reduced consumption.

Come the next elections, have enlightened candidates ready to run at all levels to throw out the bums. You're the people.

This week's news:

Women's rights and human rights leaders are calling on the Bush administration to reaffirm the historic women's rights platform at the U.N. The Platform of Action on Women's Rights was signed by 189 countries at the U.N. Fourth World Conference on Women held in Beijing ten years ago. The U.S. threatens reaffirmation of the platform by proposing the following insertion: "…while reaffirming that they do not create any new international human rights, and that they do not include the right to abortion…" The platform committed countries to protect women's human rights including freedom of political participation and access to education, employment, and health care. "The U.S. should reaffirm, not retreat from, women's rights and equality," said Eleanor Smeal, president of the Feminist Majority. "Globally, this is about saving women's lives." Women's rights and human rights leaders see the US stance against unequivocally reaffirming the platform as going against its rhetoric supporting women's rights in Afghanistan and Iraq. (truthout.org)

Major supermarkets spend millions of pounds each day keeping warehouse-sized stores overflowing with foods. But the true costs of production and transportation are far more than suggested by checkout prices. Consumers are indirectly spending billions of pounds a year more on food without realizing it. A different way of looking at prices is to calculate the savings that could be realized if things were done differently. Each person in Britain spends an average of £24.79/week for food. However, if the hidden costs of transport and the impact on the environment were included, this bill would rise by 12 per cent, a recent study has found. If all of our food came from within 12 miles of where we live, we would save £2.1 billion/year in environmental and congestion costs. If shopping by car were replaced by bus, bicycle, or walking, an additional £1.1 billion would be saved. If all farms in Britain were to use organic methods, environmental costs would drop by £1.1 billion/year. (*The Independent*)

MAPUTO, MOZAMBIQUE: The U.N. reports that beyond a devastating impact on culture, the death of a language wipes out centuries of expertise in preserving ecosystems—leading to grave consequences for biodiversity. Half of the world's 6,000 languages will disappear by 2100. A third of those are spoken in Africa and 200 already have fewer than 500 speakers. Half the world's people now use one of just eight languages: Chinese, English, Hindi, Spanish, Russian, Arabic, Portuguese, and French. Anthropologists think people whose ancestors have lived for tens of thousands of years on India's Andaman and Nicobar islands survived Asia's tsunami because of oral wisdom: signs in the wind, the sea, and the flight of birds alerted them to flee to higher ground. (AP)

Recommended reading:

Morrison, Roy, *Ecological Democracy.* Boston: South End Press, 1995.

53

THE BIGGER WE ARE, THE HARDER WE FALL

(March 20, 2005)

Population reduction, to be of any use, will have to be drastic and quick, because we've waited so long to do it. With an average age worldwide of 35 years, uncommonly old even in developed countries, we would have to go to one child per couple today to lower world population far enough, fast enough. For lower average ages, it is worse. The first chart illustrates three age distributions, with ascending ages along the left axis and number of people at each age on the bottom axis. This is the reverse of the customary graphing convention, but demographers do it this way. For simplicity of modeling, everyone dies at 70. The dashed line depicts an age distribution for a population with an average age of 35. They have a nearly uniform age distribution of around 100,000,000 people at each age level. The dotted line reflects the age distribution for an average age of 27, the current global average. There are 3–4 times as many 1-year olds as 70-year olds. The solid line is for a population with average age of 20, characteristic of the most rapidly reproducing peoples on earth. The ratio of 1- to 70-year olds is 12:1. For easy comparison, each population distribution is plotted as though it consists of all 6.5 billion people on the planet.

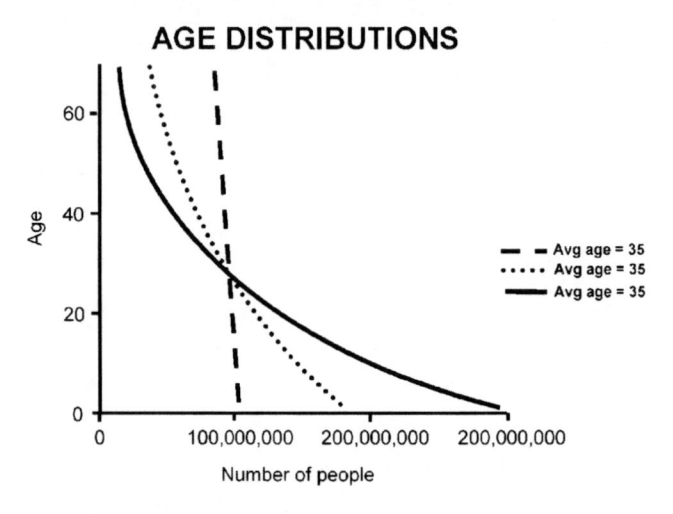

AGE DISTRIBUTIONS

The second chart shows the effect of going to a global policy of one child per couple, for the three different populations. We used the same chart in Chapter 52. The "oldest" population (dashed line) starts to shrink immediately, and reaches 3 billion about 30 years before the dashed curve, with the dotted curve in between. Unfortunately, we don't have 30 years to spare.

World Population

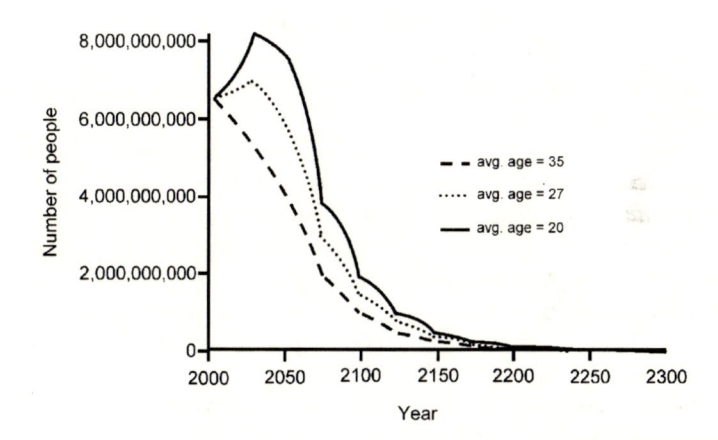

The really bad news is that the two younger populations increase in size for about 30 years because major portions of their populations are younger than the child-bearing age of 25 at the time they all shift to the one-child rule. This is disastrous because the more people there are during this period, the more damage they do to their environment, and the fewer resources will be left for the survivors, if any, one or two centuries from now. For every person, for every year that a population remains unsustainable, they inflict more damage.

To get a sense of the relative amounts of damage caused by the three population groups, draw a horizontal line across the graph at around 5 billion people. The area bounded by the left axis, the line you've drawn, and the line representing the group's population size indicate the relative damage done by a group. The area under the dotted line is at least twice as large as the area under the dashed line, and the area under the solid line is at least three times as large as the dashed-line area. In regions where there are that many children soon to start having babies, the few remaining options are very unpalatable indeed. Moreover, inaction will bring humanitarian disasters of unprecedented magnitude. Even if we go to a one couple-one child limit or less, it's too late for population reduction alone to solve our problems. A drastic reduction in population size will be indispensable, but other measures will be required as well.

This week's news:

WASHINGTON: World Resources Institute (WRI): Two thirds of Caribbean reefs are threatened by over-fishing, runoff of pollution and sediment, and bleaching from warming oceans, disease, and more frequent hurricanes. Human activity has undermined the health and vitality of reefs. Caribbean coral reefs provide goods and services with an annual net economic value in 2000 between $3.1 billion and $4.6 billion from fisheries, dive tourism, and shoreline-protection. (Common Dreams)

An article in *Nature* reports that more than 500 million people—nearly twice earlier estimates—were infected by the deadliest form of malaria in 2002. One person in three worldwide—2.2 billion people—are at risk. The disease kills 1 million/year in sub-Saharan Africa alone, mostly children under five. It has developed resistance to many drugs. In 2002, there were around 515 million cases worldwide, 365 million in Africa: 1 million new cases per day. Control requires safe water, effective public health, education, up-to-date drugs, and mosquito nets. (*The Guardian*)

UN: The US, under pressure, has withdrawn its demand that a UN declaration of women's equality disclaim a "guarantee" of the right to abortion.

WASHINGTON: The American dream of prosperity and material luxury will not work for China or India because there are not enough resources, according to the Earth Policy Institute (EPI). Chinese consumption of grain, meat, coal, and steel has overtaken that of the United States, although Americans still consume more oil than Chinese do. To reach the US 2004 meat consumption, China's total meat consumption would equal 4/5 of current world meat production. If China were to burn coal at the current US level, they would need more than the full current world production. For the Chinese to consume oil at today's US per capita level, China would need 20% more oil than the world currently produces. (Common Dreams)

SYDNEY: Australian scientists have confirmed that deforestation means less water evaporated from vegetation, resulting in less rainfall. They examined the cycle of a heavy molecular version of water common in the Amazon that evaporates more readily through plants than from lakes and rivers. They discovered a decline in heavy-molecule water since the 1970s. It turns out that the only plausible explanation is that heavy-molecule water is no longer being returned to the atmosphere to fall as rain due to less vegetation, signaling a relationship between deforestation and rainfall. "The bottom line is for the first time we can tell the difference between moisture that has been transpired through the plants, and water that has come through the rest of the water cycle. Trees play a critical role in moving heavy-water molecules through the cycle. This is the first demonstration that deforestation has an observable affect on rainfall." (Reuters)

LONDON: A photo of Mount Kilimanjaro devoid of its 11,000-year snowcap will be used as evidence of global warming at a Group of Eight meeting of rich nations. A picture of coastal defenses in the Marshall Islands threatened with swamping from rising sea levels will also be presented. The Kyoto Protocol became active in February but is still ignored by the US, the worst emitter, and imposes few restraints on China, with emissions increasing rapidly. British think-tank Institute for Public Policy Research has proposed a multi-tiered approach, calling for progressively deeper cuts in greenhouse gas emissions by rich nations but more flexible commitments from the developing world. (Reuters)

GENEVA: Himalayan glaciers are melting rapidly, threatening water shortages for hundreds of millions who depend on glacial-fed rivers in China, India, and Nepal, according to a report by WWF. The region has the greatest concentration of ice in the world after the polar regions. Glaciers are now receding at 33 feet/year. "The rapid melting of Himalayan glaciers will first increase the volume of water in rivers, causing widespread flooding," said Jennifer Morgan, director of the World Wide Fund for Nature's Global Climate Change Program. "But in a few decades this situation will change and the water level in rivers will decline, meaning massive economic and environmental problems for people in western China, Nepal, and northern India...Ministers should realize now that the world faces an economic and development catastrophe if the rate of global warming isn't reduced," Morgan said. A WWF study indicates that if nothing is done soon, the earth will warm by 3.6°F above pre-industrial levels by 2060. (spacedaily.com)

Recommended reading:

Harkavy, Oscar, *Curbing Population Growth: An Insider's Perspective on the Population Movement.* New York: Plenum Press, 1995.

54
THE DEMOGRAPHIC TRAP
(March 27, 2005)

Overpopulation has wiped out entire societies. Yet progress in curbing population growth has been negligible. Why? Not for lack of trying.

The *demographic transition theory* is an attempt to understand the economic basis for population dynamics. It originated with the observation that European populations went through a three-phased transformation. (1) The first, *undeveloped*, phase was the Dark Ages (more descriptive than "Medieval Period"), in which both birth rates and life expectancy were low, and population grew slowly if at all. (2) In the second, *transition*, phase birth rates increased and death rates declined due to an increased quality of life, and population growth accelerated. (3) The third, *developed*, phase was marked by continued improvement of the quality of life, yet birth rates declined and population growth slowed, stopped, or reversed in different developed nations.

Population policy makers in the last half of the twentieth century assumed that improvement in the quality of life would stabilize population levels as countries went from undeveloped to developed phases. Unfortunately, the transition phase has rarely been followed by the developed phase. Why?

The European experience has eluded most undeveloped countries because the circumstances were dissimilar. Europe was less heavily populated than today's undeveloped countries. As death rates declined, Europeans had a safety valve to relieve their population pressure: massive emigration. Colonization, slavery, and mercantilism allowed Europeans and their colonists to enjoy unprecedented economic growth. Europe's new wealth, extracted from the colonies, financed the Renaissance, Enlightenment, and Industrial Revolution. As European, American, and other developed countries' populations became more crowded and women acquired more reproductive rights, they opted for fewer children.

None of these advantages applied to the undeveloped world. These countries enjoyed enough economic improvement to lower death rates, so they entered the transition phase. Family sizes increased dramatically. These families, whose incomes didn't increase, were less well off. They couldn't adequately support the increased number of young and old dependents. Capital for economic growth wasn't there, and unemployment increased. Tax revenues needed to expand public utilities, education and health care didn't materialize. Tradition, declining education, and poverty prevented birth rates from declining as they had in developed countries.

These countries are falling into a demographic trap: they can't support their existing populations or slow their growth, so the standard of living can only decline further. Many regions face a runaway positive feedback loop in which poverty, lack of infrastructure, illiteracy, reproduction-oriented belief systems, unemployment, malnutrition, and disease increase from one short generation to the next. Massive famines, plagues, desperate migrations, takeovers by organized crime and warlords, anarchy, and terrorism have begun. In these regions, the gravest humanitarian disasters in the history of humankind are coming to pass and their reverberations will shake the foundations of all the civilizations of the world.

This week's news:

GENEVA: Eastern Congo is undergoing the world's worst current humanitarian crisis. U.N. emergency relief director Jan Egeland says that over the last six years, 600,000 people are dying each year. Asked if too much emphasis was being put on Darfur, Egeland responded, "The amount of focus on Darfur is correct, but there is too little on (eastern) Congo." Egeland also conveyed indignation to Khartoum that some women raped and now pregnant by Janjaweed fighters are being persecuted for violating Islamic sharia law. "That is the ultimate insult for women who have been raped." (Reuters)

Poor nations are urging the European Union to stop subsidizing farm exports within 5 years. The EU offered last July to end the practice but hasn't done so. Trade talks have made no progress since they were begun in Doha in 2001. (Al Jazeera)

GOP senators defeated by 51 to 49 a move to delete a provision that would open ANWR to drilling from the 2006 budget resolution. Democratic and Republican lawmakers acknowledge that supporters of drilling now have a better chance to succeed. Oil and gas companies have contributed $179.7 million since 1989 to federal candidates and political parties, 74 percent to Republicans. ChevronTexaco and ExxonMobil are the top contributors. In addition to campaign contributions, ChevronTexaco has spent over $38 million since 1997 on lobbying; Exxon Mobil has spent more than $62 million. The Teamsters, which have opposed many of Bush's initiatives, deserted the largely Democratic labor movement to invest large sums of money in the political fight. The 19.6-million acre wildlife refuge was established in 1960 and expanded by President Carter in 1980 under the Alaska National Interest Lands Conservation Act. The Bush administration estimates there could be 10 billion barrels within ANWR. Republicans have long pressed for drilling in ANWR on the excuse that it would reduce US dependency on foreign oil. Opponents argue that the yield would not make an impact on foreign oil use and would wreak havoc on the environment. Seven GOP senators, including John McCain (R-AZ) and Lincoln Chaffee (R-RI.) opposed this week's vote. (CapitalEye)

In several US states, Imax cinemas are refusing to show movies that mention evolution. In southern states, theatres that had test screenings of several of these films were accused of blasphemy. This decision could have a ripple effect because Imax films are expensive to produce and require special equipment, making the market very tight. (*The Observer*)

BOULDER: Even if all greenhouse emissions had been stopped in 2000, temperatures would rise one degree Fahrenheit and seas would rise another 4 inches due to thermal expansion by 2100, according to research findings at the National Center for Atmospheric Research (NCAR) published in *Science*. First author Gerald Meehl says, "The longer we wait, the more climate change we are committed to in the future." These numbers don't include effects of fresh water from melting ice sheets and glaciers, which could at least double the sea level rise. The North Atlantic thermohaline circulation, which currently warms Europe by transporting heat from the tropics, becomes weaker in the models. Nevertheless, Europe will heat up with the rest of the planet because of the overwhelming greenhouse effect. The inevitability of the climate changes is the result of thermal inertia, mainly from the oceans, and the long lifetime of carbon dioxide and other greenhouse gases in the atmosphere. The scientists also examined scenarios in which greenhouse gases continue to build in the atmosphere at low, moderate, or high rates. The worst-case scenario projects an average temperature rise of 6.3°F and sea level rise from thermal expansion of 12 inches by 2100. All scenarios analyzed in the study will be assessed by international teams of scientists for the next report by the Intergovernmental Panel on Climate Change, due out in 2007. (terradaily.com)

Recommended reading:

Department of Economic and Social Affairs, Population Division, *Review and Appraisal of the Progress Made in Achieving the Goals and Objectives of the Programme of Action of the International Conference on Population and Development: The 2004 Report*. New York: United Nations, 2004.

55

THE SOCIAL COSTS OF POPULATION GROWTH

(April 3, 2005)

If the people of the world are to bring population growth under control, we will have to be more successful than we have been over the last fifty years. At this point, the problem of population control really consists of two parts: controlling the negative effects of overpopulation and reversing population growth. They are inextricably interrelated, as illustrated by the demographic trap phenomenon, where the threshold for a positive feedback loop between population size and the breakdown of society is crossed, and both accelerate in a runaway process.

The costs of overpopulation include deterioration of the nutrition and health of mothers and children as family size becomes insupportable. In areas of explosive population growth, mothers start having children in their mid-teens, and continue having them, if they live long enough, into their forties and fifties. Mothers at both age extremes are much more vulnerable to complications. Many or most of these pregnancies are unwanted by the mothers. Causes of maternal deaths due to undesired pregnancies include hemorrhaging, blood poisoning, and other birth complications, as well as back-street abortions (either because they're illegal or the pregnant women can't afford legal ones). Survivors of these events have impaired health. Especially where health care is poor, the more children women have, the worse their health becomes, leading to higher disease and mortality rates in both mothers and children. Short intervals between births, nutritional deficiencies, lack of hygiene and sanitation, and shorter breast-feeding periods all stack the odds even more against these mothers and children.

With the exception of South Asia, female children—future mothers—are not discriminated against nutritionally. However, in some South Asian cultures, the dowry system requires families to pay bridegrooms to marry their daughters. The resulting nutritional discrimination against girls is not limited to the poor. In some sub-Saharan African cultures on the other hand, where bridegrooms pay a bride price, daughters receive better nutrition than sons do. There is little evidence of gender nutritional bias (I use the term *gender*, with reservations because its traditional use is for grammar, where it has nothing to do with sex—like male and female nouns or electrical connectors—to denote sociological rather than biological sex distinctions) in Latin America.

Large families incur higher costs for food, clothing, and education than small families, forcing cost cutting in the largest and poorest families. Child labor and caring for younger siblings cut into the time children should spend in school or playing (playing is important for normal cognitive and social development). The government can't provide new schools

147

fast enough to keep pace with the growing school-age population. Since over half the population is too young for full-time employment, per capita income dwindles.

Countries with rapidly growing populations have a poor success record in establishing or maintaining stable constitutional governments and civil and political rights. The balance of trade in such countries is generally negative because the ruling oligarchy pockets most of the income and the countries' natural and economic assets are sold off by the same corrupt people (who collect handsome "commissions" for selling what is not theirs) to foreign corporations at fire sale prices. Government corruption and malfeasance accelerate the impoverishment of the people.

Even where education has improved, there are not enough jobs to absorb the expanded educated work force. This, coupled with the ever-increasing income inequality in oligarchic societies, provides prime fuel for explosive unrest.

One of the few relief valves for such populations is migration. Those who can, move, in hopes of finding better conditions. As we've seen in our review of animal and human migrations, species always have migrated and they always will, in response to population pressure. All developed nations and even some underdeveloped ones are now experiencing the increased immigration, legal and illegal, which results from other countries' overpopulation.

This week's news:

EUGENE, OR: A study indicating a stormy greenhouse future has been published in the journal *Geology*. "We know the gathering greenhouse will be warm, but this new information confirms that the contrast between the rainy season and the dry season will increase dramatically," says Greg Retallack, the lead author. Retallack explored the relationship between seasonality and rainfall in contemporary soils, and then applied the same techniques to buried soils spanning a greenhouse event 55 million years ago. "This is known to have been a time of high atmospheric carbon dioxide from studies of the breathing pores in fossil leaves. At that time, Wyoming warmed from a mean annual temperature of some fifty-five degrees to a summer-like sixty-five degrees Fahrenheit. Rainfall in Utah jumped from sixteen inches per year to twenty-six inches per year. As a result, sagebrush deserts of the western U.S. were transformed into sub-humid woodlands." Retallack agrees with earlier studies indicating that the trigger for the greenhouse spike, which lasted less than half a million years, was a catastrophic release of natural gas from undersea ices and permafrost. "In a remarkable parallel to modern hydrocarbon pollution of the atmosphere, this natural methane oxidized to carbon dioxide and created a global greenhouse event. The past methane outburst dwarfed even human consumption of hydrocarbons, and there is a danger that another similar outburst could be triggered by warming due to human activities. Our little warming push could repeat the troubled times of 55 million years ago. During the greenhouse spike of 55 million years ago, tropical mangroves and rain forests spread as far north as England and Belgium and as far south as Tasmania and New Zealand. Turtles, alligators, and palm trees graced Ellesmere Island in the Canadian Arctic, which is now the treeless abode of musk oxen and polar bears." (terradaily.com)

Recommended reading:

Ehrlich, Paul, and Ehrlich, Anne, *One with Nineveh: Politics, Consumption, and the Human Future*. Washington: Island Press, 2004.

56

ENVIRONMENTAL COSTS OF OVERPOPULATION

(April 10, 2005)

Water is scarce in the most overpopulated regions: Africa, northern China, and parts of India, Mexico, and the Middle East. Water is also scarce in thinly populated regions, as we would expect. How could water-poor regions have become overpopulated? There's no single answer, but we can be certain that scarce water doesn't cause overpopulation. Can it be that overpopulation causes scarce water? You bet.

Overpopulated areas aren't the only places where demand for water will increase, but they will need still more water in coming decades because these regions will see the fastest growth of population. Leisinger et al. (2002) explain what happens next: "Wherever water is scarce, its quality deteriorates first, and eventually even the available quantity shrinks." They report that more than five million people a year die now from lack of clean water for drinking, sanitation, and hygiene. About *half* the inhabitants of underdeveloped countries have illnesses related to inadequate water stocks. This can only get worse as their populations continue to grow.

Such scarcities are most often the result of previous growth. Population increased threefold between 1900 and 1995, and worldwide water consumption increased six-fold, a ratio of two. If coming population increases are accompanied by any economic development, the ratio of increases in water consumption to population size will climb well beyond its current value of two. The useable water supply isn't adequate now, and the deficit will accelerate because the supply will shrink due to depletion and pollution at the same time that demand is increasing. Global temperature increases will make it even worse.

Just as water depletion is increasing faster than population increases, population growth will cause a disproportionate increase in the rate of depletion of other natural resources. Worldwide demand for energy is growing much faster than population, not only for development but also just to maintain the status quo in a deteriorating environment. Our vessel has sprung an entropy leak, and we need ever-increasing amounts of fuel for our entropy pumps. Overpopulation has already caused desertification of arable land due to increased erosion, salinity, and overgrazing; massive deforestation; depletion of marine food supplies; and mass extinctions due to loss of habitat, pollution, climate change, and water depletion. Further population growth will accelerate these rates of depletion at a time when increases in these resources will be desperately needed, because these disintegrative processes have now become mutually reinforcing. We have triggered a set of interlocking positive feedback loops in which even zero population growth could no longer halt the runaway processes of environmental degradation.

To halt this runaway process, we must go not just to a replacement fertility level (where population size is stable, neither growing nor shrinking), but to rapid negative growth. We've reached the point where, in many regions, only a near-zero birth rate will suffice and even then massive aid will be required. Runaway environmental degradation can only be halted by moving quickly to a period of *negative resource consumption*, in which we rebuild the world's renewable resources and repair the world's ecologies as much as possible, followed by zero net consumption of resources. Economic structures will have to align with this reality.

This week's news:

CONCORD, NH: The state Senate has voted to permit women to buy emergency contraception without a prescription after unprotected sex. Gov. John Lynch will sign the bill if it passes the House. The bill would permit specially trained pharmacists to offer the so-called "morning after" pill, which is a large dose of birth control hormones that can prevent ovulation or fertilization and can prevent a fertilized egg from implanting into the uterus. It doesn't abort pregnancy, but if taken within 72 hours of unprotected intercourse, it can cut a woman's chances of pregnancy by up to 89%. The FDA last year recommended the drug be available over the counter, but an agency director felt that not enough was known about the pill's effects on teenagers' sexual behavior. That sentiment was echoed by opponents in the Senate. Opponents also claimed the bill would encourage promiscuity. They tried without success to get the Senate to limit access to women age 18 and older. "This is saying there is no consequence for sex," said Boyce, R-Alton. "One of the consequences is conception. One of the consequences is disease." Maine voted last year to permit pharmacists to dispense the contraceptives without a prescription—becoming the sixth state to do so. (Associated Press)

NAIROBI: The UN warns that increasing poverty and urbanization may triple the people in the world's slums to three billion by mid-century. The growth of slums is a key risk to public health and development. Nairobi is the site of the Kibera ghetto, with at least 500,000 residents. UN head Kofi Annan says slum dwellers in Africa, Asia, and Latin America make up 30% of the global urban population, now higher than at any point in history. (terradaily.com)

ExxonMobil, the world's biggest oil company, is the last oil major challenging the science on global warming, opposing the Kyoto Protocol on climate change and insisting that only fossil fuels can meet the world's future energy needs. Three resolutions have been proposed for the Exxon annual shareholder meeting in May. One would require the corporation to report within six months "on how ExxonMobil will meet the greenhouse gas reduction targets of those countries in which it operates that have adopted the Kyoto Protocol." A second would require the corporation to "make available to shareholders the research data relevant to ExxonMobil's stated position on the science of climate change." The third resolution would require the corporation to appoint an expert on environmental science as a non-executive director. Exxon spends only $10 million/year on research into alternatives to fossil fuels, compared with more than $100 million/year spent by BP, and $1 billion between 2003 and the end of this year by Shell. (*The Independent*)

WASHINGTON: While hydrogen has been used for years by NASA to power rockets, it's only in the last decade that car companies have looked seriously into using it for mass transportation. The effort gained momentum in 2003 when President Bush announced a $1.2 billion Hydrogen Fuel Initiative during his State of the Union Address. US Department of Energy secretary Samuel Bodman has signed recognition agreements with several auto companies, to make hydrogen vehicles more efficient and cost-effective by 2010. According to DOE estimates, hydrogen is currently three to four times more expensive to produce than gasoline. However, hydrogen produces zero emissions and can be made domestically, potentially reducing US dependency on foreign oil. "The progress that DOE and the automotive and energy industries have made so far has us on the path to an industry commercialization decision in 2015. If our research program is successful, it is not unreasonable to think we could see the beginning of mass market penetration by 2020," Bodman said. Representatives from Ford Motor Company, Daimler-Chrysler, General Motors, and ChevronTexaco signed the recognition document. However, those corporations represent only a small fraction of the ones involved in developing hydrogen-powered vehicles, and a Toyota representative says the company was upset at being excluded. "We're in it for the long haul and we've been there from the very beginning," said Cindy Knight, environmental communications administrator for Toyota. Knight pointed out that Toyota and Honda were at the forefront of hydrogen technology since the two were the first to make vehicles using fuel cells. Toyota has a fleet of about 20 vehicles in the U.S. and Japan. Ballard Power Systems, a Canadian corporation that holds contracts with both Ford and Daimler-Chrysler, has made vehicles powered by fuel cells. Vehicles powered by fuel cells are quieter and cleaner, with less vibration than traditional gasoline cars. (spacedaily.com)

Five thousand people are dying from tuberculosis each day in Africa. Global TB incidence has dropped by more than 20 per cent since 1990, but in Africa, it has tripled since 1990 in countries with high HIV rates. Even Uganda, where HIV has been reduced, is curing fewer TB cases than four years ago. More than half of TB victims in Uganda have no access to adequate treatment. The two worst areas are Africa, with the TB/HIV co-epidemic, and Europe, with high levels of multi-drug-resistant TB and slow advances in countries of the former Soviet Union. (UN News Service)

LUANDA: Angola is struggling to contain a deadly outbreak of the Ebola-like Marburg virus that has claimed a record toll of 126—three higher than the worst outbreak in neighboring Democratic Republic of Congo. "The situation is serious. Really serious. It's a disease about which not much is known. It's worse than the SARS in Asia," said Margarida Correia, head of maternity in a Luanda hospital. "We are very worried because we are in direct contact with the ailing. We are tending to them without sufficient protection," she said. A severe form of hemorrhagic fever, Marburg was first identified in 1967. It spreads on contact with body fluids such as blood, urine, excrement, vomit, and saliva. Angola's former colonial ruler Portugal announced it was donating gloves, protective eyewear, boots, and masks. (terradaily.com)

Recommended reading:
Barney, Gerald O., et al., *The Global 2000 Report to the President of the United States. Entering the 21st Century.* New York: Pergamon Press, 1980.

57

THE PSYCHOLOGY OF FERTILITY

(April 17, 2005)

We've already learned that the demographic transition theory is wrong. That is, improving people's quality of life doesn't suffice to cure overpopulation, and a poor quality of life doesn't guarantee overpopulation. Nevertheless, this theory still influences government policy. Partly because of this, nearly every country on Earth is overpopulated and population growth is totally out of control in many countries. The fact remains, however, that improvements in education, health, and reproductive rights are necessary if not sufficient for reversal of population growth.

Abernethy (2000) amply documents data that suggest the most powerful determinant of fertility (number of children per woman) is desired family size, regardless of education or availability of contraception. Regardless of government policy and other pressures, family size approaches the desired size. Even women who use contraceptives have large families if that's their goal. Conversely, lacking modern contraception, women limit the number of children they have by draconian measures if that's their goal. To do so, they resort to crude contraceptive practices that can lead to sterility, abortions, and infanticide (even in the US today). In some societies, men resort to equally drastic approaches. Their societies condone these practices, and support smaller families by traditions such as late marriages, murdering women (including suttee—the killing of widows—and virgin sacrifice), sexual taboos, primogeniture (estate inheritance only by the oldest son), male and female genital mutilation, a large celibate priesthood, and large dowries.

All societies that survived a great length of time had to come to grips with the need to live in balance with nature, consuming resources only as fast as they could be replenished. They came to believe in the concept of *finite resources*. They adjusted their populations and lifestyles to fit their environments, a practice of *sustainability*. Societies that didn't do this eventually failed.

A Western aberration was the notion of *limitless resources*, born of the benefits of technology, emigration, colonization (usually accompanied by genocide), and empire. Our economic system is built on the absurd concept of *endless growth* rather than sustainability. There are two principal substitutes for sustainability in this system, both centered on family structure. One was the model of the nuclear family, in which parents assumed the responsibility to limit the number of children to the number they could provide for. A similar concept was the model of the extended family, in which a wider group of relatives took collective responsibility. Either can work, in principle.

However, in such a system supportable family size largely depends on wealth, not sustainability *per se*. The pursuit of wealth can become a means to larger families, or an end in its own right. Wealth confers power, advantage, and privilege. In a classic positive feedback loop, the power of wealth makes it easier to acquire more wealth, and poverty (lack of power) makes it harder, separating society into a few wealthy and many poor people. This system was rationalized by the ideals of hard work, delayed gratification, savings, and upward mobility: the Horatio Alger paradigm. Wealth became a sign of virtue and of God's favor, especially among Protestants and most explicitly, Calvinists. Social Darwinism provided a pseudoscientific rationale, disguised today as "meritocracy." Just below the surface of meritocracy lies the positive feedback system of privilege and exclusion.

Unfortunately, wealth can be achieved in very unsustainable ways, such as plundering our environment and our global neighbors, and borrowing so heavily from other developed countries that we, and they, face economic ruin. We react to this threat by plundering more recklessly, rattling our military sabers, and going further in debt.

If actual family size is most influenced by desired family size, what determines desired family size? If you're guessing that it's determined by supportable family size, you're close. Remember that all "knowledge" is subject to error. Desired family size, according to Abernethy, is determined by the *perceived* supportable family size. How could it be otherwise? But perceptions can be seriously inaccurate. For example, the mere expectation of upward mobility leads to unrealistically high fertility.

This brings us to the most important question we've ever faced. **How can we shape perceived supportable family size to match actual supportable family size?**

This week's news:

BEIJING: China plans to build 2–3 1-gigawatt reactors/year over 15 years. "We are speeding up development of nuclear power because it is clean and green energy," said Zhang Fubao, deputy department director of China Atomic Energy Authority. China uses coal for 70% of its energy supply. China plans to increase nuclear power from 2.4 to 4% of total power. The average among countries with nuclear power is 17%. (spacewar.com)

The World Health Organization has traced an Ebola-like disease to Yambio, Sudan. Fifteen cases have been confirmed in two weeks and four people have died. More than 100 people are being kept under surveillance. Four are in isolation at Yambio Hospital. The virus is hard to diagnose in its early stages because some of the symptoms mimic malaria. Many victims develop internal bleeding. The last major outbreak of Ebola was in Uganda 2 years ago, in which hundreds of people died. (Al Jazeera)

UíGE, ANGOLA: World Health Organization workers so far are losing the race to control the Marburg epidemic in eastern Sudan. The virus killed at least 193, mostly in the past month—the largest number on record. Nine out of 10 infected people die, quickly. Bodily fluids, and even stray drops of spittle spread the virus or beads of sweat can lead to death. Health officials fear a disaster that could spread rapidly. (*New York Times*)

PETERBOROUGH, NH: Senator Lamar Alexander (R-TN) has proposed legislation to cushion the tight natural gas market with a 5-year 30% investment tax credit for both residential and commercial solar technology. That 30% credit is taken after any existing state credits. The tax credit would be capped at $7,500 for residential solar PV and solar thermal and would be unlimited for commercial projects. A 10 percent investment tax credit currently exists for solar, but that's not enough to stir the residential market. (truthout.org)

TOKYO: Japan plans to kill two more species of whale—humpback and fin whales, considered endangered by the World Conservation Union—in the Antarctic and double the catch of minke whales, currently set at 440. Countries can get around the commercial whaling limits using exemptions for research. Japan claims their research shows whale populations are thriving and provides data showing whales are consuming fish stocks—points disputed by environmentalists. "The humpback and fin whales have been pushed to the brink of extinction because of commercial whaling," said Junichi Sato, a senior campaigner for the Japan branch of Greenpeace. "If Japan goes ahead to hunt them again, it will become clearer still that its research whaling is just commercial whaling in disguise." The meat from the research cull—about 2,000 tonnes annually—goes to supermarkets and restaurants across Japan. Japan has also hunted whales in the northern Pacific since 1994 under the research program. The number totaled 330 last year, including 220 minke whales, 50 Bryde's whales, 50 sei whales, and 10 sperm whales. Japan argues there is a need to increase its target species to analyze the ecosystem of the Antarctic and develop a method to manage whale resources. (terradaily.com)

Recommended reading:

Abernethy, Virginia, *Population Politics*. New Brunswick, NJ: Transaction Publishers, 2000.

58

POPULATION AND ITS MISCONCEPTIONS

(April 24, 2005)

Assuming that the world's new leaders are committed to a humane, rational, and effective policy to reduce population density to a sustainable level, do we have enough information to describe the general outline of such a policy?

Surprise. We do. Building on the information in previous chapters, we know the biggest determinant of family size is the perceived supportable family size, and that errors of such perceptions are the main causes of inappropriate family sizes. Our highest priority must be to eliminate these errors, so that people accurately perceive the safe limits to population growth as reflected in their own quality of life. Then they'll decide on their own to reduce family size and will accept assistance doing so.

First, what are common forms of error? In roughly descending order of importance:

- **Over-optimism about the environment and overpopulation—including overconfidence in technological fixes.** Most people are poorly informed about the state of the environment and the population explosion. Many still view the planet's resources as essentially inexhaustible or believe that technological solutions will always come to our rescue, in spite of evidence all around us that neither is true. In fact, most people avoid these issues, an irrational but natural psychological defense. Governments, corporations, some religions, and the corporate news media are happy to help them avoid reality. If there's no obvious relationship between large families and quality of life, environmental impact has no effect on family size.
- **Belief that one, or one's offspring, can migrate to some place better.** Some overpopulated areas don't feel much impact because many of their inhabitants emigrate. About 50% of the Indian state of Kerala's labor force works in the Persian Gulf. In contrast, only 0.7% of Mexico's workforce emigrates to the US annually, mostly to regions we seized from Mexico. Although these population shifts are largely benign in the short term, they encourage rather than discourage population growth through large families.
- **Dependence on others.** In many places where the population is exploding, an unhealthy dependence on foreign aid, government assistance, and NGO inter-

vention has contributed to unrealistic family sizes. Over-reliance on credit is another form of unrealistic dependence on others. The illusion of getting something for nothing or putting off payment to an indefinite future tempts people into having families that are too large.

- **Trust in one's ability to survive while others do not.** The Social Darwinist concept of survival of the fittest in society is built on the view that unrestrained competition is normal and just, and that this process will allow winners ("superior individuals") to flourish. Aside from its immorality, this stance ignores three important facts: quality of life in non-sustainable cultures is based more on privilege than ability; the illusion of upward mobility encourages the poor to have families that are too large; and there are no winners on a ruined planet.

- **Confidence that good times will continue or that better times are coming.** People ignore danger signs when they are doing well. We also forget that tough times have utterly destroyed many once-flourishing civilizations. We also illogically argue that worldwide ecological collapse hasn't happened yet (ignoring the many cases of local collapse and global mass extinctions), so it never will. We need to suffer bad consequences of large families during the good times in order to avoid the bad times.

- **Belief that a larger family, tribe, or nation is more secure.** Most animal populations grow only as long as their environments can support them. Then negative feedback mechanisms, internal and external, slow or reverse their population growth. When these mechanisms fail, the species crashes, taking other segments of the ecosystem with it. Sustainable cultures never rely on the delusion that bigger is better when it comes to family size.

- **Faith, dogma, and resignation.** These are perhaps the most extreme forms of irrational behavior we indulge in. Blind *faith* usually involves a selective rejection of the processes we use successfully every day to learn about and cope with the challenges that face us. Many religions encourage large families (except once overpopulation has already brought famine and disease). *Dogma* is an unquestioning adherence to an approach that may once have been rational and useful but often isn't any longer. The overthrow of a dogma can take too long to rescue a society. *Resignation* is the abandonment of hope. In the form of acceptance that a cause is lost, it's appropriate in the last stage of a final defeat. The most pernicious form of resignation is the "It's God's will" variety.

This week's news:

A UN report, compiled by 1,360 experts from 95 countries, says humans have changed the world's ecosystems more in the last 50 years than in any other period, and 10 to 30% of mammal, bird, and amphibian species now face extinction. In the last 50 years, we have lost 20% of the world's coral reefs, with another 20% seriously degraded, and 35% of the world's mangroves. Fishing stocks have collapsed. "Any progress achieved in addressing the goals of poverty and hunger eradication, improved health, and environmental protection is unlikely to be sustained if most of the ecosystem services on which humanity relies continue to be degraded," the study said. The report offered several strategies including one that involves a global economy where the sharing of education, skills, technology, and resources leads to a reduction in poverty and pressures on local environments. The worst possible approach the report called "Order from Strength," which results in "a regionalized and fragmented world, concerned with security and protection, emphasizing primarily regional markets, paying little attention to public goods, and taking a reactive approach to ecosystem problems." That describes the U.S. today. We gobble 25% of the world's energy and emit the most greenhouse gases. There has been no response to the Millennium Ecosystem Assessment by the administration. The study was co-chaired by the World Bank's chief scientist, Robert Watson, who says that the study reinforces his belief that climate change "may become the most dominant threat to ecological systems over the next hundred years." (*Boston Globe*)

CORVALLIS, OR: If increased precipitation and sea surface heating from global warming disrupts the Atlantic Conveyer current, as many scientists fear, phytoplankton in the North Atlantic could drop 50%. A simulation study by Andreas Schmittner, at Oregon State University, indicates that such a disruption would prevent the normal upwelling of nutrient-rich North Atlantic and Pacific waters that nurtures phytoplankton growth. A growing body of evidence indicates that the Atlantic Conveyer current switched on and off 20–25 times during the last ice age. "During the last ice age, from about 100,000 years before present to 20,000 years B.P., thick ice sheets over Canada sporadically dropped armadas of icebergs into the North Atlantic where they melted, sufficiently freshening the water to disrupt the conveyer," Schmittner says. "Deep ocean sediment core samples show pebbles from land delivered by the floating icebergs." Ice cores from Greenland show temperature changes that correlate with changes in ocean nutrient concentrations revealed by deep-sea sediment cores. "One full oscillation of these switches took 1,500 years," Schmittner said, "but the individual transitions happened surprisingly fast. The climate went from a cold state to a warm state in as little as 20 to 50 years. Surface temperatures in Greenland increased 20 to 30 degrees Fahrenheit and water temperatures increased 10 to 20 degrees." Schmittner said the impact of the current on the Pacific Ocean generally isn't as great, even though the system is a global one. Still, he added, plankton production would also decrease in the Pacific if the current were reduced. (terradaily.com)

Recommended reading:

Diamond, Jared M., *Collapse: How Societies Choose to Fail or Succeed*. New York: Viking, 2005.

59
DEMOGRAPHIC MISCONCEPTIONS

(May 1, 2005)

We know the biggest determinant of family size is the perceived supportable family size, and that errors of such perceptions (which collectively I call *demographic misconceptions*) are the main causes of inappropriate family size.

We've listed some causes of demographic misconceptions:

- Excessive optimism about the environment and overpopulation—including overconfidence in technological fixes.
- Belief that one can migrate, or one's offspring can migrate, to some place better.
- Dependence on others.
- Trust in one's ability to survive while others don't.
- Confidence that good times will continue or that better times are coming.
- Belief that a larger family, tribe, or nation is more secure.
- Faith, dogma, and resignation.

Here are some (certainly not all) of the principal social forces behind these proximate causes:

- **Pseudo-economics based on perpetual growth.** Investment, as opposed to simple thrift, requires that invested wealth grow over time. Invested wealth can grow three ways: by taking wealth away from other people, by being lucky, or through real growth of wealth. The first is theft, the second is gambling, and the third, if it is to last, requires perpetual growth. The myth of unlimited growth reinforces demographic misconceptions. The drive for Social Security personal investment accounts was a cynical attempt to keep this myth alive a little longer so the wealthy could steal the last of the not-so-wealthy's wealth (see privilege, below).
- **Global trade.** If people only use local resources, the limiting resources, those that start running out first, determine the supportable population size. Global trade allows peoples to supplement their limiting resources by trading *excess resources* with other peoples. This encourages demographic misconceptions, because few people are aware of how much of anything there is if it comes from elsewhere.

- **Pseudo-charity.** Charity that provides symptomatic relief from the consequences of overpopulation boosts the quality of life of the recipients, but can't make an unsustainable population sustainable. It perpetuates demographic misconceptions because it leads to over-optimism about supportable family sizes.
- **Foreign workers.** Workers who earn better incomes outside their home countries lower the apparent cost of large families. In the host country cheap labor artificially lowers the cost of goods. In the home country, revenues from the absent labor pool artificially raise family income. The same effect comes from outsourcing work: the only difference is that goods or services are transported instead of people.
- **Privilege.** Privilege is advantage conferred upon a group (e.g., the wealthy, government leaders, clergy, and corporations) over other groups. The privileged can use the pseudo-merit system to more advantage than the less privileged. This sustains demographic misconceptions among the privileged, and among the underprivileged who hope to join them.
- **Credit.** Credit is fake money issued by the consumer in the expectation of redeeming it with real money, plus interest, in the future. It fuels demographic misconceptions through an illusion of wealth, and promotes usury.
- **Tribalism.** It's natural that we try to protect our best interests and those of our families, tribes, ethnic groups, and nations over others. It's part of our biology, and recognized in law. Unfortunately, it promotes demographic misconceptions through a sense of solidarity.
- **Pseudo-information.** Misinformation is an obvious cause of misconceptions. Deliberate misinformation, presented as news, moral principles, history, theology, or science, bombards us every day. Detecting and resisting it is tricky.
- **Superstition.** This is an incorrect inference about how the world works, based on personal experience, anecdotes, or trust in authority. We all have superstitions, which we have to keep testing against reality.
- **Pseudo-morality.** Humans invent moral rules to inculcate behavioral wisdom. Many perpetuate demographic misconceptions.
- **Tradition.** Our ancestors' family sizes haven't been sustainable for some time. If we don't recognize the bad effects of large families, we perpetuate the tradition.
- **Misplaced incentives.** Many societies today reward people for having babies, through tax breaks, workplace policies, and other incentives. Some societies are even trying to reverse the shrinking of their populations in a misguided effort to cope with their aging populations. Dumb. There are sustainable ways to adjust to such consequences of reversed overpopulation. Putting off population reduction only makes it more difficult.

This week's news:

GREENBELT, MD: According to a study appearing in *Science*, a global warming-induced loss of winter and spring snow cover over Southwest Asia and the Himalayan mountain range has led to decreased albedo and more heating of the exposed land in the summer. This increases the temperature difference between the Indian subcontinent and the Arabian Sea, which in turn creates a lower pressure system over the land and a higher-pressure system over the Arabian Sea than in previous years. The increased pressure difference generates stronger monsoon winds that mix the ocean water in the Western Arabian Sea more than in the past. This brings more nutrients from deeper layers to the surface, causing better growing conditions for phytoplankton, the base of the ocean food chain. Since 1997, a reduction in snow has led to wider temperature differences between the land and ocean during summer, and phytoplankton concentrations in the Western Arabian Sea have increased by over 350 percent over the past 7 years along the coasts of Somalia, Yemen, and Oman. While large blooms of phytoplankton can enhance fisheries, exceptionally large blooms can lead to oxygen depletion and a decline in fish populations. (terradaily.com)

Excerpts from "Some Like It Hot"
by Chris Mooney
Mother Jones, May/June 2005
(reproduced with permission)

Michael Crichton's latest science fiction novel, *State of Fear,* is an anti-environmentalist page-turner in which shady eco-terrorists plot catastrophic weather disruptions to stoke unfounded fears about global climate change. The author, who is not an environmental scientist, was recently greeted as an authority by policy makers at the Wohlstetter Conference Center of the American Enterprise Institute for Public Policy Research (AEI). In his introduction, AEI president Christopher DeMuth congratulated Crichton for presenting "serious science with a sense of drama to a popular audience."

Conservative think tanks like AEI are trying to undermine the sound science of global warming with a disinformation campaign using, among other tactics, events like Crichton's talk. They provide pseudo-intellectual talking points for their point men to contest the legitimate science of climate researchers, at the beck and call of politicians like Senator Inhofe (R-OK), chair of the Environment and Public Works Committee, who calls global warming "a hoax." This effort reflects the convictions of antiregulatory politicians who serve the wealthy fossil-fuel industry.

Groups like AEI are funded by ExxonMobil, the world's largest oil company. *Mother Jones* has tallied 40 ExxonMobil-funded organizations that seek to undermine mainstream scientific findings on global climate change or are affiliated with a handful of professional "skeptic scientists" who do so. These front organizations also include quasi-journalistic outlets like TechCentralStation.com (a website that received $95,000 from ExxonMobil in 2003), a *FoxNews.com* columnist, and even religious and civil rights groups. These organizations received a total of more than $8 million between 2000 and 2003 (figures below are for that period unless noted otherwise). ExxonMobil chairman and CEO Lee Raymond is vice chairman of the board of trustees for the AEI, which received $960,000 in funding from ExxonMobil. The AEI-Brookings Institution Joint Center for Regulatory Studies, which hosted Crichton, received another $55,000. Crichton drew an analogy between believers in global warming and Nazi eugenicists. "Auschwitz exists because of politicized science," Crichton asserted, to gasps from some in the crowd, who were there, after all, to politicize science. Attending was Myron Ebell, censured by the British House of Commons for "unfounded and insulting criticism of Sir David King, the Government's Chief Scientist." Ebell is the global warming and international policy director of the Competitive Enterprise Institute (CEI), which has received $1,380,000 from ExxonMobil. Also attending was Christopher Horner, counsel to the Cooler Heads Coalition who's also a CEI senior fellow. Also attending: Paul Driessen, a senior fellow with the Committee for a Constructive Tomorrow ($252,000) and the Center for the Defense of Free Enterprise ($40,000 in 2003). ExxonMobil's funding of think tanks is small change compared to its lobbying expenditures of $55 million over the past six years.

Consider attacks on the Arctic Climate Impact Assessment (ACIA), an international collaboration of 300 scientists over a period of 4 years, commissioned by the Arctic Council. Their report warned that the Arctic is warming "at almost twice the rate as that of the rest of the world," and that melting sea ice and glaciers are already apparent and "will drastically shrink marine habitat for polar bears, ice-inhabiting seals, and some seabirds, pushing some species toward extinction." Industry defenders attacked the study, and, with no science to marshal to their side, used opinion pieces and press releases instead. "Polar Bear Scare on Thin Ice," tooted *FoxNews.com* columnist Steven Milloy, an adjunct scholar at the libertarian Cato Institute ($75,000 from ExxonMobil) who also publishes the website *JunkScience.com*. The conservative *Washington Times* published the same column. Neither disclosed that Milloy, who debunks global warming routinely, runs two organizations that receive money from ExxonMobil. Between 2000 and 2003, the company gave $40,000 to the Advancement of Sound Science Center, which is registered to Milloy's home address in Potomac, Maryland. ExxonMobil gave another $50,000 to the Free Enterprise Action Institute—also registered to Milloy's residence. Under the auspices of the Free Enterprise Education Institute, Milloy publishes *CSRWatch.com*, a site that attacks the corporate social responsibility movement. Setting aside any questions about Milloy's journalistic ethics, at the scientific level his attack on the ACIA was inept. Citing a single graph from a 146-page overview of a 1,200-plus page, fully referenced report, Milloy claimed that the document "pretty much debunks itself" because high Arctic temperatures "around 1940" suggest that the current temperature spike could be chalked up to natural variability. "In order to take that position," counters Harvard biological oceanographer James McCarthy, a lead author of the report, "you have to refute what are hundreds of scientific papers that reconstruct various pieces of this climate puzzle."

Milloy's charges were taken up by other groups. TechCentralStation.com published a letter to Senator McCain from 11 "climate experts," who maintained that recent Arctic warming was not at all unusual compared to "natural variability in centuries past." The conservative George C. Marshall Institute ($310,000) issued a press release asserting that the Arctic report was based on "unvalidated climate models and scenarios…that bear little resemblance to reality and how the future is likely to evolve." In response, McCain said, "General Marshall was a great American. I think he might be very embarrassed to know that his name was being used in this disgraceful fashion."

The day of McCain's hearing, the Competitive Enterprise Institute put out its own press release, citing the aforementioned critiques as if they should be considered on a par with the massive, exhaustively reviewed Arctic report: "The Arctic Climate Impact Assessment, despite its recent release, has already generated analysis pointing out numerous flaws and distortions." The Vancouver-based Fraser Institute ($60,000 from ExxonMobil in 2003) labeled the Arctic warming report "an excellent example of the favored scare technique of the anti-energy activists: pumping largely unjustifiable assumptions about the future into simplified computer models to conjure up a laundry list of scary projections...2004 has been one of the cooler years in recent history." Never mind that the UN World Meteorological Organization would later declare 2004 to be "the fourth warmest year in the temperature record since 1861."

Senator Inhofe—who received nearly $290,000 from oil and gas companies, including ExxonMobil, for his 2002 reelection campaign—cited the Marshall Institute's work in his own critique of the report.

According to a memo uncovered by the *New York Times* in 1988, a plan was developed to sow uncertainty about the first Intergovernmental Panel on Climate Change (IPCC) report confirming global warming. Framers of the plan included Jeffrey Salmon, then executive director of the George C. Marshall Institute; Steven Milloy, now a *FoxNews.com* columnist; David Rothbard of the Committee for a Constructive Tomorrow ($252,000); the Competitive Enterprise Institute's Myron Ebell, then with Frontiers of Freedom ($612,000); and ExxonMobil lobbyist Randy Randol.

Less than three weeks after Cheney assumed the vice presidency, he met with ExxonMobil CEO Lee Raymond for a half-hour. Corporation officials also met with Cheney's energy task force. A memo forwarded by Randy Randol recommended that Harlan Watson, a Republican staffer with the House Committee on Science, help the United States' diplomatic efforts regarding climate change. Watson is now the State Department's "senior climate negotiator." The Bush administration also appointed former American Petroleum Institute attorney Philip Cooney—who headed the institute's "climate team" and opposed the Kyoto Protocol—as chief of staff of the White House Council on Environmental Quality. In June 2003 the *New York Times* reported that the CEQ had watered down an Environmental Protection Agency report's discussion of climate change, leading EPA scientists to charge that the document "no longer accurately represents scientific consensus."

Larisa Dobriansky, the deputy assistant secretary for national energy policy at the Department of Energy—in which capacity she manages the department's Office of Climate Change Policy—was previously a lobbyist with the firm Akin Gump, where she worked on climate change for ExxonMobil. Her sister, Paula Dobriansky, is undersecretary for global affairs in the State Department. She headed the U.S. delegation to a UN meeting on the Kyoto Protocol in Buenos Aires, where she charged that "science tells us that we cannot say with any certainty what constitutes a dangerous level of warming, and therefore what level must be avoided." At a November 2003 panel sponsored by the AEI, she declared, "the extent to which the man-made portion of greenhouse gases is causing temperatures to rise is still unknown, as are the long-term effects of this trend. Predicting what will happen

50 or 100 years in the future is difficult." Memos uncovered by Greenpeace show that in 2001, within months of being confirmed by the Senate, Dobriansky met with ExxonMobil lobbyist Randy Randol and the Global Climate Coalition. She also met with ExxonMobil executives to discuss climate policy just days after September 11, 2001.

Naomi Oreskes, a science historian at UC San Diego, reviewed nearly 1000 scientific papers on global climate change published between 1993 and 2003, and couldn't to find any that explicitly disagreed with the view that humans are contributing to the phenomenon. As she hastens to add, that doesn't mean no such studies exist. But given the size of her sample, about 10 percent of the papers published on the topic, she thinks it's safe to assume that the number is insignificant. Conservative think tanks deal with this unanimity by magnifying debates beyond their scientific significance. For example, drawing upon several independent studies including one by Michael Mann and colleagues, the IPCC's 2001 report stated, "the increase in temperature in the 20th century is likely to have been the largest of any century during the past 1,000 years." This statement was followed by a graph, based on one of the Mann group's studies, showing relatively modest temperature variations over the past thousand years and a dramatic spike upward in the 20th century. It looks like a hockey stick.

During his talk at the AEI, Michael Crichton attacked the "hockey stick," calling it "sloppy work." In fact, a whole movement has formed to criticize this analysis, much of it linked to ExxonMobil-funded think tanks. At a congressional briefing sponsored by the Marshall Institute, Senator Inhofe described Mann's work as the "primary scientific data" on which the IPCC's 2001 conclusions were based. Wrong. Mann points out that he's not the only scientist to produce a "hockey stick" graph—other teams of scientists have come up with similar reconstructions of past temperatures. And even if all of the studies that served as the basis for the IPCC's statement on the temperature record were wrong, that would not in any way invalidate the conclusion that humans are *currently* causing rising temperatures. "There's a whole independent line of evidence, some of it very basic physics," explains Mann. Nevertheless, the ideological allies of ExxonMobil virulently attack Mann's work, as if discrediting him would somehow put global warming concerns to rest. This seems to have started with Willie Soon and Sallie Baliunas of the Harvard-Smithsonian Center for Astrophysics, both senior scientists with the Marshall Institute. Soon serves as science director to *TechCentralStation.com*, is an adjunct scholar with Frontiers of Freedom, and wrote (with Baliunas) the Fraser Institute's pamphlet *Global Warming: A Guide to the Science*. Baliunas, meanwhile, is "enviro-sci host" of *TechCentral*, and is on science advisory boards of the Committee for a Constructive Tomorrow and the Annapolis Center for Science-based Public Policy ($427,500 from ExxonMobil), and has given speeches on climate science before the AEI and the Heritage Foundation ($340,000). In 2003, Soon and Baliunas published an article, partly funded by the American Petroleum Institute, in a small journal called *Climate Research*. Presenting a review of existing literature, the two concluded "the 20th century is probably not the warmest nor a uniquely extreme climatic period of the last millennium." Another version of the paper was quickly published with three additional authors: David Legates of the University of Delaware, and longtime paid skeptics Craig and Sherwood Idso of the Center for the Study of Carbon Dioxide and

Global Change in Tempe, Arizona. The Idsos received $40,000 from ExxonMobil for their center in the year the study was published, and Legates is an adjunct scholar at the Dallas-based National Center for Policy Analysis (which got $205,000 between 2000 and 2003).

Calling the paper "a powerful new work of science" that would "shiver the timbers of the adrift Chicken Little crowd," Senator Inhofe devoted half a Senate hearing to it, bringing in both Soon and Legates to testify against Mann. The day before, Hans Von Storch, editor-in-chief of *Climate Research* resigned to protest the slack review process that led to its publication; two other editors soon joined him. Von Storch later told the *Chronicle of Higher Education* that climate science skeptics "had identified *Climate Research* as a journal where some editors were not as rigorous in the review process as is otherwise common." Meanwhile, Mann and 12 other leading climate scientists wrote a blistering critique of Soon and Baliunas' paper in the American Geophysical Union publication *Eos*, noting, among other flaws, that they'd used historic precipitation records to reconstruct past temperatures—an approach Mann told Congress was "fundamentally unsound."

On February 16, 2005, 140 nations celebrated the ratification of the Kyoto Protocol. In the weeks before the event, as friends of ExxonMobil rushed to protect the Bush administration from the bad press, a congressional briefing was organized. Sponsored by the George C. Marshall Institute and the Cooler Heads Coalition, the briefing's panel of experts featured Myron Ebell, attorney Christopher Horner, and Marshall's CEO William O'Keefe, formerly an executive at the American Petroleum Institute and chairman of the Global Climate Coalition. The emcee, Senator Inhofe, best represented the spirit of the event. Stating that Crichton's novel should be "required reading," the senator asked for a show of hands to see who had finished it. He attacked the "hockey stick" graph and damned the Arctic Climate Impact Assessment for having "no footnotes or citations," as indeed the ACIA "overview" report—designed to be a "plain language synthesis" of the fully referenced scientific report—does not. But never mind, Inhofe had done his own research. He pulled out a 1974 issue of *Time* magazine and, in mocking tones, read from a 30-year-old article that expressed concerns over cooler global temperatures. Inhofe again called the notion that humans are causing global warming "a hoax," and said that those who believe otherwise are "hysterical people, they love hysteria. We're dealing with religion." Having thus dismissed some 2,000 scientists, their data sets and temperature records, and evidence of melting glaciers, shrinking islands, and vanishing habitats as so many hysterics, totems, and myths, Inhofe vowed to stick up for the truth, as he sees it, and "fight the battle out on the Senate floor."

Seated in the front row of the audience, former ExxonMobil lobbyist Randy Randol looked on approvingly.

(Excerpted, with permission, from Chris Mooney, "Some Like it Hot", *Mother Jones*, May/June 2005, http://www.motherjones.com/news/feature/2005/05/some_like_it_hot.html)

Recommended reading:

Diamond, Jared M., *Guns, Germs, and Steel: The Fates of Human Societies*. New York: W. W. Norton & Co., 1997.

60

ABORTION GAG RULE, by Frank Carpenter, DMin

(May 8, 2005)

The great challenge confronting humanity today is overpopulation. Solutions for global warming or global epidemics cannot develop without education about population growth and its causes. Controversies over abortion and sex education prevent this.

Population growth will reverse, whether we choose it voluntarily, have it regulated by government, as is the case in China, or have it imposed upon us by Mother Nature in the form of disease and starvation as is happening in parts of Africa and Asia. The only proven voluntary approach anywhere in the world is through family planning, including sex education, contraception, sterilization, and abortion—as well as abstinence.

The greatest obstacle to family planning is attitudes about abortion, the major wedge in American culture wars. The most significant date is Jan. 22, 1973, when the U.S. Supreme Court decided Roe v. Wade. The court's decision held that the right of privacy embraced in the due process clause of the fourteenth Amendment includes "a woman's decision whether or not to terminate the pregnancy." Privacy is well grounded in the Constitution. The Supreme Court considers it basic in cases protecting the privacy of personal papers and the right of individuals to choose whom to marry, how to raise their children, and whether to use contraception.

The Court sought to balance potentially conflicting concerns. A woman's health and well-being are essential to a woman's self-determination. They held a woman has the right to decide, in consultation with her physician, whether to have an abortion taking into account factors such as her own health, fetal abnormalities, and whether the pregnancy result from rape or incest. As pregnancy advances, the Court recognized the state's interest to protect the fetus as well as a woman's health. The Court has consistently found against antiabortion laws that do not take into account the health of the mother.

In the wider debate over abortion are conflicting views about life. In Roe v. Wade, the Court stated that a fetus has legal protections when it becomes viable—can survive outside the womb. Other views on the beginning of human life suggest criteria such as fetal brain development, or the ability to feel pain. Others suggest the stage of development when the fetus is as behaviorally and cognitively different from other species as would be an adult human.

Although there is an arbitrary quality to any attempt to define when a life 'begins', absolute definitions have been put forward. These include theological views, for example, where Aquinas follows Aristotle in saying that the soul, created by God, is infused into the body at 40 days for males, and at 90 days for females. Given the ambiguity as to what a person is, it's not surprising there is ambiguity as to when a human being may be said to 'begin.' Basing an embryonic right to life on such views appears even more arbitrary when many 'pro-life' people see no contradiction with advocating the death penalty.

Abortion is a battle cry in the rhetoric of the religious right. The battle over sex education textbooks in schools has intensified, leaving it an open question whether young people today are receiving clear information as to how babies are made. The use of all forms of contraception is condemned, if even mentioned, and information about the morning after pill is no longer considered suitable to be given to a rape victim. Religious arguments are made that the solution to the HIV/AIDS epidemic and unwanted pregnancy is the same: abstinence. Unwanted pregnancy and HIV/AIDS would thus be signs of moral weakness.

Given the intensity and inconsistency of arguments against abortion, some inquiry into why is in order. The second most important date in the abortion war in the United States may well be July of 1991 in Wichita, Kansas. The Summer of Mercy, as its organizers—Operation Rescue—called it, coalesced the Christian Right as the rising political force in Kansas.

The founder of Operation Rescue, a self-described Christian group, is Randall Terry. Terry uses the most powerful of organizing techniques: hate. In 1993, Terry urged, "I want you to let a wave of intolerance wash over you. Yes, hate is good…Our goal is a Christian nation. We have a Biblical duty; we are called by God, to conquer this country. We don't want equal time, we don't want pluralism." Terry changed the name of his organization to Operation Save America. As with Jerry Falwell, who saw abortion as one of the reasons God let the terrorist attack of 9/11 happen, anxieties over American supremacy may be more important in abortion debates than the question of when a human life begins.

The hostility of the American Christian Right has forced a gag upon the honest and open global discussion about overpopulation so urgently needed.

This week's news:

OSLO: A new study shows injecting CO_2 into oil fields boosts production and helps reduce emissions of the gas, but is for now too costly and too risky. "CO_2 injection is technically feasible, and the potential for increased recovery is substantial," said the authors of the report, who have studied 20 Norwegian oil platforms where the method could help extract an additional 150 to 300 million cubic meters of oil in total. "However, the threshold costs for establishing a delivery chain for injection of CO_2 are so high that other methods of improving recovery emerge as being more attractive for the licensees at this time," the authors said. With CO_2 injection, production costs jump to $30–33/barrel. (spacedaily.com)

As reported at a Royal Society meeting on Food Crops in a Changing Climate, experiments have now shown that essential crops such as wheat, rice, maize, and soybeans will be affected much more negatively by global warming than previously predicted. The benefits of higher CO_2 will be outweighed by climate change. Rising temperatures, longer droughts and higher levels of ground-level ozone gas are likely to cause a substantial reduction in crop yields. Just a few days of hot temperatures can greatly reduce yields of wheat, soybeans, rice, and peanuts, if they occur during flowering periods. (*Independent*)

BOULDER, CO: Paleontologist and conservationist Richard Leakey, along with other scientists and policy-makers, are meeting at SUNY Stony Brook to discuss how to protect biodiversity in wildlife parks. Leakey hopes to interest governments, companies and international agencies in establishing a $100 million fund to help those efforts. "I have spent quite a bit of my time looking at conservation in Africa. Climate change is upon us, and I'm not sure that anyone is giving any thought to maintaining the national parks systems. Will the systems within the parks survive a 3°C change at the poles, a different pattern of monsoons? We should be rethinking the boundaries to give the parks a better chance in the future." In the past, Leakey said, animals could adapt to changing climate by expanding or contracting their ranges. Now this option is not always available, for species hemmed in by human settlement and activity. In his book "The Sixth Extinction," Leakey wrote migration "illustrates the need of species to move when climate changes, if they can. Sometimes the climate change is too extensive for species to be able to respond; and geographical barriers such as rivers and mountains can block the only route available. When this happens, extinction is the most likely outcome." The signs of warming are discernible in Africa. "It's obvious in the loss of the ice caps in the high mountains. Kilimanjaro has lost its glaciers. The glaciers on Mount Kenya have lessened over the last decade. The permanent ice has been there for more than 10,000 years. I don't think you'll see the elephants peel off and die, but what impact will it have on the food chain, the sources of water? Huge numbers of creatures will have no water, then what do you do?...Change climate dramatically and conservation, in any context, is adversely affected...We are in another period of great extinction." (terradaily)

US corporations intend to make it impossible to save seed for planting anywhere on earth. The "life-science industry" is patenting and introducing "terminator seed" technology that prevents plants from making viable seeds, forcing farmers to buy seed for planting. As patent holder, the Department of Agriculture expects to license this technology worldwide for maize, wheat, and rice, staples of the developing world. Corporate lobbyists have bought regulations, laws and court decisions that require registration of every crop variety, impose heavy fines for planting and distribution of unlicensed seeds, and require licensing and extensive record keeping by anybody selling or giving seeds away. US patent laws were modified to permit corporations to claim genetic material as their intellectual property. It is now illegal for American farmers to save and plant their own seeds for some plant varieties. American corporations have filed patents on food and medicinal plants grown for centuries in India, Mexico, Africa, South and Southeast Asia, Amazonia, and elsewhere. Corporate geneticists have also learned to transfer genes from one kind of organism into another, creating the first transgenic organisms. Human genes were spliced into animals, animal and bacterial genes into plants, commonly using infectious viruses as "vectors" to carry genes into recipient cells. Risks such as transgenic plants accidentally pollinating native ones, or virus genes infecting people who eat genetically modified (GM) food, are ignored or glossed over. Europe, Africa, and Asia have had some success resisting importation of American "Frankenfood" or planting of genetically GM crops. But transgenic pollution has already occurred in Mexico, contaminating native varieties of corn. Monsanto recently sued a Canadian farmer for non-payment of royalties after his crop was polluted by transgenic Monsanto pollen, and other companies are doing the same in Argentina. (*Black Commentator*)

HONG KONG: Greenpeace and other activists confronted shareholders in the billion dollar China Light and Power (CLP) company with the real cost of burning fossil fuels. For every dollar of profit made by CLP last year, it cost communities across the region nearly $4 in health and environmental impacts. Representatives from affected communities in Hong Kong, mainland China, Thailand, Philippines, and India traveled to the company's annual general meeting. CLP Group made profits of $1.1 billion in 2004, their highest for a decade, mostly from burning coal. Greenpeace demands that CLP invests these profits in clean renewable energy projects like wind power in southern China's Guangdong province. (Common Dreams)

Six years of drought have cut the Missouri river's flow by a third, hampering agriculture, draining reservoirs, and limiting shipping. By next year, the river is expected to become unnavigable. Thirty-nine of Montana's 52 rivers are listed as "very dry" because of the drought. This year is expected to be even drier than the past six. More than 60 of 231 miles of the Oahe reservoir have dried up, with just a narrow channel left of a lake that was up to five miles wide. Boat ramps have been stranded a mile from the nearest water. Hydroelectric power from the dams on the river has been cut by a third, and riverside coal and nuclear plants are running out of cooling water. Farmers don't have water to irrigate their crops. The largest barge operator has not run a vessel up the river for two years. Governor Schweitzer has asked the Pentagon to bring home some of the state's National Guard troops from Iraq so they'll be available to fight forest fires. The most serious threat to the Missouri is the lack of snow in the Rockies, where the snowpack provides 70% of the river's water. In January, golfers were playing on courses that are normally covered by snow until April. Global warming is being blamed for the lack of snow, and may bring even harsher conditions in future. Studies of tree-rings in the area, stretching back 1,200 years, show that in previous warm periods, the area was hit by mega-droughts lasting 20 or 30 years at a time.. (*The Independent*)

Recommended reading:

Luker, Kristin, *Abortion and the Politics of Motherhood.* Berkeley: University of California Press, 1984.

Dorsen, Norman, "Rights in Theory, Rights in Practice," in Berlowitz, Leslie (ed.), *America in Theory.* New York: Oxford University Press, 1988.

Tribe, Laurence H., *Abortion: the Clash of Absolutes.* New York: Norton, 1990.

Schwartz, Lewis M., *Arguing about Abortion.* Belmont, CA: Wadsworth Pub. Co., 1993.

Kagin, Edwin, "The Gathering Storm," in Blaker, Kimberley, et al., *The Fundamentals of Extremism: The Christian Right in America.* New Boston, MI: New Boston Books, 2003.

Frank, Thomas, *What's the Matter with Kansas? How Conservatives Won the Heart of America.* New York: Metropolitan Books, 2004.

Micklethwait, John, and Wooldridge, Adrian, *The Right Nation: Conservative Power in America.* New York: Penguin Press, 2004.

Energy

61

ENERGY—WHAT IT IS

(May 15, 2005)

Just as we need to understand evolution to deal with mass extinctions and human behavior, we need to understand our energy options to attain a sustainable energy economy. It's discouraging that most Americans don't know what energy is, or what the physical limits are to energy production and consumption.

In a nutshell, energy is what's needed to move things and to change things. There's a fixed amount of matter and energy in the universe, although they can be converted from one to the other (e.g., in nuclear reactions). Energy comes in many forms, including kinetic (moving matter), thermal (random molecular movement), electromagnetic (light, radio, X rays), and potential ("stored") energy. Potential energy comes in many forms, including the energy stored in mass itself (nuclear), separation of charge (e.g., lightning), thermal gradients (magma, hot springs), chemical bonds, and even the locations of objects in a gravitational field (snow can release potential energy in an avalanche).

Energy can be converted from one form to another. Whenever such a conversion occurs, some is wasted, lost in the form of heat that is dissipated throughout the environment. This wasted energy is called *entropy*, which inevitably increases over time. For this reason, we can't keep re-using energy. Therefore, we use up available energy sources (e.g., combustion of petroleum) and heat up our environment. Besides heat pollution, we also produce other pollutants as we use energy (e.g., CO_2 from combustion). Cleaning up pollution requires still more energy.

Some energy sources are essentially non-renewable. Fossil fuels (petroleum, coal, and natural gas) took hundreds of millions of years to form during an era when a particularly large fraction of the biosphere's carbon was in plants. They would take much longer to form under today's conditions. Other sources, like peat, are renewable, but they form slowly. Wood is more rapidly renewed.

There are three very long-term sources of energy on Earth: solar (light and heat from the sun), tidal (water moved by the sun's and moon's gravity) and geothermal (heat from the earth's interior).

As long as we use an energy source no faster than it is renewed and we repair the harm it does (e.g., capture and detoxify pollutants), that use is said to be *sustainable*. We can continue that use as long as the sun and the Earth hold out—essentially forever. If we don't use an energy source sustainably, we use it up or we accumulate unacceptable levels of damage.

These are the basic physical operating rules of energy. There are no exceptions to these rules, although with human ingenuity we're learning how to reduce the harmful effects (pollution, resource depletion) and increase the efficiency of energy use.

At present, global energy use is not sustainable: we're using up our resources and wrecking our environment, both at an accelerating rate. We've probably used up half the world's readily retrievable petroleum, a smaller fraction of natural gas, and a yet smaller fraction of coal, about a million times faster than they took to make. While we could switch from petroleum to gas to coal as primary energy sources as we use them up, we aren't removing the main pollutant produced by their combustion: CO_2. That pollutant is causing global warming, and if not stopped soon, it will kill us.

Some scientists fear we've already entered a positive feedback loop caused by fossil fuel combustion, in which runaway climate change can't be stopped. We can only hope there's still time to head off such a calamity.

This week's news:

The world's biggest solar energy power station, capable of sustaining 130,000 households, will be built at an abandoned mine in Portugal. A German manufacturer of solar panels has said it also plans to build a factory at the site, bringing 250 permanent jobs and 3 jobs in ancillary areas for every employee working directly on the solar power plant or in the factory. With a potential output of 116 megawatts, the new station would be several times the size of what is now the world's largest solar energy plant. Last year a solar power plant near Leipzig pronounced itself to be the world's largest, producing 5 megawatts. A 15-megawatt solar power station is being built in South Korea, and Israel is reportedly planning a 100-megawatt station for the Negev desert. (*The Guardian*)

The IEA forecasts both consumption of and emissions from fossil fuels will grow 60% by 2030. The growth will be limited to the most industrialised, richer nations, and "1.4 billion people would still have no access to electricity in 2030." The IEA admitted that its members, the major oil consuming nations, needed to "curb our growing energy import dependence as world reserves narrow to fewer sources." They also noted that desire to "reduce the environmental impact of the world's growing reliance on fossil fuels…[The] IEA estimates that $16 trillion in investment will be needed in the energy sector by 2030." This comes to $640 billion/year for a quarter century, 7–8 times the norm. "We are witnessing under-investment in power generation and transmission and up and down stream, along the oil and gas value chains," the IAE said. (Al Jazeera)

A leak of 20 tonnes of uranium and plutonium dissolved in concentrated nitric acid has forced the closure of Sellafield's Thorp reprocessing plant. The highly dangerous mixture leaked through a fractured pipe into a huge stainless steel chamber so radioactive it's impossible to enter. Recovering the liquids and fixing the pipes will take months and may require special equipment and techniques. The leak is not a danger to the public but is likely to be a financial disaster for the taxpayer since income from the Thorp plant, calculated to be more than £1m a day, was supposed to pay for cleanup of redundant nuclear facilities. (*The Guardian*)

Recommended reading:

Bent, Robert, et al., *Energy: Science, Policy, and the Pursuit of Sustainability*. Washington: Island Press, 2002.

62

OIL WARS

(May 22, 2005)

Our dependence on fossil fuels is the main obstacle to reversing global climate change, mass extinctions, overpopulation, and resource wars. We can't afford to wait any longer to switch to sustainable fuels (think wind, sun, and hydrogen). Instead, we're increasing our dependence on carbon energy, particularly foreign oil. The following is an important time line of U.S. geopolitical maneuvers and military exercises aimed at prolonging and expanding the use of oil. It goes back to the beginning of the 20th century, but we pick up the story during the Carter administration.

July 3, 1979: Carter doctrine: any interference with U.S. access to Persian Gulf oil was declared an assault on U.S. vital interests that will be repelled if necessary with military force. Carter signs secret order to help Mujahedeen fight pro-Soviet regime in Kabul. Soviets invade Afghanistan, as anticipated.

September, 1986: Reagan supplies Mujahedeen with Stinger missiles.

1989: Soviets withdraw from Afghanistan, USSR soon collapses.

Late 1980s: U.S. helps Saudis bring foreign companies into Afghanistan, including Saudi royal family associate Osama bin Laden. Bin Laden recruits fundamentalists to form al Qaeda, brings the Taliban to power.

1990s: American Unocal seeks trans-Afghan pipeline for Caspian and Caucasus oil.

1997: Congress declares Caspian and Caucasus zones of vital American interests. Taliban meet with Unocal in Texas, Clinton officials in DC to discuss pipeline.

1998: Cheney, CEO of Halliburton, states, "I cannot think of a time when we have had a region emerge as suddenly to become as strategically significant as the Caspian." Unocal spokesman tells House of Representatives Taliban should be replaced, arguing trans-Afghanistan pipeline would increase profit by 500% by 2015. Unocal announces delay in finalizing project due to Afghanistan civil war. Al Qaeda bombs U.S. embassies in Kenya and Tanzania. Clinton attacks Afghan and Sudanese targets with cruise missiles. Unocal suspends the pipeline project.

1999: Pakistan, Turkmenistan, and Taliban reach agreement without U.S. Clinton freezes Taliban assets, blocks trade. U.N. imposes sanctions on Taliban, demands bin Laden.

2000: Project for a new American Century (PNAC) advocates U.S. world domination, acknowledges this would take "some catastrophe and catalyzing event—like a new Pearl Harbor." Dubya is illegally appointed president, Supreme Court loses trust of many Americans.

2001: **January:** The first National Security Council meeting after W's inauguration discusses military action against Iraq. Cheney's Energy Task Force formed. **February:** NSC instructs its officials to cooperate with Cheney Task Force, "melding…the review of operational policies towards rogue states, *capture of new and existing oil and gas fields*." Bush administration negotiates with Taliban. **March:** Energy Task Force Policy Report. **April:** Judicial Watch seeks task force documents, Cheney refuses. **May 15:** U.S. official tells Taliban, "…accept our offer…, or we bury you [with] bombs." **July:** Former Pakistani Foreign Secretary advised by U.S. of planned military action against Afghanistan by mid-October. **September 11:** U.S. fails to intercept hijacked airliners in spite of ample warnings before attacks, and alerts during them. Congress grants Bush war powers with little investigation or debate. **October:** State Department plans for post-Saddam. Congress funds studies to restore electricity and water, demobilize Iraq's military. **October 7:** U.S. attacks Afghanistan. **December 5:** U.S-selected committee appoints Hamid Karzai, former Unocal employee, leader of interim Afghan government.

2002: **May:** CIA war-games postwar Iraq. Pentagon officials ordered not to participate. **September:** USAID plans postwar work. Relief organizations warn of serious postwar unrest. **October:** War College lists postwar tasks, warns long-term gratitude unlikely, predicts suspicions of U.S. motives. **December:** Karzai signs oil pipeline agreement.

2003: **January:** CIA war-games postwar issues again, defense officials again prohibited from participating. Problems identified: political reconstruction, public order, humanitarian relief. **March:** Bush diverts military from al Qaeda for invasion of Iraq, ignoring warnings of agencies, task forces, nongovernmental organizations, members of Congress. Administration claims troops will be welcomed as liberators and postwar problems will be benign. **April:** Iraq's social order disintegrates. AID Administrator insists U.S. rebuilding cost will only be $1.7 billion. **July 17:** Judicial Watch publishes Energy Task Force documents obtained through Supreme Court, including maps of Saudi Arabian, United Arab Emirates', and Iraqi oil fields and list of oil companies who want oil contracts with Iraq. U.S. media bury the story. Congress does nothing.

2004: **January:** U.S. weapons inspector David Kay concludes Iraq had no weapons of mass destruction. **June:** 9/11 Commission concludes there was no alliance between al Qaeda and Saddam Hussein. **October:** Bush administration admits there was no solid evidence for WMD or Saddam-al Qaeda links, but continues to refer to both. U.S. food-for-oil profiteers exposed. Billions in Iraqi oil revenues missing. U.S. platoon mutinies over 'suicide mission' in Iraq. Budget gap reaches record $413 billion.

U.S. military now guarding oil pipelines around the world. US oil output at 50-year low.

This week's news:

The Bush Administration is repealing a rule that protects 58.5 million acres of untouched National Forests from logging, mining, and other commercial interests. Some 60 million Americans get their clean drinking water from pristine forest areas. Under the new rule, 34.3 million acres are immediately made available for road construction. The land can be protected only if state governors successfully petition the U.S. Forest Service within 18 months. To develop on the remaining 24.2 million acres, governors can request that the U.S. Forest Service write new management plans that allow construction. "The rule…establishes a meaningless process for governors to petition," said U.S. PIRG Executive Director Gene Karpinski in a news release: "Governors will be used as window dressing while the Secretary of Agriculture will retain control." During the 18-month petition period, the National Forest land remains vulnerable to development, since there are no longer any federal protections. Once a petition is complete it goes to an advisory committee that will make recommendations to the Secretary of Agriculture. The petition may be rejected or accepted, based on criteria that are not yet clear. The 386,000 miles of roads already built in America's national forests have generated $10 billion in maintenance costs for the Forest Service. While spending nearly $49 million on logging programs and roads last year, the Forest Service received only $800,000 from timber sales—a $48 million loss to the American taxpayer. (BushGreenwatch)

Recommended reading:

Roberts, Paul, *The End of Oil: On the Edge of a Perilous New World*. Boston: Houghton Mifflin, 2004.\

63

A BRIEF HISTORY OF HUMAN ENERGY USE

(May 29, 2005)

Organisms use energy contained in chemical bonds to power biological processes such as growth, locomotion, and reproduction. Most of this energy comes from the sun in the form of light, used by plants for the photosynthesis of energy-rich carbon compounds from the CO_2 in the air. Other organisms eat the plants or the plant-eaters for their energy supplies.

Like other animals, prehominids did all the work of food-gathering, shelter construction, and other daily tasks with their own muscle power. Without external energy supplies, they had only modest effects on their environments. Hominids began to change their environments more drastically, using fire to harden their wooden spear tips, clear land, cook food, and stampede herds of prey over cliffs. Hominid numbers grew, and they began driving other species to extinction at an unprecedented rate.

In principle, if all our energy came from combustion of contemporary biological materials (biomass) like wood and the methane produced by decomposition, and we only used these materials as fast as they could be replaced, we would have a sustainable energy economy in which light from the sun produces chemical energy and oxygen through the photosynthetic conversion of CO_2 and we release that energy through oxidation, producing CO_2. We passed the point of biomass sustainability long ago.

Combustion of biomass, mostly wood, peat, and dung, is still the main source of thermal energy for cooking and heating for about 30% of the world's population today. Use of wood for fuel and paper, as well as clearing forests for farmland, have deforested large areas of the planet, including rain forests that are our main repositories of biodiversity. At current rates of deforestation, the rain forests will be gone in a few decades. This has already increased atmospheric CO_2 and decreased capacity for CO_2 clearance by photosynthesis. Deforestation is accelerating, as human population continues to grow and per capita consumption increases.

With the advent of agriculture, *Homo sapiens* began clearing land for crops and domesticating animals. By harnessing the muscle power of animals, we could cultivate fields, grind seeds for flour, haul and lift larger loads, and travel faster. We also harnessed the muscle power of slaves, and many nations, including our own, were built on slavery.

We also learned in prehistoric times to convert potential energy to kinetic energy through the downhill flow of water, redistributing water for irrigation. Later, the kinetic energy of wind and water were used to provide mechanical energy to grind grains into flour, and later on to power tools in factories and generate electricity. These sources of kinetic energy are relatively "clean" environmentally (although they can have their problems), and are important sources of renewable energy, derived from solar heat through meteorological processes.

In the Copper, Bronze, and Iron Ages, metal smelting and pottery and ceramic kilns used wood. By the Dark Ages, small towns consumed hundreds of acres of forests per year for these purposes alone.

Fossil fuels, predominantly coal, petroleum, and natural gas, came into use at different times around the world. They have been used for heat for a long time. Coal came into massive use for steam engines in the eighteenth century, oil for internal combustion engines in the nineteenth century. Discoveries in physics and chemistry led to even greater dependence on fossil fuels for electricity generation and for greatly expanded metallurgy and manufacturing industries.

Nuclear energy has been produced since the 1940s by the conversion of small amounts of matter into large amounts of energy. Although we're currently limited to fission reactors, researchers have been working for decades to derive usable energy from fusion.

Wind, solar, geothermal, wave, and tidal power are just now coming into more prominent use. The combustion of hydrogen shows promise, but only under the proper circumstances.

This week's news:

The modern food system relies heavily on cheap oil. As food undergoes more processing and travels farther, the food system consumes more energy per calorie of food energy. The US food system uses over 10 million billion BTU of energy each year, as much as France's total energy consumption. Twenty-eight percent of energy for agricultural production is used to manufacture fertilizer, 7% for irrigation, 34% for farm vehicles and 31% for pesticide production, grain drying, and facility operations. World grain production has tripled in the last 50 years: 80% due to population growth, 20% due to more people eating higher up the food chain. This has required increasing land productivity with more oil-intensive mechanization, irrigation, and fertilizer use. Farming now depends heavily on fertilizers and on the oil needed to mine, manufacture, and transport them. Nitrogen fertilizer requires natural gas for synthesis. The use of pumps for irrigation crops makes farming in the desert possible, at the cost of aquifer depletion. As water tables drop, more powerful pumps have to be used, increasing the oil requirements. Only 21% of overall energy use is for actual agricultural production. Of the rest, 14% is for transport, 16% for processing, 7% for to packaging, 4% for retailing, 7% for restaurants and caterers, and 32% for home refrigeration and preparation. Food travels farther than ever: fruits and vegetables in western industrial countries often travel 2,500–4,000 kilometers from farm to store. Trucking accounts for the most food transport, though it's almost 10 times as energy-intensive as rail or barge. Rather than propping up fossil-fuel-intensive, long-distance food systems through oil, irrigation, and transport subsidies, governments could promote sustainable agriculture, locally grown foods, and energy-efficient transportation.

Incentives to use environmentally friendly farming methods such as conservation tillage, organic fertilizer application, and integrated pest management could reduce farm energy use significantly. Rebate programs for energy-efficient appliances and machinery for homes, retail establishments, processors, and farms would cut energy use throughout the food system. Legislation to minimize unnecessary packaging and promote recycling would decrease energy use and waste going to landfills. Direct farmer-to-consumer marketing, such as farmers' markets, bypasses centralized distribution systems, cutting out unnecessary food travel and reducing packaging needs while improving local food security. Preferentially buying local foods in season cuts transport and farm energy costs and improves food safety and security. Buying fewer processed, heavily packaged, and frozen foods cuts energy and marketing costs, and using smaller refrigerators reduces electricity bills. Eating lower on the food chain reduces pressure on land, water, and energy supplies. Decoupling the food system from the oil industry is key to improving food security. (Earth Policy Institute)

FORT MCMURRAY, ALBERTA: This is the largest known petroleum deposit in the world outside Saudi Arabia. Monstrous shovel loaders gouge tons of tar-like sands from strip mines 250 feet deep and covering many square miles. Nearby refineries burn natural gas to steam oil from the sands. The oil sands are the world's most expensive, most polluting source of oil under large-scale production. For every four barrels of crude, it's necessary to burn the equivalent of a fifth barrel. It takes 2 tons of sand to yield one barrel. The mines and refineries release vast amounts of greenhouse gases, more than a third of California's car emissions. These oil sands will be the main foreign oil supply for the US for at least a century. The sands cover an area the size of Florida. The oil sands industry now consumes about 400 billion cubic feet of natural gas per year, an amount that could triple by 2015 as oil production rises by the same amount. (*San Francisco Chronicle*)

Recommended reading:

Blumrosen, Alfred W., and Blumrosen, Ruth G., *Slave Nation: How Slavery United the Colonies and Sparked the American Revolution.* Naperville, IL: Sourcebooks, 2005.

64

CARBON FUELS

(June 5, 2005)

Presently the world's human population gets more energy by burning various carbon compounds than from any other source: about 8% biomass, 20% natural gas, 25% coal, and 36% oil—a total of 89%. Of the other 11%, we get about 3% from hydropower, 5% from nuclear, and 3% from renewable sources.

Biomass refers to animal and vegetable material: agricultural, land-clearing and forestry residues, animal wastes, and dedicated energy crops such as wood and plant materials that can be burned directly or used to make gaseous (e.g., methane or hydrogen) or liquid (e.g., ethanol, methanol, biodiesel) fuels. Biomass is potentially renewable in that the plants and animals can be replaced, but at current rates of use they aren't sustainable because we're using up the world's forests faster than we replace them. Burning wood produces CO_2, contributing to global warming. Wood burning is a significant source of particulate air pollution as well.

Natural gas consists of hydrocarbons, strings of carbon atoms with attached hydrogen atoms—mostly methane, CH_4. Natural gas has virtually no impurities, and produces less CO_2 per unit of energy than other carbon fuels because of the high proportion of hydrogen, which burns to produce heat and water vapor. Nevertheless, burning gas does produce CO_2, contributing to global warming.

At the opposite extreme, coal is the dirtiest fossil fuel. Its combustion dumps more CO_2 into the air per unit of energy (there's almost no hydrogen in coal, which is mostly carbon) than any other fuel. Coal also contains nitrogen and sulfur, whose combustion products cause acid rain that kills vegetation and fish, and the deadly smog that killed so many people in London until burning coal was prohibited. Coal-fired power plants are poisoning our air, our water, and our fish with mercury, causing nervous system damage to all of us, especially our children. Thousands of miners' lives have been cut short by black lung disease. The process of mining coal has devastated hundreds of thousands of acres worldwide, causing subsidence, long-smoldering underground fires, disastrous floods from collapsed slag heaps used as dams, and major water pollution. It is not unfair to say that everyone in the coal power industrial complex is complicit in mass murders; all of them are understandably in denial about it. We need to find jobs for them that don't involve maiming and killing innocent victims.

Oil (petroleum) consists largely of longer-chain hydrocarbons, with plenty of impurities. It's second only to coal in the amount of CO_2 and other pollutants per unit of energy produced by its acquisition, refining, and combustion. Gasoline, diesel, and aviation fuel made from petroleum are practically the only fuels used for transportation. We have now tapped into almost all of the least expensive (economically and environmentally) deposits. For this reason, future oil drilling will be more expensive and will cause more environmental and geopolitical disruption, at a time when we should be retooling for sustainable energy. Today, the carbon lobby, crooked politicians, and the journalistically bankrupt media have suppressed the truth about oil, global warming, and alternate energy sources. They have destabilized global geopolitics and caused untold human misery. Today's coal and oil barons and their cronies are a far worse threat than any other terrorists. Only we the people can stop them, and only if we act soon.

This week's news:

DUBLIN: Ireland's first offshore wind farm, the 25-megawatt Arklow Bank Offshore Wind Park, is located on a sandbank about 10 kilometers off County Wicklow in the Irish Sea, just south of the capital Dublin. The seven wind turbines are capable of meeting the annual electricity requirements of 16,000 homes. Under a 2001 EU directive, Ireland is required to increase green electricity to 13.2% of total electricity by 2010. (terradaily)

ROME: The UN Food and Agriculture Organization reports that sub-Saharan African nations' food security is the most at risk from global warming, being the least able to adapt to global warming because of reduction of already scarce farm land, or to compensate through increased food imports. Climate change is also expected to worsen animal diseases and plant pests. The very existence of global warming was denied for decades by government officials in countries such as the U.S. (terradaily.com)

SYDNEY: Australia's largest city announced unprecedented water restrictions as the country's worst drought on record left dam levels at less than 40%. More than 4 million residents will be limited to watering their gardens just twice a week. The New South Wales government quadrupled the penalty for stealing water to a $1,660 on-the-spot fine. Water restrictions were first introduced in Sydney in October 2003 and have been tightened as the drought continues. Even stricter rules could be introduced if dam levels fall below 30 percent. Australia's federal government has increased its drought aid for farmers by $250 million to $1.25 billion. (terradaily.com)

Recommended reading:

Heinberg, *The Party's Over: Oil, War and the Fate of Industrial Societies*. Gabriola Island, BC: New Society Publishers, 2005.

65

THE END OF OIL

(June 12, 2005)

Fossil fuels were produced very slowly, under special conditions not present today. This means that they constitute an essentially non-renewable resource. Whenever we use a million tons of coal, oil, or natural gas, that's one million tons less that's available for future use. Until the twentieth century, no one would have thought that was a pressing problem, because we believed the supply was vastly greater than our needs.

But in the mid-twentieth century a geologist named Hubbard suggested that given a finite supply of a resource, consumption over time would follow a bell-shaped curve, rising from nothing at the time of its first use to a peak when use matched the highest production rate we could achieve, followed by a decline as the resource became more scarce and production became more expensive. This wasn't a particularly revolutionary idea, but the prediction he made was based on the history of oil production and use. He forecast a peak of US oil production in the 1970s, followed by a decline. Sure enough, his prediction came true.

From that time on, we've become progressively more dependent on foreign oil, and rather than switch to other sources of energy we've deepened our commitment to oil. The story of this deepening commitment is one of the tawdriest chapters of our nation's history, revealing the lack of values of our richest corporations and individuals, our news media and our government, the same ones who claim the noblest of values. We're now committed by these same unprincipled companies and people to continued oil wars in order to continue enriching our deeply overcommitted fossil energy industry. This is being done with the uninformed consent of US voters, while other countries have begun their switch to alternative energy sources.

The consequences for the US will be a disaster, quite possibly the collapse of our nation. The consequences for the rest of the world will be little better because of their close economic links to us. We're now the world's largest debtor, for example, and they're the lenders who have underwritten our debt. Also, the world's oil producers know that we're at or near peak *global* oil production, after which the cost of oil will rise catastrophically, regardless of where it comes from. Given our nation's rogue behavior in recent years, we'll most likely overextend ourselves militarily, estrange our few remaining allies, and in the end, it will be US against the rest of the world. Our creditors will call in their debts, costs of US goods will skyrocket due to increased energy costs, outsourcing will evaporate because of prohibitive transportation costs, and we'll be left with an inadequate manufacturing infrastructure (little things like factories). Nations that have invested in sustainable energy will rule. Third world nations enslaved by globalization will be on their own.

It's urgent that we build the new infrastructure for a sustainable energy economy in order to stop global warming, halt population growth, and end mass extinctions. We'll need lots of energy to solve those problems. But the well will have run dry. There will be a great temptation to turn to coal, natural gas, oil sands and shale, and nuclear energy, which would only make things worse.

This prospect probably would have been much the same regardless of the outcome of our recent election. Energy wasn't mentioned much because both parties and their candidates are in the pockets of our energy industries. We don't have another four years in which to await the miraculous arrival of honest and effective leaders. We'll have to salvage what we can through our own initiatives.

This week's news:

GENEVA: In 10 years, Borneo could lose most of its forests to logging, fires, and plantations. In a report entitled "Treasure Island at Risk," the WWF says deforestation on the world's third-largest island will seriously endanger orang-utans and pygmy elephants, as well as Borneo's future economic potential. Today, 50% of Borneo's forest cover remains, down from 75% in the 80s. 1.3 million hectares of forest are destroyed every year. "The consequences of this scale of deforestation will not only result in a major loss of species but also disrupt water supplies and reduce future economic opportunities, such as tourism, and subsistence for local communities," said Chris Elliott, head of the WWF Global Forest Programme. More than 210 mammals, including 44 that are found nowhere else in the world, live on Borneo. Between 1994 and 2004, at least 361 new species were discovered and new ones are constantly being found. (terradaily.com)

Easter Island's inhabitants committed mass suicide by systematically destroying their own habitat. They cut down their forests much faster than they could grow back. At first, the island was plunged into resource wars. Then they turned on their leaders in rage. Then they were left with nothing. They committed mass cannibalism and almost completely died out. In his chilling new book *Collapse—How Societies Choose to Fail or Survive*, Jared Diamond describes how some of the most advanced civilizations in history, like the Maya, committed ecocide without realizing it. According to the world's leading climatologists, we are now on-course for the most rapid increase in global temperatures since the last Ice Age. The biggest common factor in past ecocides has been the pursuit of short-term "rational bad behavior," that ranks short-term interests above the long-term interests of everyone. One solution is "true cost economics": instead of only paying the market price for something, we would have to pay the environmental cost as well. (*The Independent*)

Recommended reading:

Kleveman, Lutz, *The New Great Game: Blood and Oil in Central Asia.* (2003).

Klare, Michael T., *Resource Wars: The New Landscape of Global Conflict.* New York: Henry Holt, 2001.

66

RENEWABLE ENERGY

(June 19, 2005)

We need a sustainable energy economy, i.e., renewable energy, means of energy transmission, and a portable energy source to propel vehicles, with zero net environmental damage. Power grids provide means of transmission. There are several affordable, sustainable candidates for generation of electricity. Nuclear and fossil fuel power are not among them.

Wind

Modern, quiet, and non-polluting wind turbines are already producing cost-competitive electricity. Wind could replace all carbon-based power in a decade or two, at lower cost. Turbine manufacture and wind farm construction and maintenance would employ at least as many people as fossil fuel and nuclear electrical utilities. Wind farms would take up about 0.7% of our land. A properly dispersed system would be impervious to local catastrophes. Wind could power much of the inhabited world, raising standards of living enough to make population control possible. We should give away the means to make wind turbines so every nation can build its own wind farms, helping to break the shackles of globalization and develop steady-state national economies. With no remaining reason for military and economic competition over fossil fuels, there would also be less motivation for terrorism or economic or military domination by developed nations.

Solar

Solar energy is ideal for sunny areas and it can be installed on rooftops in urban and industrial areas. Like wind, its distributed nature is attractive. Like wind, solar power could replace all carbon-based power in a decade or two, at a comparable price. Solar power would employ as many people as those displaced by it. Solar electricity for homes currently pays for purchase costs in ten to twenty years, and solar panels have a lifetime of about forty years, making solar power less than half as expensive as carbon. As fossil power becomes more expensive and solar becomes less so, and if governments offer financial incentives presently provided the fossil and nuclear power industries, amortization times will become even more attractive. California's tax incentives cut amortization time roughly in half. We should give away means of production so every nation would be energy independent. It would cost less than Star Wars and would be far more effective as a deterrent to resource wars.

Wind and solar power are both intermittent, and the demand for energy is subject to large fluctuations. I'll discuss means for storing power during periods of excess supply, in order to generate electricity during periods of excess demand.

Hydroelectric

This energy comes from water flowing over waterfalls or dams. In most developed nations, this resource is being used to full capacity already, but it is potentially a major source of power in many developing countries where it hasn't yet been fully exploited. There are generally social and ecological costs, such as displacing people, drowning archeologically important sites, destroying habitats, and blocking fish migration routes. In addition, siltation and redistribution of water can cause serious problems.

Tidal

Tidal power is available anywhere significant amounts of water flow under the influence of the sun and moon's gravity, such as fjords and tidal bores. It's available about one quarter of the day, and subject to some monthly variation. Essentially the same technology is used as for hydroelectric power.

Wave

The kinetic energy of ocean waves can be converted to electricity, and England is considering wave power stations.

Geothermal

Intense sources of heat can be used to generate steam, which in turn can drive turbines for power generation. Such intense sources are available for little cost in areas where hot magma comes close enough to the surface of the earth, e.g., where volcanic activity or hot springs are found. Such areas include the Philippines, Iceland, and other sites at the edges of tectonic plates. The supply is very long-term.

If sustainable sources could replace fossil fuels for electricity generation in a decade or two at no additional cost and great environmental benefit, why haven't we humans done it? The answer is that we are doing it, and successfully, but not nearly as fast as we should.

This week's news:

UNITED NATIONS ENVIRONMENT PROGRAMME: Urban Environmental Accords Green Cities Declaration

RECOGNIZING for the first time in history, the majority of the planet's population now lives in cities and that continued urbanization will result in one million people moving to cities each week, thus creating a new set of environmental challenges and opportunities; and

BELIEVING that as Mayors of cities around the globe, we have a unique opportunity to provide leadership to develop truly sustainable urban centers based on culturally and economically appropriate local actions; and

RECALLING that in 1945 the leaders of 50 nations gathered in San Francisco to develop and sign the Charter of the United Nations; and

ACKNOWLEDGING the importance of the obligations and spirit of the 1972 Stockholm Conference on the Human Environment, the 1992 Rio Earth Summit (UNCED), the 1996 Istanbul Conference on Human Settlements, the 2000 Millennium Development Goals, and the 2002 Johannesburg World Summit on Sustainable Development, we see the Urban Environmental Accords described below as a synergistic extension of the efforts to advance sustainability, foster vibrant economies, promote social equity, and protect the planet's natural systems.

THEREFORE, BE IT RESOLVED, today on World Environment Day 2005 in San Francisco, we the signatory Mayors have come together to write a new chapter in the history of global cooperation. We commit to promote this collaborative platform and to build an ecologically sustainable, economically dynamic, and socially equitable future for our urban citizens; and

BE IT FURTHER RESOLVED that we call to action our fellow Mayors around the world to sign the Urban Environmental Accords and collaborate with us to implement the Accords; and

BE IT FURTHER RESOLVED that by signing these Urban Environmental Accords, we commit ourselves to moving vital issues of sustainability to the top of our legislative agendas. By implementing the Urban Environmental Accords, we aim to realize the right to a clean, healthy, and safe environment for all members of our society.

[Signatures & Dates]

Issues:
Energy Renewable Energy | Energy Efficiency | Climate Change
Waste Reduction Zero Waste | Manufacturer Responsibility | Consumer Responsibility
Urban Design Green Building | Urban Planning | Slums
Urban Nature Parks | Habitat Restoration | Wildlife
Transportation Public Transportation | Clean Vehicles | Reducing Congestion
Environmental Health Toxics Reduction | Healthy Food Systems | Clean Air
Water Water Access and Efficiency | Source Water Conservation | Waste Water Reduction

Energy
Action 1 Adopt and implement a policy to increase the use of renewable energy to meet ten percent of the city's peak electric load within seven years.

Action 2 Adopt and implement a policy to reduce the city's peak electric load by ten per cent within seven years through energy efficiency, shifting the timing of energy demands, and conservation measures.

Action 3 Adopt a citywide greenhouse gas reduction plan that reduces the jurisdiction's emissions by twenty-five per cent by 2030, and that includes a system for accounting and auditing greenhouse gas emissions.

Waste Reduction

Action 4 Establish a policy to achieve zero waste to landfills and incinerators by 2040.

Action 5 Adopt a citywide law that reduces the use of a disposable, toxic, or non-renewable product category by at least fifty percent in seven years.

Action 6 Implement "user-friendly" recycling and composting programs, with the goal of reducing by twenty per cent per capita solid waste disposal to landfill and incineration in seven years.

Urban Design

Action 7 Adopt a policy that mandates a green building rating system standard that applies to all new municipal buildings.

Action 8 Adopt urban planning principles and practices that advance higher density, mixed use, walkable, bikeable and disabled-accessible neighborhoods that coordinate land use and transportation with open space systems for recreation and ecological restoration.

Action 9 Adopt a policy or implement a program that creates environmentally beneficial jobs in slums and/or low-income neighborhoods.

Urban Nature

Action 10 Ensure that there is an accessible public park or recreational open space within half-a kilometer of every city resident by 2015.

Action 11 Conduct an inventory of existing canopy coverage in the city; and, then establish a goal based on ecological and community considerations to plant and maintain canopy coverage in not less than fifty per cent of all available sidewalk planting sites.

Action 12 Pass legislation that protects critical habitat corridors and other key habitat characteristics (e.g., water features, food-bearing plants, shelter for wildlife, use of native species, etc.) from unsustainable development.

Transportation

Action 13 Develop and implement a policy that expands affordable public transportation coverage to within half-a-kilometer of all city residents in ten years.

Action 14 Pass a law or implement a program that eliminates leaded gasoline (where it is still used); phases down sulfur levels in diesel and gasoline fuels, concurrent with using advanced emission controls on all buses, taxis, and public fleets to reduce particulate matter and smog-forming emissions from those fleets by fifty per cent in seven years.

Action 15 Implement a policy to reduce the percentage of commute trips by single occupancy vehicles by ten per cent in seven years.

Environmental Health
Action 16 Every year, identify one product, chemical, or compound that is used within the city that represents the greatest risk to human health and adopt a law and provide incentives to reduce or eliminate its use by the municipal government.
Action 17 Promote the public health and environmental benefits of supporting locally grown organic foods. Ensure that twenty per cent of all city facilities (including schools) serve locally grown and organic food within seven years.
Action 18 Establish an Air Quality Index (AQI) to measure the level of air pollution and set the goal of reducing by ten per cent in seven years the number of days categorized in the AQI range as "unhealthy" or "hazardous."

Water
Action 19 Develop policies to increase adequate access to safe drinking water, aiming at access for all by 2015. For cities with potable water consumption greater than 100 liters per capita per day, adopt and implement policies to reduce consumption by ten per cent by 2015.
Action 20 Protect the ecological integrity of the city's primary drinking water sources (i.e., aquifers, rivers, lakes, wetlands and associated ecosystems).
Action 21 Adopt municipal wastewater management guidelines and reduce the volume of untreated wastewater discharges by ten per cent in seven years through the expanded use of recycled water and the implementation of a sustainable urban watershed planning process that includes participants of all affected communities and is based on sound economic, social, and environmental principles.

DHAKA, BANGLADESH: River currents strengthened by rising sea levels have devoured half of Bangladesh's biggest island over the last forty years, leaving half a million people homeless. "If the erosion continues at the same rate, it will completely disappear over the next four decades," said Mohammad Shamsuddoha, who carried out the research for The Coast Trust. Shamsuddoha said rising sea levels were responsible for the erosion of coastal islands such as Bhola that were not previously vulnerable to the problem. "The erosion of Bhola Island only started in the 1960s. Before that the size was stable and only a small amount of erosion took place on one side...But from the mid-1960s the erosion began and the rate has accelerated over the years." The government estimates that 6 million people out of the country's 140 million population are displaced each year due to river erosion. (terradaily.com)

Recommended reading:
Geller, Howard S., *Energy Revolution: Policies for a Sustainable Future*. Washington: Island Press, 2003.

67
THE NUCLEAR OPTION
(June 26, 2005)

Now that global warming and higher oil prices are upon us, the nuclear power industry is advertising their electricity as cheap and emission-free. But the nuclear materials used in bombs and reactors, and the waste materials produced by bombs and reactors, are highly radioactive. The only benefactors of radioactive materials have been some patients and consumers of irradiated foods, and nuclear power and nuclear weapons are not needed for either purpose. Oh, there's another group of benefactors: those who get filthy rich off the very lucrative nuclear power and weapons industry.

In early 1896, only a few months after Roentgen's discovery of x rays, French physicist Henri Becquerel reported that uranium could cloud photographic film. In 1898, Marie Curie deduced that radioactivity was a consequence of fundamental atomic properties of certain elements. She went on to discover radium and polonium, and died of radiation sickness. So did many of the women who licked the tips of the brushes they used to paint the radium dials on wristwatches and clocks. Since then, many uranium miners, nuclear fuel refiners, physicists who worked on the Bomb, soldiers and civilians who served as guinea pigs near nuclear blasts, prisoners who were unwitting subjects in experiments on the effects of radioactivity, people living near nuclear facilities and nuclear test sites, sailors on nuclear submarines, and nuclear armaments and power plant workers died. Untold numbers of people have sickened or died because of fallout from nuclear testing and the nuclear accident at Chernobyl. There's no telling how many more thousands of people will die from nuclear wastes, especially from corroding drums dumped into the ocean and from contaminated drinking water.

The nuclear attacks on Hiroshima and Nagasaki, which killed and sickened hundreds of thousands of civilians, were war crimes on a par with the fire bombings of German and Japanese cities, exceeded only by the genocides in Uganda, the Soviet Union, Germany and the Columbian New World. Our use of nuclear weapons opened the Pandora's Box of nuclear weapons proliferation, and every time we rattle our nuclear saber, threatened nations scramble to join the nuclear club. We now face the real threat of nuclear terrorism. More recently, we've spread tons of depleted uranium dust from bombs and bullets around the Balkans, Afghanistan, and Iraq. Civilians and combatants in these countries have suffered toxic effects including organ damage, cancers, birth defects, and stillbirths.

Nuclear power is currently used to provide about 20% of U.S. electricity. France has the highest level of use, about 75%.

There is one simple argument against ever building another nuclear power plant, and against using any of the existing ones any longer, which trumps all arguments in favor of the nuclear "option." They produce high-level nuclear waste in the form of spent fuel. This waste no longer has enough fuel material to be used in reactors, but it is still very radioactive, and its radioactivity will last for thousands of years. Nuclear power plants have short lifetimes: only a few decades. Their radioactive carcasses are accumulating. No one has come up with a nuclear reaction pathway that could be used to change the radioactive material into harmless non-radioactive materials. No one has come up with a foolproof way of storing the stuff for thousands of years until it decays enough to be negligibly dangerous.

Our government, and the governments of other nations, have provided massive subsidies to the nuclear power industry for decades. Nuclear power is more expensive than is reflected in our electricity bills but like so many of the hidden costs of our corrupt energy industry, subsidies from our tax dollars continue to fatten the coffers of the nuclear con artists. This squandering of tax money, which could be used to develop wind, solar and hydrogen energy continues today. This madness has to stop, soon, if we are to reverse global warming, mass extinctions, and the population explosion.

This week's news:

OSLO: Inuit hunters claim the U.S. is violating their human rights by causing global warming. Sheila Watt-Cloutier, chair of the Inuit Circumpolar Conference (ICC), also says the US is hampering a follow-up to a 2004 report by 250 scientists that said the thaw could make the Arctic Ocean ice-free in summer by 2100. She says a planned petition to the Organization of American States (OAS) could exert pressure on the United States to do more to cut greenhouse emissions. The U.S. claims to be investing heavily in energy research and clean hydrogen fuel but has not joined almost all its allies in signing up for the United Nations' Kyoto protocol, which sets caps on carbon dioxide emissions. There are 155,000 Inuits in Canada, Alaska, Greenland, and Russia. (planetark.com)

Documents obtained by *The Observer* reveal Bush administration efforts to undermine completely the science of climate change, and show that the U.S. position has hardened during the G8 negotiations. They also show the U.S. has abandoned a critical U.N. effort to stabilize greenhouse gas emissions. The documents show that Washington officials removed from a draft statement by the G8 all reference to the fact that climate change is a "serious threat to human health and to ecosystems"; deleted any suggestion that global warming has already started; and expunged any suggestion that human activity was to blame for climate change. One of the sentences removed was: "Unless urgent action is taken, there will be a growing risk of adverse effects on economic development, human health and the natural environment, and of irreversible long-term changes to our climate and oceans." Another section erased by the White House said, "Our world is warming. Climate change is a serious threat that has the potential to affect every part of the globe. And we know that…mankind's activities are contributing to this warming. This is an issue we must address urgently." Earlier this month, the top science academies of the

G8 nations, including the U.S. National Academy of Science, issued a statement saying that evidence of climate change was clear enough to compel their leaders to take action. "There is now strong evidence that significant global warming is occurring," they said. It is now clear that this advice has been completely ignored by Bush and his advisers. "Every year, it (local air pollution) causes millions of premature deaths, and suffering to millions more through respiratory disease," reads another statement removed by the U.S. Washington is also uncooperative regarding the plight of Africa. The documents show the Bush administration has withdrawn from earlier pledges to fund a network of climate monitoring centers. Another section deleted by the U.S. is, "Africa, Asia-Pacific, and the Arctic are particularly vulnerable to climate variability and are starting to experience the impacts." Other components scuttled by the U.S. include the Clean Development Mechanism to help nations develop while controlling greenhouse gas emissions. (*The Observer*)

A naturally occurring bacterium that historically hadn't been a problem in Alaska's Prince William Sound is contaminating oysters and making people sick. Waters were previously too cold for *Vibrio parahaemolyticus*, but a warmer summer in 2004 heated coastal waters, causing a bacterial bloom. Health officials tested the bacterium strains found in Alaska oysters and in sick people and found they matched. The bacteria's virulence was unexpected. Environmental samples had the highest proportion of a gene known to cause sickness in people ever reported. The flourish of vibrio bacteria in 2004 illustrates how climate change can spread infectious diseases. Vibrio has been more of a problem farther south, where water temperatures are warmer. The bacterium needs a temperature of about 62 degrees to start multiplying. Median summertime water temperatures at one Prince William Sound oyster farm have been increasing and exceeded 62 degrees last summer. According to the Arctic Climate Impact Assessment published by the International Arctic Council, the average temperature in the Arctic has climbed at twice the rate in the rest of the world. (*Anchorage Daily News*)

Recommended reading:

Heintzman, Andrew, and Solomon, Evan, *Fueling the Future: How the Battle over Energy is Changing Everything.* Toronto: House of Anansi Press, 2003.

68

HYDROGEN

(July 3, 2005)

Fossil fuels can be replaced in a decade or two by renewable (wind, solar, geothermal, tidal, and wave) sources to generate electric power, if we start now. And we have to start now in order to stop global warming and mass extinctions and to provide the decent standard of living needed to reduce population. The one remaining ingredient for eliminating carbon fuels is an environmentally friendly fuel to propel vehicles and store energy. There's really only one option: hydrogen, the most common element in our biosphere.

When hydrogen is combined with oxygen, it yields power and water. Don't settle for hydrogen made from fossil fuels—that's just fossil fuel (generally natural gas) power by another name. Go for the clean stuff, produced by using electricity from renewable sources to convert water into two gases: hydrogen (the fuel) and oxygen (released into the atmosphere to combine with our fuel). The process is called *hydrolysis*. What could be more elegant? The idea has been around at least since Jules Verne: apply electric power to water to produce oxygen and a fuel from which we can regenerate power and water (and some entropy).

There are two ways to use hydrogen for transportation: by burning in an internal combustion engine similar to current gasoline-powered ones, or by combining it with oxygen in fuel cells to produce electricity to run an electric motor. Both can have negligible environmental impact if managed properly. Working hydrogen-powered cars were first developed last century, and modern versions are in use today in some locations. To put enough in your car's gas tank to drive 100 miles, it has to be pressurized, like propane or butane.

The truly revolutionary aspect of hydrogen as a fuel is that it can be produced anywhere there is water and enough electricity. In principle everyone could make their own, although the corner gas station ("gas station" acquires a new meaning) is still probably the most cost effective approach. Companies and other entities with vehicle fleets would have their own hydrogen fueling points, as they do today for gasoline. Distributed hydrogen production, like distributed electricity production, minimizes the dangers from local disasters, be they natural, accidental, or premeditated.

Hydrogen does not store well, though. It doesn't spoil, but it leaks. It's such a small molecule it can slip through the intermolecular spaces in container walls in significant amounts. Although new materials may be devised to contain it better, at least for now on-demand, distributed production and supply will meet our needs. Leaked hydrogen is harmless if not allowed to accumulate in unventilated spaces, where it could be ignited by a spark or flame—like natural gas or gasoline fumes, but worse.

Like natural gas, hydrogen can be used for cooking and heating, but it would be more cost effective to use electricity, since that will be the source of our hydrogen. One less energy transformation means less energy lost to entropy. It also means hydrogen could be banned from indoor locations, reducing explosion hazards.

Think of it. No coal mining, or gas or petroleum drilling. No polluting, expensive and wasteful refining or transport of fuels, no power plant pollution (particulates, sulfur and nitrogen oxides, heavy metals), no greenhouse gases, and a real prospect of reversing global warming, mass extinctions, and overpopulation if we act now. And we'd have enough fossil carbon left to meet the needs of the worldwide organic chemical industry, e.g. for plastics, artificial rubber, and pharmaceuticals.

Two last notes: (1) the hydrogen industry is developing fast and one line of thought is to pipe hydrogen to fuel cell electricity generators. That's dumb: power lines are more cost-effective than pipelines and leakage from pipelines would be significant. (2) The Bush administration has promised to push hydrogen development, but Bush and Congress are funding hydrogen extraction from fossil fuels, which produces CO_2, rather than electrolysis of water, which produces only hydrogen and oxygen. They're giving big bucks to fossil fuel industries rather than to the wind and solar industries.

This week's news:

STANFORD, CA: Stanford researchers say in an article in the journal *Science* that a conversion to hydrogen-fueled cars would improve air quality, health, and climate—especially if wind were used to generate the electricity needed to split water and make hydrogen and fuel cells were used to react it with oxygen to produce water and energy. Associate Professor Mark Z. Jacobson and colleagues report that the conversion could prevent millions of cases of respiratory illness and tens of thousands of hospitalizations and save 3,000–6,000 lives each year. The *Science* study compared emissions that would be produced in five cases—if all vehicles on the road were powered by 1) conventional internal-combustion engines, 2) a combination of electricity and internal combustion of gasoline, as in hybrid vehicles, 3) hydrogen generated from wind electrolysis, 4) hydrogen generated from natural gas, and 5) hydrogen generated from coal gasification. Wind is the most promising means of generating hydrogen. The cost of making hydrogen from wind is $1.12 to $3.20 per gallon of gasoline or diesel equivalent ($3 to $7.40 per kilogram of molecular hydrogen)—on par with the current price of gas. But gasoline has a hidden cost of 29 cents to $1.80 per gallon in societal costs such as reduced health, lost productivity, hospitalization, and death, as well as cleanup of polluted sites. So gasoline's true cost in March 2005, for example, was $2.35 to $3.99 per gallon, which exceeds the estimated mean cost of hydrogen from wind ($2.16 equivalent per gallon of gasoline). Jacobson envisions wind turbines generating electricity fed to the power grid. At fuelling stations, it would power an electrolyzer, splitting water into oxygen, which would be released into the air, and hydrogen, which would be compressed and stored. Next, the group plans to look at the effects of converting all power plants to hydrogen fuel cell power plants. They also plan to explore the long-term effects of switching to a hydrogen economy on global climate change and the ozone layer. (terradaily.com)

GUAM: International officials at the Pacific Islands Environmental Conference yesterday warned of what global climate change may bring in coming decades: having no tuna because they've left for cooler waters; a dead reef that provides no protection from tsunamis; droughts that kill nearly all the vegetation on the island; 10 inches of rain in one hour, wiping out roads, homes and schools. Global warming is already taking a toll on smaller Pacific islands. Slight changes have driven tuna from Yap's waters, closer to Hawaii. It's devastated one of Yap's few sources of income: commercial fishing. (*Pacific Daily News*)

PAMPLONA, SPAIN: A research team from the Public University of Navarra has started a study of absorbent materials for storage of hydrogen, an alternative to fossil fuels. Storage is a key process in the changeover from internal combustion engines to fuel cells. Hydrogen gas normally has to be stored at very high pressure or very low temperature. These two factors cause technical difficulties, apart from safety ones. Hydrogen can also be absorbed into metals (as hydrides) and physiadsorbed in suitable materials. This last method, involving the "physical adsorption onto porous materials," is what's being developed in this research project, using nanoporous materials: activated carbons, zeolites and stacked clays. These materials have mechanical resistance and are safe, light, and cheap. (spacedaily.com)

ROME: The heat wave that hit much of Europe in 2003 killed almost 20,000 people throughout Italy, the national statistics institute said on Monday, more than doubling the previous official estimate of the toll and taking it above that recorded in France. In Italy, where temperatures are expected to rise for the next three days throughout the country, four cities—Milan, Turin, Genoa, and Rome—have activated emergency plans that place 10,000 people at risk under automatic surveillance. Italian civil protection forces are using a forecasting system in eight cities that combines weather and health data to predict and prevent heat-related incidents three days ahead. In Rome, a refuge network will provide cool places for the city's elderly and a volunteer force will check at-risk people in their homes. Rome has 550,000 citizens over 65, 20% of the population. Four thousand Romans will also use a heat monitoring system that links an electronic bracelet wearer to health authorities via telephone. When the bracelet detects bodily signals related to heat stress, it automatically sends an alarm signal to a central control center. (terradaily.com)

The U.S. still has the highest incarceration rate in the world, with 714 prisoners per 100,000 inhabitants, ahead of Russia and Belarus, according to London University King's College International Centre for Prison Studies. The United States has held first place since 2000. Of nine million people imprisoned in the whole world, more than two million (22% of the total) are behind American bars. Russia has the highest incarceration rate in Europe, with 550 prisoners per 100,000 inhabitants. South Africa has the highest rate in Africa (413/100,000) and Surinam the highest in South America (437/100,000). (AP)

Recommended reading:

Rifkin, Jeremy, *The Hydrogen Economy: The Creation of the Worldwide Energy Web and the Redistribution of Power on Earth*. New York: J. P. Tarcher/Putnam, 2002.

69
ENERGY POLICY AND POWER
(July 10, 2005)

This will be a very brief overview of current U.S. energy policy, to set the stage for proposing a sane energy policy.

Current policy simply escalates the worst features of past policy. In the face of an urgent need for renewable energy and conservation, most U.S. investment still goes into fossil fuels, nuclear energy, and energy-wasting vehicles and transport systems.

The robber barons of the nineteenth century, through monopolies, trusts, and interlocking directorships between the oil, coal, railroad, and auto industries, developed a powerful, essentially unrestrained system for plundering the peoples' wealth. They were finally brought under some control by federal antitrust laws and by the labor unions, but this was not to last. Unions and regulatory mechanisms have been eroded by Reagan-Bush-Clinton-Bush, to the point where conditions are as bad as a century ago, maybe worse.

The power of the nuclear sector arose from the gargantuan nuclear arms industry, paid for by our tax dollars.

The extremely wealthy fossil and nuclear industries have developed powerful lobbies that are difficult to oppose, as politicians have discovered when it comes time to finance their campaigns. The automobile, airline, highway construction, and trucking industries have flourished under this system, while ironically, the more economical railroads have not. Our country is unique among developed nations in this regard.

US foreign policy is grossly distorted by our corporation-imposed dependence on oil. Official national policy dictates that if we need it, we will take it. The evolution of this policy dates back at least to the Great Game, a contest among the Western powers before and after the First World War to secure shares of the oil, primarily in the Middle East and central Asia. The Great Game was, of course, a continuation of similar contests going back to the first chiefdoms around 10,000 years ago—soon after the origin of agriculture.

Even the geopolitical outcomes of the Second World War were partly determined by the contest for oil. The U.S. and USSR maneuvered their Cold War confrontations partly around the control of oil. Carter, during his oil crisis, notched up our militarism. The Carter doctrine declared that any interference with our access to oil would be viewed as an assault on our national interests and would be opposed if necessary with military force. Most of our military actions and geopolitical maneuvers in South America, Europe, North Africa, Indonesia, Asia, and the Middle East had to do with oil. The Dubya Doctrine went further, to declare that the US reserved the option of preemptive war against any nation perceived to threaten our national security—read access to oil.

This means our foreign relations are largely determined by our oil dependence. We support despots who provide us oil like the Saudis, and undermine legitimate governments if they even threaten to put a political price on their oil, like Venezuela. The Department of Defense, not the State Department, is our primary instrument of international relations.

As a result of government support of the oil industry, gasoline prices at the pump are far lower than actual costs. The difference is paid through our taxes and through horrific military actions resulting in the deaths or severe hardship of millions of people—so we can drive our gas-guzzling, polluting SUVs and consume, consume, consume.

The nuclear power industry is particularly ruinous. The cost of nuclear electricity has little relation to its actual cost: it doesn't reflect the true cost of tax-supported construction, fuel production, spent fuel processing and disposal (storage), and decommissioning of reactors, not to mention the enormous ecological costs.

Our country is no longer the wealthy nation it used to be, thanks to plundering by government-assisted corporations. We never could afford present energy policy, but it is especially onerous now, when it impedes development of sustainable energy.

This week's news:

ASHLAND, OR: Scientists and conservationists say the Endangered Species Act fails to protect hundreds of species headed for extinction. Property rights advocates say it harms farmers, ranchers, and developers. Western governors want more influence over how the law is defined and enforced. And congressional critics say it's enforced by judges who don't considering their decisions' economic or social impact. Lawmakers want to change it. Religious groups get involved on both sides. Some species have done very well, among them the peregrine falcon, the American alligator, the bald eagle, and the California condor. But of the more than 1,200 species listed as endangered or threatened since passage of the ESA in 1973, very few have recovered to the point where they no longer need special protections such as limiting activity in a designated habitat. Rep. Richard Pombo (R–CA), chair of the House Resources Committee, wants fewer species on the list because so few are ever de-listed. Environmentalists argue that without ESA protections such as designation of critical habitat, many listed species might have gone extinct, and none would have improved. Pombo and like-minded lawmakers are pushing for more rigorous (and expensive and time-consuming) scientific studies before a plant or animal can be listed and therefore require protections. They also want to provide more financial incentives to property owners—at least three-quarters of all listed species reside on private land—and to involve state and local governments more in decisions to list species. Not surprisingly, land owners agree. Given the current Congress and administration, such initiatives have a good chance of succeeding. Scientists might agree to minor changes to the act but they take a much longer view than politicians. "Earth is faced with a mounting loss of species that equals or exceeds any mass extinction in the geological record," ten prominent scientists headed by Harvard's E. O. Wilson recently wrote to US Senators. "Habitat destruction is widely recognized as the primary cause of species loss. In the face of this crisis, we must strengthen the [Endangered Species Act] and broaden its protections, not weaken them." (*Christian Science Monitor*)

Renewable power, particularly schemes where thousands of homes have their own microgenerators for heat and electricity, are a far cheaper way of meeting the UK's energy needs and combating climate change than nuclear stations, says a report by the New Economics Foundation. Renewable energy is abundant and cheap to harvest, and the technology is quick to build. It is also flexible, safe, secure and climate friendly. "The opposite conclusion is only possible if renewable energy technologies are negatively misrepresented and if the numerous weaknesses, high costs, and unsolved problems of nuclear power are glossed over." The report also says the industry would create more jobs, with cheaper and faster results than nuclear energy. "Renewables also do not leave a legacy of radioactive waste that endures in the environment for tens of thousands of years." The other advantage of micro-power, which uses solar, wind, hydropower, and tides, depending on location, is that it provides security of supply, since it uses such a variety of sources, the report says. Surplus electricity generated can be put into the local grid. The report estimates that the probable net benefit to the UK of micro-generation would be £35m a year, and calls for withdrawal of subsidies to nuclear power. So that renewables can reach their full potential, public support for renewables should match the levels historically enjoyed by nuclear power. (*The Guardian*)

Since 1996, there's been a 60 percent increase in the average number of thunderstorms in the Anchorage area. The farther east you go, toward the Chugach Mountains, the more thunder you hear. Air temperatures, the water temperature in Cook Inlet, and other factors are creating the right conditions for thunderclouds. A fire manager for the state Division of Forestry says that so far this summer, the number of lightning strikes throughout coastal South-central Alaska, particularly on the Kenai Peninsula, has been unprecedented. Nearly 100 strikes were recorded in a single day recently, he said. (*Anchorage Daily News*)

Americans overwhelmingly support the US joining other members of the Group of Eight leading industrialized nations in limiting greenhouse gas emissions. A poll by the Program on International Policy Attitudes, a Washington-based research group, found that 94 per cent of respondents said the US should make efforts to limit greenhouse gas emissions, in line with other developed nations. (*Financial Times*)

Recommended reading:

Phillips, Kevin, *Wealth and Democracy: A Political History of the American Rich*. New York: Broadway Books, 2002.

Johnson, Chalmers, *The Sorrows of Empire: Militarism, Secrecy, and the End of the Republic*. New York: Henry Holt, 2004.

Global Climate Change

70

GROWING A DESERT

(July 17, 2005)

The Mesopotamians, some of the earliest farmers, earned the dubious distinction of being one of the first known civilizations to turn forests into farms, and then into deserts. More than 5000 years ago, they cleared forests in an area extending from the Levantine shores of the Mediterranean to the region now known as Jordan. They used the wood for a variety of purposes, especially for fuel, and instead of reforesting the land, they planted crops to feed their growing population.

There was enough rainfall to sustain the forest before the trees were cut. But much of the atmospheric moisture that produced the rain came from ground water drawn from underground by the trees roots and evaporated from their leaves (a process called *transpiration*). Without the trees, the ground water didn't get recycled into the atmosphere, rain didn't fall, and the land became too dry for agriculture. The Mesopotamians used irrigation to replace the rainwater.

Rainwater is distilled through the process of evaporation and condensation, and it is very pure (at least in the absence of atmospheric pollution). Rain can fall essentially forever on a land area through the transpiration-condensation cycle without changing the mineral content of the land except through very gradual leaching of mineral salts by runoff into rivers, lakes, and seas. Over millennia, the seas and some lakes with no outlets become salty from the accumulation of mineral salts.

Unfortunately, irrigation water contains the dissolved salts leached from the soil and fine suspended particles of dirt (*silt*). When irrigation water is distributed over farmland the water evaporates, leaving salts behind, and over time the land becomes salty, a process called *salination* (or *salinization*). Eventually, nothing can grow there and the land becomes a desert. Under some circumstances, the groundwater becomes salty too, making well water undrinkable. At the same time, silt settles out of the slowly flowing irrigation water, clogging the irrigation channels. This process is called *siltation*.

After a few hundred years, the land of Mesopotamia was exhausted of nutrients such as nitrogen compounds, and became too salty to grow anything. It had become desert, never to be reclaimed. This process is called *desertification*. The Mesopotamians no longer had a wood supply, because their forests were gone. They had to invade adjoining lands for wood and fertile land. They deforested large stretches of land along the Tigris and Euphrates rivers, making more farmland that eventually became desert too. After a millennium or so, the empire collapsed from widespread famine.

What have people learned from this? Not much, apparently. Other civilizations that independently developed agriculture also independently developed desertification. Greece and Rome destroyed their forests through deforestation and much of their cropland through irrigation. Centuries later the English deforested their island for fuel and ships but their moist climate allowed them to farm. Even today, deforestation continues all over the world and irrigation-caused salination continues in the Middle East, Africa, Asia, and the Americas—including the U.S. Salination of groundwater is rendering it undrinkable in many places—including parts of the U.S.

Clearly, any civilization that cuts down its forests or irrigates its land is not living sustainably. If it has to do these things to support its population, it must be overpopulated. This means we've had overpopulation in some parts of the world since the development of agriculture—including parts of the U.S. today. There were probably pockets of overpopulation even before that, and overpopulation probably forced some hunter-gatherers to take up agriculture in the first place (*viz.*, Tudge, 1999). Once farming settlements took hold, much larger population densities could be supported (although not sustainably), and the fuse of the population bomb was well and truly lit.

This week's news:

ALTAMONT, CA: The Altamont Pass wind farm is one of the leading producers of wind power in the U.S., generating about 820 million kilowatt-hours of pollution-free electricity a year, enough to power 120,000 homes. But the turbine blades are killing 1,700–4,700 birds annually as they fly through or hunt for prey. Eight hundred eighty to thirteen hundred are federally protected raptors. Wildlife advocates have taken legal action. The case could go to trial late this year or early next year. The wind farm owners agree that something must be done to protect the birds. FPL Energy, which runs about half of the Altamont's turbines, has already taken down about 100 of its most deadly windmills and replaced another 169 with 31 larger, high-tech towers, but they fear more stringent actions will force them to go out of business. (*San Diego Union-Tribune*)

In a letter to three scientists, Rep. Joe Barton (R-TX) charges that other researchers have found methodological flaws and data errors in their study, known as the hockey-stick paper, and that the researchers failed to share their raw data and the computer code used in the analysis. Scientists familiar with the research say the paper has stood up to intense scrutiny and the raw data are available. They say the request is meant to intimidate climate scientists from linking global warming to human activity. The hockey-stick paper (*Nature* **1998**, *329*, 779–787) was a fundamental part of the UN Intergovernmental Panel on Climate Change (IPCC) third assessment report, *Climate Change 2001*. The article and report synthesize 12 data sets—such as the width of tree rings and the isotopic composition of ice cores—to generate a chart of temperature variation in the Northern Hemisphere. Although many scientists point to this analysis and numerous other studies to argue that global warming is real and is caused by human activity, a small group of self-styled skeptics continue to pick away at it. A growing number of scientists charge that Bush has been attempting to cloud the science on global warming. Rick Piltz, in the U.S. Climate Change Science Program, recently resigned because the White House was altering the program's reports. According to Barton's spokesperson, the key arguments in his letter are found in a paper by McIntyre and McKitrick. McKitrick is a senior fellow at the Fraser Institute, a conservative Canadian think tank that received $60,000 from oil giant ExxonMobil in 2003. In June, the presidents of 11 national academies of science signed a statement that climate change is real. "Action taken now to reduce significantly the buildup of greenhouse gases in the atmosphere will lessen the magnitude and rate of climate change," wrote the presidents. Signatories included scientific leaders from the U.S., U.K., China, India, and Russia. This statement was largely ignored by major media outlets in the US. (*Environmental Science and Technology*)

The strain of bird flu responsible for the deaths of tens of millions of chickens and 54 people in East Asia over the past 2 years is now circulating in long-distance migratory birds, opening a way for the deadly virus to reach India, Australia, and Europe. Spread of the virus beyond its current home in China and neighboring countries could cause billions of dollars in losses to poultry farmers around the world. It could also the virus further opportunity to adapt to human as well as avian hosts, a development that could lead to an epidemic. Until now, the H5N1 virus has chiefly attacked chickens and ducks in farms and markets, killed a few birds in Hong Kong parks, and has been found in some hawks, herons, and swans. Those birds presumably acquired it from direct contact with poultry. Now it appears the virus is being transmitted among wild birds that have had no known contact with domesticated birds. The species most affected is the bar-headed goose, whose migration over the Himalayas to Burma, India, and Pakistan starts in about a month. Illness and death were also recorded in brown-headed gulls, black-headed gulls, and great cormorants. The virus has also infected 108 people (most of them in Vietnam), of whom 54 have died. Most human victims had direct contact with dead or dying chickens, but in a few cases, it appears the virus was acquired directly from an infected person. The more the virus circulates, the greater its chance of acquiring genetic changes that permit easy human transmission. If that occurs, the virus would have "pandemic potential"; it could travel quickly and infect much of the world's population, which has no immunity to it. (*Washington Post*)

ITHACA, NY: Turning plants such as corn, soybeans, and sunflowers into fuel uses much more energy than the resulting ethanol or biodiesel generates, according to a new Cornell University and University of California/Berkeley study. Their report is published in *Natural Resources Research* (Vol. 14:1, 65–76). Corn requires 29%, switch grass requires 45%, wood biomass requires 57%, soybean plants require 27%, and sunflower plants require 118% more fossil energy than the fuel produces. The researchers considered the energy used to produce the crop (including production of pesticides and fertilizer, running farm machinery and irrigating, grinding and transporting the crop) and in fermenting/distilling the ethanol from the water mix. "The government spends more than $3 billion a year to subsidize ethanol production when it does not provide a net energy balance or gain, is not a renewable energy source or an economical fuel. Further, its production and use contribute to air, water, and soil pollution and global warming," David Pimentel, professor of ecology and agriculture at Cornell says. He points out that the vast majority of the subsidies do not go to farmers but to large ethanol-producing corporations. "Ethanol production in the United States does not benefit the nation's energy security, its agriculture, economy, or the environment," says Pimentel. "Ethanol production requires large fossil energy input, and therefore, it is contributing to oil and natural gas imports and U.S. deficits." He says the country should instead focus its efforts on producing electrical energy from photovoltaic cells, wind power, and burning biomass and producing hydrogen fuel. (terradaily.com)

Recommended reading:

Tudge, Colin, *Neanderthals, Bandits, and Farmers: How Agriculture Really Began*. New Haven: Yale University Press, 1999.

71
GLOBAL WARMING
(July 24, 2005)

The most urgent problems associated with carbon fuels are not the particulate air pollution, acid rain, mercury pollution, or damage to the environment caused by coal mining and oil drilling and refining—although those are certainly very serious problems. In principle, they can be alleviated, although at increasing cost as we're forced to tap oil deposits that are less accessible, involve more environmental damage or have more harmful impurities. Although we'll run out of economically practical sources of oil soon, there's enough natural gas and coal to last for centuries.

The really serious problems caused by our dependence on carbon are global warming and mass extinctions, caused primarily by human-produced increases in atmospheric CO_2 and the rapid destruction of the world's forests (which remove CO_2 from the air). The increased atmospheric CO_2 is primarily due to fossil fuel combustion, in which carbon reacts with atmospheric oxygen. To a lesser degree, the production of Portland cement, used to make concrete, contributes to the problem: limestone is heated to produce cement and CO_2.

The evidence for global warming is all around us. Most chapters in this book mention news reports of record-breaking heat waves; melting ice caps and glaciers; droughts and famines; violent weather; die-offs, overpopulation and migrations of species; and rising ocean levels. There is no longer any doubt that the principal cause is human combustion of fossil fuels. We're producing CO_2 faster than it could be removed from the atmosphere even if we had not reduced the world's photosynthetic capacity through forest destruction.

Global warming is caused by a greenhouse effect, in which solar light passes through the atmosphere and that portion that is not reflected warms the surface of the globe. In the recent past, enough of that heat was radiated back into space in the form of infrared (IR) light so the Earth didn't heat up. However, CO_2 absorbs IR, trapping some of the heat in the atmosphere. It acts just like the glass in a greenhouse, which passes visible light but not IR. That's why CO_2 is referred to as a greenhouse gas. Atmospheric CO_2 has gone from 280 parts per million (ppm), a level that had remained much the same for the previous 10,000 years, to 360 ppm in 1998. Scientists project that the concentration will reach 560 ppm by 2050, and average global temperature will increase by 3–7°F.

There's a very real danger that a positive feedback loop will be (or already has been) initiated, in which rising temperatures will decrease reflectivity due to ice melting; atmospheric water vapor concentration will go up because warmer air can hold more water (and water's a greenhouse gas too); methane will be released from thawing tundra, swamps, and undersea deposits (yes, methane's a greenhouse gas too) because warm water can't hold as much gas as ice or cold water; and CO_2 will be released from rocks, water and ice. This will cause global warming to skyrocket far faster than would be caused by our increased CO_2 production alone. If that happens, drastic temperature shifts could take place in just a few years. New scientific evidence suggests that just such drastic temperature shifts have occurred in the past. If that were to happen, there would be nothing we could do, and our ecosystem would collapse. We would go with it. ⟩

What the people of the world do not need to face this challenge is continued dependence on fossil fuels, "clean coal," crackpot schemes to store massive amounts of CO_2 underground or at the bottom of the ocean or in oil deposits for future generations to deal with, or a CO_2-intensive hydrogen economy: making hydrogen with fossil fuels is no help. We don't need more public disinformation or delaying tactics. We do need a crash program to implement sustainable carbon-free sources of energy, using existing technology, with the same total commitment as for a world war. This *is* a world war.

This week's news:

ZURICH: After Denmark, Switzerland has the highest rates of testicular and prostate cancer for men aged 15–34 in Europe. Scientists have suspected that pesticides, cosmetics, and some drugs cause this and fertility problems. The Swiss environment agency concluded in 1999 that such chemicals had already affected wildlife. Researchers at Lausanne University Hospital hope to analyze sperm samples from 3,000 army recruits, for standard parameters such as concentration, sperm count, mobility, and morphology. An American study found that men in areas where there was extensive use of agricultural chemicals had reduced semen quality. Retrospective analysis of 20,000 spermiograms conducted at the hospital over the past 30 years has already shown that the average sperm count is twice as high among Swiss men born in 1945 compared with those born in 1975. "Exposure to estrogens or related compounds in the environment has been suspected of inducing detrimental effects on male reproductive development and of being responsible for a decline in spermatozoa concentrations in humans," said Marc Germond, a professor of reproductive medicine, who is leading the study. (*Neue Zürcher Zeitung*)

France is fighting a plague of hundreds of thousands of locusts that are devouring everything from crops to flowers in village window boxes. The Aveyron Chamber of Agriculture says the locusts have hatched because of a drought that began in 2003. The chamber's development director, Patrice Lemoux, said, 'There is nothing we can do for the 700 or 800 farmers affected. The locust has no known predator and the only insecticides that might make a difference are banned.' Although the problem in France is not yet as serious as in Spain, Portugal, and Italy, by last night most départements had introduced water restrictions. The environment ministry says the effects of the drought are being 'felt across most of France'. Experts fear forest fires in high-risk areas. Jean-Marc Billac, a woodlands campaigner in Aquitaine, said, 'All it takes is a sudden change in the weather—a mistral or a temperature rise—for the situation to become critical.' Forecasters say the heat, with temperatures of between 32 and 37°C, will continue well into this week. Prefects in all major cities have announced level three—out of four—alerts that allow them to reduce speed limits, cutting pollution, and raise medical staffing to levels used in winter flu epidemics. (*The Guardian*)

Recommended Reading:

Tennesen, Michael, *The Complete Idiot's Guide to Global Warming*. Indianapolis: Alpha, 2004.

72

AEROSOLS AND GLOBAL WARMING

(July 31, 2005)

Aerosols are suspensions of fine particles in air. They have an important impact on global temperature. They also have an impact on human health, because many aerosols are harmful. Unfortunately, the needs of human health may conflict with reducing global warming.

Net heat gain or loss by our planet depends primarily on the amount of solar radiation (mostly in the form of light) penetrating our atmosphere and retained as heat. A smaller contribution comes from planetary sources: heat from the hot core.

The light from the sun is partly reflected back into space. That which is not reflected is mostly absorbed as heat.

An important factor determining planetary temperature is the amount of sunlight reflected by the Earth. Bright objects reflect more than dark ones. The total reflectivity or "brightness" of the planet is referred to as its *albedo*. Important contributors to albedo include planetary ice and cloud cover. Another important contributor to albedo is aerosols.

Aerosols include natural ones such as smoke from fires, and volcanic smoke and ash. Large volcanic eruptions can have major effects on weather and climate, because their aerosol releases can block sunlight for long periods, causing global cooling enough to bring about little ice ages, famine due to crop losses, and even mass extinctions. Meteoroid and asteroid impacts, which can also cause major aerosol release, can have similar effects. So can a nuclear holocaust—hence concerns about a "nuclear winter."

Man-made aerosol releases include smoke and the crap dumped into the air by chemical plants and coal-fired power plants. These are very harmful to human health, and it's urgent that we eliminate them. However, these aerosols increase the earth's albedo, and the resulting cooling effect slows global warming. As we eliminate them, we'll eliminate their braking effect on global warning. There may even be a positive feedback effect involved, because with increased atmospheric temperature another aerosol—clouds—will be reduced, further reducing albedo. So reducing hazardous aerosols may speed global warming.

This week's news:

TOURS: Population growth, poor countries getting richer and the failure of wealthy countries to reduce greenhouse gases means there will be massive increase in CO_2 emissions, Tim

Dyson, professor of population studies at the London School of Economics, told a world population conference. "We're on a toboggan and we've gone over the edge...It [global warming] will screw everyone up, no matter where you are," he said at the start of the four-day conference of 2,000 demographers, economists, geographers, and sociologists from 110 countries. Scientists predict global warming, caused mainly by increasing carbon dioxide emissions from the burning of coal, oil, and petrol in motor vehicles and power stations, will increase the frequency and severity of droughts, flooding, and storms, threatening global agricultural production. The Intergovernmental Panel on Climate Change (IPCC) predicted in its 2001 report that rising levels of greenhouse gases like carbon dioxide will increase temperatures by 2.1–10°F by the end of the century and raise sea levels by 3.5–35 inches. The IPCC, set to produce its next report in 2007, is likely to "increase its temperature estimates by 0.4 degrees at both the low and high end." If per capita CO_2 emissions remained at their 2000 levels, which is unlikely, population increases would raise world emissions by 27%. World population is expected to reach 9 billion in the next 50 years. Even a 40% reduction in per capita emissions in the developed world would be outweighed solely by the effects of demographic growth elsewhere in the world. At the worst, where emissions in the developing world double but remain constant in the industrialized countries, the increase in CO_2 emissions would be 90% above 2000 levels by 2050. Developed countries have so far been unable to reduce emissions, even in Europe where population is expected to fall in the next fifty years. The US, responsible for 25 percent of the world's CO_2 emissions, has refused to ratify the Kyoto protocol, which came into force this year and commits industrialized nations to cut emissions to 5 percent below 1990 levels by 2012. Even those countries that have ratified the Kyoto protocol are unlikely to meet its modest goals, Dyson said. Between 1990 and 2002 Canada's CO_2 emissions rose by 22% and Japan's by 13% while EU emissions have risen by 3.4%. The world's poor response so far is reminiscent to that for other long-term threats such as HIV/AIDS, with the early development of a scientific consensus followed by "avoidance, denial and recrimination" with little behavioral change, said Dyson. The US and Canada have the highest per capita CO_2 emissions, 20 times sub-Saharan Africa's, and are expected to increase their population by 132 million during the next 50 years, due largely to immigration. Economic development in poor countries will also increase emissions. From 1990–99 emissions in North Africa and West Asia rose by 20%, South America 22.5%. Population growth is also likely to put more people at greater risk from climate change. "The continuing process of urbanization will mean that extremely large numbers of people, probably several billion, will be living in low-lying, densely populated coastal areas of the developing world, and their situation is likely to be particularly exposed. Flooding of coastal areas, which might result partly from sea level rise and partly from increase rainfall, could lead to the simultaneous loss of cropland and urban infrastructure, producing food price rises, large scale migration and possibly significant socio-political disruption," the professor said. (terradaily.com)

Recommended reading:

Mooney, Chris, *The Republican War on Science*. New York: Basic Books, 2005.

73

THE CLIMATE TIPPING POINT

(August 7, 2005)

Recall that positive feedback loops accelerate (speed up) global warming, countering the negative feedback loops that previously kept temperature stable. The excessive CO_2 in the atmosphere caused by burning carbon fuels is producing a greenhouse effect in which the ecosphere absorbs more solar energy than is radiated back into space. Our environment is heating up. The warmer air can hold more water, another greenhouse gas, further speeding the warming process. Global ice is now melting faster than it's produced, diminishing the Earth's albedo, or brightness, so less sunlight is reflected, again speeding up warming. As the oceans get warmer, dissolved CO_2 is coming out of solution into the air. As the land heats up, carbonates in rock are becoming CO_2 and oxides, releasing more CO_2 into the air. The increased atmospheric CO_2 in the air is speeding up warming. Warming is releasing methane as well. Methane is a greenhouse gas also, so yet again warming causes more warming.

As population continues to grow, people are burning more carbon fuels, increasing the rate of human CO_2 production, speeding warming. Forests are vanishing due to logging and settlement, decreasing the rate at which CO_2 is converted to living matter and oxygen. Consequently, CO_2 is building up faster in the air, speeding warming. The warming oceans are speeding up oxygen consumption by algae, lowering oxygen levels to the point where enormous dead zones are developing in which plankton die, slowing removal of CO_2 and replenishment of oxygen.

As forests vanish, rainfall is decreasing and erosion is increasing, producing more desert land and decreasing arable land. The forests are being killed faster for a short-term increase in arable land.

As ecospheric heating accelerates the positive feedback loops get worse, and new ones come into play. The negative feedback processes, like photosynthesis, are weakened, further enhancing the effects of positive feedback loops. If left alone these positive feedback processes will kill us. Long before that, they will have reached the point where there's nothing we can do to stop them.

The only hope of reversing these positive feedback processes is to intervene aggressively, immediately, damping the positive feedbacks and strengthening the negative feedbacks.

Our intervention must take the form of rapidly decreased population, rapidly switching to renewable energy, and ecological repair. Even if we devote all our wealth and energy to this one endeavor, living at a subsistence level worldwide, it will take decades before temperature can be stabilized and centuries before we can return to pre-industrial temperatures. This will require a coordinated international crash program.

This week's news:

NEW YORK: Heat advisories are posted across all or parts of nine states, with higher-level excessive heat warnings in Philadelphia, Baltimore, and Washington. Raleigh, North Carolina, recorded 101 degrees, a record. The two-week heat wave has been blamed for nearly sixty deaths nationwide. In Arizona 21 immigrants trying to cross the US-Mexico border died; twenty people in Phoenix died, most of them homeless. New York City set a new record for power usage, 12,551 megawatts. New York State set another record of more than 32,000 megawatts. New York Mayor Michael Bloomberg announced the opening of "cooling centers"—public, air-conditioned facilities—across the city's five boroughs. As well as urging people to conserve power where possible, Bloomberg asked New Yorkers to keep a close check on elderly neighbors, relatives, and friends. In Nebraska, the heat caused 1,200 cattle deaths, and truck farms were hit in the cornbelt states. (terradaily.com)

WOODS HOLE, MA: From Cape Cod to Cape Town, and coastal regions on at least three other continents, algal blooms are being carefully examined. Scientists are preparing to study blooms that occur where nutrient-rich waters well up from the deep ocean along the Pacific coasts of North and South America and the Atlantic coasts of Africa and Spain. Nutrients in upwelling waters support some of the most productive fisheries in the world, but they also are prone to algae blooms that threaten those fisheries. This year, blooms have struck sections of the New England, California, and Oregon coasts and Florida's Gulf Coast. Texas had a nontoxic bloom. Some blooms are caused by human activities. In Southeast Asia and the Mississippi Delta, algae feed on nutrients from agricultural runoff that originates far inland. But blooms in the Gulf of Mexico often coincide with dust plumes from Africa. The dust carries iron and other nutrients. In the Gulf of Maine, nutrient sources are largely natural. (*Christian Science Monitor*)

Half of fish species in the sea have disappeared, according to a study of ocean life over the past 50 years, due to overfishing. The study examined fishing logbooks dating back to the 1950s, and found that the size of ocean "hot spots", traditionally rich in a diversity of fish, have shrunk significantly. The most important predators such as sharks, tuna, swordfish, and marlin have suffered most. (*The Independent*)

Recommended reading:

Harvey, Danny, *Global Warming: The Hard Science.* Upper Saddle River, NJ: Prentice Hall, 2000.

74

RUNAWAY GLOBAL WARMING

(August 14, 2005)

Our planet is now approaching the stage of runaway global warming that climatologists have feared for decades. We can describe stages of global temperature as: (A) Unperturbed stable state. (B) Perturbed stable state: something forces the temperature away from the operating point, but if the perturbing force is removed, the temperature is brought back to normal by negative feedback. (C) Initial positive feedback stage, where global warming or cooling in itself causes more of the same. In this initial stage of positive feedback, it may be possible with drastic measures to quench the positive feedback by massive human intervention. (D) Runaway global warming or cooling, which we can't stop.

We have now reached stage C, possibly stage D. It's too early to say which. If we wait and see as Bush commands, it will definitely be stage D.

It's safe to say that we had no effect on global temperature until the dawn of agriculture. That was stages A and B. There were perturbations that caused climate changes such as the glaciation cycle, but we weren't one of them. We reached continuous stage B with increased population density, deforestation, and carbon fuel use, roughly 10,000 years ago. Prevailing negative feedback mechanisms could accommodate such perturbations. However, we were eroding negative feedback mechanisms and increasing the greenhouse load. We reached stage C in the mid-twentieth century, when global temperature began to rise due to our activities. If we had reversed our pattern of deforestation and increasing greenhouse gas production, the negative feedback mechanisms would have prevailed in time to return us to stage B. Our comprehension of the situation wasn't adequate to the task initially, but by the late twentieth century, our scientists had a good enough grasp of the situation to recommend remedies that would have worked. However, human societies hadn't evolved culturally to the point where they could respond rationally to scientists' warnings. If they had, sufficient reductions of population size, deforestation, and carbon combustion starting in the 1960s or 70s would have put us on track for a return to stage B. Now we're in late stage C or even stage D.

The measures needed to reverse global warming take decades. Even with stringent birth control, including widespread sterilizations and abortions, population doesn't decline rapidly unless the death rate increases dramatically. Massive reforestation and wildlife habitat restoration still would require decades before their effects could be felt. Carbon combustion can be greatly curtailed through rationing of fuel and electricity, but the amount of energy required to sustain life and convert to a non-carbon energy economy involve continued CO_2 emissions that will continue to increase the greenhouse effect. And it takes

100–200 years to clear CO_2 from the air, so even if we stopped using fossil fuels today, it would be decades before the greenhouse effect would begin to abate.

This leaves little time, if any, before we enter stage D, runaway global warming. In this short time, the only chance of heading off global warming is a drastic global effort that must start now and last for one or two centuries. It may not succeed, but it's the only remaining hope.

This week's news:

HOUSTON: Commander Eileen Collins said astronauts on shuttle Discovery had seen widespread environmental destruction on Earth and warned Thursday that greater care is needed to protect natural resources. "Sometimes you can see how there is erosion, and you can see how there is deforestation. It's very widespread in some parts of the world… We would like to see, from the astronauts' point of view, people take good care of the Earth and replace the resources that have been used…The atmosphere almost looks like an eggshell on an egg, it's so very thin. We know that we don't have much air, we need to protect what we have…I think in the old days we would not have worried about this so much." (Environmental News Network)

As Bush prepares to sign the new energy bill, groups report that it allocates at least $4 billion in subsidies and tax breaks for the oil industry. The "Exxpose Exxon" campaign includes twelve of the nation's largest environmental public interest groups, with combined membership of 6.4 million. ExxonMobil has already reported a record-breaking $15 billion in profits in the first half of 2005, $7.85 billion in the second quarter. Last year the company brought in a record $24 billion. From April to June 2005, BP reported profits of $5 billion, ConocoPhillips $3.1 billion. The bill also loosens environmental protections, and limits states' controls on liquified natural gas (LNG) facilities and pipelines and offshore oil drilling. The law provides $1.7 billion in tax breaks, millions in "royalty relief," and up to $1.5 billion in new subsidies for ultra-deepwater oil drilling and exploration. (BushGreenwatch)

Recommended reading:

McQuaig, Linda, *It's the Crude, Dude: War, Big Oil and the Fight for the Planet.* Toronto: Doubleday Canada, 2004.

The End of Life

75
EXTINCTION
(August 21, 2005)

The disappearance (extinction) of species has been going on since life began. It's been estimated that approximately 99.9% of all species have gone extinct. This results from changes in their environments to which they cannot adapt, e.g., changes in climate, availability of food, predators, parasites, or diseases, or species competing for the same resources. Such changes may pose little threat if they occur slowly enough, but they can cause extinction if they're too sudden. For a species to survive it's necessary for it to maintain a large enough population. Once population size falls too low, individuals won't produce enough offspring to maintain their numbers due to factors like inability to find mates, inadequate reserve numbers or genetic diversity to survive ecological challenges, or genetic diseases due to inbreeding, and the species will die out. This is why the population size of a species is an important determinant of whether it should be declared endangered, and should be protected. Since a species only evolves adaptations sufficient for its environment, new environmental challenges need not constitute major changes for an endangered species to go extinct. It appears that all species have a limited capacity for adaptation, and are doomed to eventual extinction.

For these reasons, there is a normal, ongoing (background) rate of extinction. Superimposed upon that, there have been major extinction events, which have killed off large percentages of existing species. The major extinctions we know about so far are summarized in this table:

Millions of years ago	percent of species lost
439	35
367	76
245	53
208	45
187	53
145	80
90	95
65	83
35	85

It's likely that some major extinctions were caused by volcanic activity. Sharp temperature drops such as the "year without a summer" (1816) occurred after major volcanic eruptions. After the eruption of Mt. Pinatubo, a reduction of 0.73°C in global temperature occurred in August 1992. The eruption is believed to have influenced such events as the disastrous 1993 floods along the Mississippi River and the drought in the Sahel region of Africa. The United States experienced its third coldest and third wettest summer in 77 years during 1992. These changes resulted from the massive amounts of dust and sulfur oxides released into the atmosphere by the eruption of one volcano.

The event 65 million years ago was caused (or perhaps climaxed) by an asteroid impact off the Yucatan Peninsula that produced a worldwide cloud of dust, probably widespread forest fires, and possibly extensive volcanic activity, cutting off so much sunlight that there was a sudden drop in temperature. The global cooling spelled extinction for the majority of species on Earth, including the dinosaurs. This provided the opportunity for explosive diversification of small mammals that spread throughout the niches vacated by the dinosaurs and other extinct species.

Four other major *mass extinctions* wiped out 75% or more of species. They were all caused by major changes of climate or atmospheric composition. Lest you draw hope for humans from the fact that biodiversity was restored after these events, our ancestors (if we have any) would not be helped by the prospect of waiting a few million years for the Earth to heal. Anyway, it's unlikely we would be one of the few species to survive the mass extinction.

Our ecosphere is now undergoing the sixth mass extinction. The reason is straightforward: Humans are changing the environment so quickly that species are going extinct much faster than they can adapt or be replaced by evolution.

This week's news:

A frozen peat bog in western Siberia the size of France and Germany combined is thawing for the first time since it formed 11,000 years ago at the end of the last ice age. Scientists fear that it will release billions of tonnes of methane, a greenhouse gas twenty times more potent than carbon dioxide, into the atmosphere. The discovery was made by Sergei Kirpotin at Tomsk State University in western Siberia and Judith Marquand at Oxford University and is reported in *New Scientist*. Dr Kirpotin says it's an "ecological landslide that is probably irreversible and is undoubtedly connected to climatic warming." Western Siberia is heating up faster than anywhere else in the world, 6°F in 40 years. Tony Juniper, director of Friends of the Earth, says, "If we don't take action very soon, we could unleash runaway global warming that will be beyond our control and it will lead to social, economic, and environmental devastation worldwide," he said. "There's still time to take action, but not much." (*The Guardian*)

Germany intends to become nuclear-free, and for renewable-energy suppliers to double their output to provide one-fifth of Germany's power, in fifteen years. By mid-century, the country expects to get more than half of its power from renewables. Five years ago, the government negotiated a Nuclear Exit Law with the power industry, requiring all nineteen of its atomic power stations to shut down by 2020. Physicist Wolfgang Neumann of Intac, a waste-management organization based in Hannover, argues, "The risk is too great for a terrible accident" on the scale of Chernobyl. "Germany should get rid of nuclear as fast as possible because, at the moment, there is no solution for the waste," warns Peter Hennicke, president of the Wuppertal Institute for Climate, Environment, and Energy. Between now and 2020, when the last plant is scheduled to close, Germany's nuclear power stations expect to produce about 6,000 additional tons of spent fuel. Were the plants' lifetimes extended until 2040, as the conservative Christian Democrats have hinted at proposing should they triumph in September, the amount of unsecured waste could reach 10,000 tons. Thus far, Germany has failed to find a safe place to store this waste. Now the phaseout's main resistance is coming from Germany's four main power producers, who control more than 80% of the energy market and all of the nuclear production, and who balk at sharing the market with small renewables producers. Thanks to the Renewable Energy Sources Act, Germany has doubled its production of renewable fuels to more than 10% of the total energy supply. Using free-market principles, Germany will be running on at least 65% renewables by mid-century. Already, Germany is the leading producer of wind power, controlling 40% of the global market. The country is second behind Japan in solar-energy production, and is boosting ties to growing solar-power production markets in places like Spain and the Middle East. Alternative-energy companies in Germany employ around 130,000 people, three times as many as nuclear.

Germans are disgusted with America's energy position. While the binding Kyoto agreement requires countries to cut emissions by 8% from 1990 levels by 2012, Germany has already cut its levels by 19%. (*Grist Magazine*)

HOOPPOLE, IL: Twenty-nine of the thirty-three corn states will see lower yields, according to the USDA. It projects the harvest will fall 12%. With extreme drought like this—the worst since 1988—those producers hit hardest will need several good years to recoup their losses. (*Christian Science Monitor*)

LISBON: Nearly 2,200 firefighters fought wildfires in Portugal on Tuesday, including ten major blazes raging out of control, which forced the evacuation of several mountain villages. The fire has already destroyed at least five homes and burned 49,500 acres. Strong winds and the steep terrain were making it difficult to control the fire. Wildfires have destroyed up to 292,000 acres of forest and scrubland so far this year in Portugal. (terradaily.com)

CLEARWATER, FL: Divers say they have documented a dead zone 20 miles offshore in the Gulf waters from Johns Pass to Clearwater. An unprecedented number of dead turtles has also washed up on Pinellas County beaches. Divers, fishermen, and scientists are worried that red tide, which has hung stubbornly to Florida's west coast for close to three months, is killing more efficiently. Scientists at the Fish & Wildlife Research Institute in St. Petersburg think the toxic algae wiped out sea life, creating the dead zone divers have discovered. They suspect a zone of cold water formed above the warmer water at the bottom, holding the algal bloom there longer than normal. Dead organisms such as crabs and shellfish decomposed, consuming dissolved oxygen in the water. (*Tampa Tribune*)

Recommended reading:

Ehrlich, Paul, and Ehrlich, Anne, *Extinction: the Causes and Consequences of the Disappearance of Species*. New York: Random House, 1981.

76

MASS EXTINCTIONS

(August 28, 2005)

We humans are part of the Earth's ecosystem. Our activities are seriously endangering other species. The fates of other species will inevitably affect our own fate.

Life on Earth seems to many to be so robust it's invincible to climate change or the mere activities of human beings, but a look into deep history shows this is not so. Approximately 30 billion species have existed since the *Cambrian explosion*, when biological diversity went through its greatest expansion. At present, there are about 30 million species, only 0.1% of all that have evolved. In other words, 99.9% of species have gone extinct. In the 540 million years since the Cambrian explosion, there have been around 20 occasions when the extinction of existing species significantly outpaced the evolution of new species, causing a net decline of species diversity. There were 5 massive die-offs in which at least 65% of species disappeared in a short time. Two of them were really, really bad ones. On one occasion, fewer than 5% of species survived. Close call.

The Earth condensed into a planet during the formation of the solar system from interstellar debris about 4.6 billion years ago. The earliest life forms we know of, single-celled organisms without nuclei, lived 3.75 billion years ago, only 250 million years after the planet had cooled enough for larger organic molecules. These were the only living things for 2 billion years—a long time with no change other than variations on the same theme. Cells with nuclei evolved "only" 1.8 billion years ago. But it wasn't until only 580 million years ago that complex multicellular organisms evolved. Apparently, their appearance depended on modern levels of oxygen in the atmosphere, made possible by photosynthesis in their single-celled predecessors. This was the *Ediacaran expansion*, during which the preceding single-celled critters were decimated—perhaps eaten by the Ediacaran organisms. Then the Ediacarans died out in a mass extinction only 20 million years before the Cambrian explosion, leaving no descendants, maybe partly because they ran out of food.

Cambrian creatures originated, like the Ediacarans, from early multicellular forms, but didn't begin their expansion until after the demise of the Ediacarans. During the Cambrian period, in only 5–10 million years, all the newer phyla (major divisions of life forms) on Earth evolved. Most of them are extinct today, yet they haven't been replaced by new phyla. They disappeared in a devastating mass extinction.

One possible explanation for the never-repeated development of so many new phyla is that this completely new complex multicellular way of life was the first that could expand into numerous previously unoccupied ecological niches. These new phyla in turn created an abundance of new niches that had never existed before. Racing in tandem like immigrants in a land rush, many could establish independent homesteads before all the territory was claimed. In contrast, after the biggest mass extinction of all at the end of the Permian period 225 million years ago, the subsequent diversification occurred mostly at the species level. It's possible that even though 95% of species died, no niches were left completely vacant and there was competition among surviving settlers, so no new phyla could develop.

This is an important finding. Today, there is more species diversity than ever before, yet fewer phyla. How might that affect the survivors, if any, of the current mass extinction? Before we can guess at an answer, we'll need to look in further detail at past mass extinctions to see what other factors can come into play.

This week's news:

Melting glaciers is only one of Alaska's problems. As Kate Troll, writing in the Anchorage Daily News, put it: "Besides retreating glaciers, insect infestations and more intense forest fires, Alaska is experiencing melting permafrost, flooded villages, warming oceans, coastal erosion, shifts in bird and wildlife populations, and shorter seasons for ice roads. And there is more to come, as Alaska is heating up at twice the rate of the rest of the world." Last year was the warmest summer on record for much of Alaska. The current warming in the Arctic was without precedent since the last ice age. All of which has prompted a mini tourist boom, a "catch-it-while-you-can" attitude among visitors eager to see the glaciers while they are still there. Muir Glacier has retreated 5 miles in the past 30 years. Portage Glacier is retreating at a rate of 50 meters a year and is no longer visible from its visitor centre. In December, a geologist with the US Geological Survey presented a series of photographs of glaciers taken in the first four years of this century, alongside pictures taken up to a century before. The result showed not just the retreat of glaciers but also the spread of vegetation where once there was ice. (*The Guardian*)

PARIS: The best way to save the planet's large wild mammals facing extinction this century is the creation of a huge nature preserve in the U.S. Midwest, leading biologists argue in *Nature*. Using the end of the Pleistocene 13,000 years ago as a benchmark, the scientists call for the "re-wilding" of great swathes of sparsely populated land. "It would take many, many hundreds of square miles," said Harry Greene, one of the authors and a professor at Cornell. "We are talking about an American Serengeti," he added, referring to the 5,800 square mile wildlife preserve in northern Tanzania. There are three reasons—biological, ethical, and economic—to do this. Repopulating the Great Plains with descendents of species that disappeared from that habitat more than 10,000 years ago is "an alternative conservation strategy for the 21st century," says Josh Donlan, a professor of ecology and evolutionary biology at Cornell. "We want to reinvigorate wild places as widely and rapidly as is prudently possible." Without major proactive steps, he suggests, many big mammals will disappear by 2100. "Africa's large mammals are dying, stranded on a continent where wars are waged over scarce resources." Humans "bear an ethical responsibility to redress these problems," Greene added. Donlan et al. argue that the creation of a network of nature preserves would generate tourism and boost depressed local economies, pointing to the successful restoration of wolves to the Yellowstone National Park, and the popularity of existing wild animal parks. Greene, 59, does not expect to see camels and cheetahs roaming the Midwest in his lifetime, but is nonetheless optimistic that things are moving in the right direction. He pointed out the recreation of "buffalo commons," where wild bison live in a network of preserves connected by corridors. Less than 300 years ago, there were tens of millions of bison roaming the American plains. "Maybe by mid-century," he said, wistfully. "I'd love to see that." (terradaily.com)

Recommended reading:

Boulter, *Extinction: Evolution and the End of Man*. New York: Columbia University Press, 2002.

77

PLATE TECTONICS AND MASS EXTINCTIONS

(September 4, 2005)

I've described some of the effects of continental drift on evolution. Some of Earth's mass extinctions also resulted from shifting tectonic plates.

The shifting of tectonic plates is caused by convection of *magma*, molten rock, which moves heat from the planet's interior to the surface and returns cooled material back to the mantle. Hot magma rises in *plumes* and melts thin spots in the crust, producing oceanic or continental spreading. Oceanic plates *spread* where magma rises from the mantle and forms new crust (*oceanic lithosphere*). The *subduction* of oceanic lithosphere occurs where it slides under the margins of continental plates, returning to the molten mantle. The subduction process causes stresses in continental margins, inland upwelling of magma, and the formation of mountain ranges parallel to coastlines.

Continents are also subject to erosion, which deposits layers of sediment offshore.

Oceans can disappear if the continents bordering them move together and collide, and an ocean can appear where a continent fractures and divides (a *rift*). The shrinking of oceans and collision of continents results from cooling of oceanic lithosphere by ocean water and continued subduction at continental margins, drawing continental plates together. These processes cause continents to combine into *supercontinents* at cooler spots, followed by formation of intracontinental plumes and breakup of the supercontinents.

Geologists have been able to trace these processes by matching up the layers of fossils along the edges of continents that once were joined, through study of the orientation of magnetically polarized minerals, and the use of radioisotope dating methods.

The formation and breakup of supercontinents takes hundreds of millions of years. They affect ocean currents, wind patterns, climate, volcanic activity, and glaciation cycles. It's no surprise, then, that they can bring about extinctions, even some mass extinctions.

The first mass extinction, 450 million years ago, wiped out a large percentage of sea life. It occurred when the Gondwana supercontinent (now South America, Africa, Arabia, India, Australia, and part of Antarctica) drifted across the South Pole, causing extensive glaciation. This sequestration of water in glaciers caused a lowering of sea levels, reducing the ecological niches available to marine organisms.

The second mass extinction, about 360 million years ago, was again mostly limited to ocean life, and was associated with a fall in oxygen concentration. The cause isn't yet known.

About 250 million years ago, 90% of marine species and 70% of vertebrate land species went extinct: the third mass extinction. At this time, the supercontinent Pangea was at its *maximum packing* (most land mass in smallest area). This produced extensive glaciation in the Gondwana part of Pangea (Pangea consisted of Gondwana plus what are now North America and Eurasia), warm global climate, arid conditions in the center of Pangea, and oceanic regression with low marine oxygen concentration. An enormous volcanic eruption in what is now Siberia contaminated the atmosphere and blocked sunlight for a long time. These processes were probably central to this mass extinction.

There was another mass extinction (the fourth) around 220 million years ago, in which the oceans lost 10% of their animal families. This was associated with oceanic transgression (coastal flooding) and regression, and low oxygen. There was no significant tectonic movement, but volcanism may have contributed.

The fifth mass extinction occurred 65 million years ago. It spelled the end of the dinosaurs and many other species, and recovery of species diversity took 5–10 million years. This was the most recent, and it was caused by the Chicxulub meteorite impact that filled the skies with ash and kindled forest fires worldwide.

We are now undergoing the sixth mass extinction.

This week's news:

WASHINGTON: Rapid growth in developing countries, combined with declining birth rates in some industrialized nations, could affect the ability of the wealthy to aid the poor, says Carl Haub, a demographer for the Population Reference Bureau. "The countries of today's developing world are growing almost three times faster than the developed countries. Almost ninety-nine percent of population growth today and for the foreseeable future will be in those developing countries," he said. Haub said the decline in the birth rate in some industrialized countries could put them "in less of a position to help developing countries. These are countries that traditionally have been quite generous in terms of foreign aid." In a study of developing countries from 1990–2003, the Population Council found some countries hadn't yet experienced fertility decline while others had "stalled" in their transition from high to low fertility rates. World population growth will continue, the study said, reaching 6.5 billion in 2005 and going to 7 billion in about seven years. The US will remain the third most populous nation behind India and China through 2050, with population increasing from 296 million to 420 million. While China has the world's largest population in 2005 at 1.3 billion, India, now number two, will overtake China by 2050 with 1.6 billion. (*San Francisco Chronicle*)

Experts predict the majority of Peru's glaciers will disappear by 2015. In the last 30 years, Peru has already lost more than 20% of its glaciers. The majority of the population lives in a narrow strip of desert between the Andes and the Pacific. Melting glaciers provide all the water for their homes, cities, hydroelectricity, industry, and farming. Pressure on water resources will continue to grow as people move to coastal cities like the capital Lima and industry expands. Once the glaciers are gone, there won't be enough water. (BBC News)

Over the past three months, bad weather has affected two million Bulgarians and claimed the lives of at least 20, while 10,000 have lost their homes. The economic damage is estimated at $625 million, with huge amounts of farmland and vital infrastructure destroyed. Much of the railway system is disabled. The army has been called in to help with rebuilding. (BBC News)

JOHANNESBURG: The World Health Organisation has declared an emergency in Africa after cases of tuberculosis tripled in countries with high rates of HIV and doubled on the continent as a whole relative to 1990. TB is the world's second largest infectious killer after HIV/AIDS, killing 2 million people every year—more than a quarter of them in Africa. It is also the most common cause of death among people with HIV/AIDS on the continent. (*Forbes*)

ZURICH: Biotech company Syngenta has filed 15 global patent applications on several thousand gene sequences from rice and other highly important crop plants. This will enable the company to set prices and control access, research and re-use of seeds. At a meeting with non-governmental organisations (NGOs) this week, Syngenta refused to drop its so-called 'megagenomic' patents. "With these patents Syngenta is claiming the work of breeders and farmers from the past centuries as the company's own invention. The attempt to monopolize thousands of gene sequences from most important crop plants in one rush is nothing less than a theft of common goods," says Tina Goethe from Swissaid. "Not to the mention the fact that these patents could block future research to a large extent." The company will claim all gene sequences that could be of commercial interest. The company aims to monopolize all similar gene sequences in any other useful plants. The company is also trying to patent the use of the plants in food and animal feed. Syngenta agreed not to pursue this kind of patent in less developed countries. "These patents must never be granted. If the company follows its claims, they should expect public protests and legal actions against it. Politicians should initiate a legal framework to stop companies such as Syngenta, Monsanto, DuPont, and Bayer to gain control on genetic resources," says Francois Meienberg from Berne Declaration. According to Adrian Dubock, head of biotech ventures in Syngenta, the patent on the GE rice will not be dropped because "Our shareholders wouldn't thank us if we had forgone that possibility." (organicconsumers.org)

Sen. Mary Landrieu (D-La.) says the federal government must spend billions of dollars to restore her state's wetlands. Intentional rerouting of the Mississippi River and rising sea levels due to climate change have eroded Louisiana's natural buffer against massive storms. Several experts and officials say Katrina highlights the need to return the coastline to its natural state. Politicians have pushed since the 20s to straighten the Mississippi for easier ship passage and better flood control, but federal officials now back a $14 billion plan to restore meandering waterways, which would deposit new silt on the shrinking wetlands. The Mississippi used to deposit sediment that built up the Delta, but now the silt is carried out to sea. As a result, Louisiana loses 25 square miles of coast a year. Global warming leads to higher seas and more extreme weather. Warmer water temperatures may lead to more violent hurricanes. In addition, global sea level has risen more than an inch over the past decade, which increases the danger from storm surges. (*Washington Post*)

Michael Mann's "hockey-stick paper" is often cited as evidence for global warming. The analysis plotted Northern Hemisphere temperatures since 1400 and found that after 1900 temperatures increased rapidly, giving the graph its shape. Subsequent studies by others gave similar results. When semi-retired businessman Stephen McIntyre's attack on Mann's paper was featured on the front page of the *Wall Street Journal (WSJ)*, *Environmental Science and Technology (ES&T)* contacted leading scientists, many of whom criticised the *WSJ* story. McIntyre's success suggests why Congress and Bush have ignored the Intergovernmental Panel on Climate Change (IPCC) report and other mainstream, peer-reviewed global climate studies. McIntyre got his start in 2003 with a paper in *Energy & Environment*, a social science journal that doesn't use peer review. McIntyre and economist Ross McKitrick claimed the hockey-stick analysis had serious errors. As a result of his paper, McIntyre was flown to Washington to brief business leaders and the staff of Sen. James Inhofe (R-OK), chair of the committee on Environment and Public Works. He also presented his findings that year at the Marshall Institute, a nonprofit whose CEO is ExxonMobil lobbyist William O' Keefe. In early January of this year, he finally had a paper accepted into a real science journal—*Geophysical Research Letters (GRL)*. Decades of research have created a massive body of scientific literature on climate change, and thousands of new studies on the subject appear every year in different science journals. Yet, within weeks of publishing his only peer-reviewed article, reporter Antonio Regalado profiled McIntyre on the front page of the *WSJ* in an article. *WSJ* praised Regalado and attacked the hockey stick, the IPCC, and the science of global warming. *ES&T* performed a Factiva search of *WSJ* articles with the terms "global warming" or "climate change" from August 1, 2004–July 31, 2005. Three stories about new research in science journals were found: one about director of NASA's Goddard Institute for Space Studies Jim Hansen's *Science* article, on page A4; Regalado's front-page feature; and one on a press conference about scientific research on page D2. No other newspaper reported on the McIntyre and McKitrick article. Max Boykoff, a graduate student at the University of California, Santa Cruz, examined US "prestige press" coverage of global warming in 2004, using stories on global warming in *New York Times, Los Angeles Times, Washington Post,* and *WSJ* from 1988–2002. He found 3543 articles: 41% from *New York Times*, 29% from *Washington Post*, 25% from *Los Angeles Times*, and 5% from *WSJ*. *ES&T* 's search found *WSJ* frequently covers business aspects of global warming, but most material on climate science was in the opinion sections. In one book review Russell Seitz opines, "Billions of dollars are spent annually on understanding aspects of climate change too ephemeral to elicit consensus." Another review presents a positive look at Michael Crichton's novel *State of Fear*, in which environmentalists try to promote a trumped-up global-climate-change scare. The *WSJ*'s op-ed page has dozens of editorials and opinion pieces that skewer the scientists and science of climate change. Professional climate-change skeptic Fred Singer is the only (former) scientist who has written a piece on global warming. The harshest critic of their coverage is former *WSJ* page-one editor, Frank Allen. He describes the front-page article as a "public disservice" littered with "snide comments" and "unsupported assumptions." "It…had this bizarre undertone of being investigative but it didn't investigate," says Allen. "…it purported to be authoritative, and it's just full of holes." *ES&T* asked Regalado and his immediate editor to respond on-the-record to the criticisms of the story and the paper's coverage of climate-change science, and were eventually told to submit their questions. They emailed 19 questions and never received a response. Rep. Joe Barton (R-TX), chair of the House Committee on Energy and Commerce, sent out letters to Mann, his colleagues, and two scientific groups. The letter to Mann begins: "Questions have been raised, according to a February 14, 2005, article in The *Wall Street Journal*, about the significance of methodological flaws and data errors in your studies of the historical records of temperatures and climate change." The same letter makes extensive requests for raw data. Mann and his colleagues have complied with Barton's demands, and the investigation is apparently still open. Jim Hansen of NASA wrote in an email, "Although I have been carrying out research in the atmospheric science and climate field for more than four decades, I have never heard of [McIntyre and McKitrick]. That perhaps tells you something." Michael Oppenheimer, professor of geoscience and international affairs at Princeton University: "All I can say is that story gave an undeserved amount of attention to a controversy that most scientists regard as ludicrous." While scientists have essentially dismissed McIntyre's research, professional societies have gone after Rep. Barton and his letters. The American Association for the Advancement of Science and the AGU, for example, has protested Barton's intrusion into the scientific process. The National Research Council has even offered to perform an independent review of the controversy for Barton. Bill Colglazier, the council's executive director, declares, "It was a sincere good-faith offer, but [the congressman] didn't seem too positive on this." McIntyre's blog has received more than 500,000 hits. In late July, Sen. Inhofe referenced McIntyre's work during a Senate debate on climate change, declaring, "We have the *Energy & Environment* report that came out in 2003 that says the original Mann papers contain collation errors, unjustifiable truncations of extrapolations of source data, obsolete data…" (*Environmental Science and Technology*)

Recommended reading:

Rogers, John J. W., and Santosh, M., *Continents and Supercontinents*. New York: Oxford University Press, 2004.

78

THE GREAT PLEISTOCENE EXTINCTIONS

(September 11, 2005)

At widely differing times, ranging from around 50,000 to 13,000 years ago, large land animals disappeared throughout much of the world. The cause wasn't climate change, nor was it likely to be disease. No large kill sites, in large numbers, which would indicate overkill by humans, have been found. What happened?

There was one event that correlated very well with the times of all these extinctions: the arrival of humans. In Australia, giant lizards, birds, and kangaroos disappeared soon after humans arrived 50,000 years ago. In North America giant mammals like saber-tooth cats, ground sloths, camels, mammoths, mastodons, and many others were extinct soon after humans crossed the Bering land bridge from Siberia. The same things happened, at different times, in Europe, South America, Hawaii, New Zealand, Madagascar, and Asia—in fact, everywhere we went. This happened regardless of the stage of cultural development of the immigrants: Paleolithic, Neolithic, Bronze or Iron Age.

Perhaps a few species were victims of overkill. Humans may have brought diseases of their domesticated livestock that killed off related species. There's a good chance that tuberculosis killed off the mammoths. We destroyed habitats of indigenous species big time. Species that had never experienced human predation didn't recognize us as predators, and made no attempt to escape from us. We may have introduced competitor species, as we have in modern times. We may have killed off the saber-tooth cats indirectly, by eliminating their prey. Recent research has shown that animal species with the slowest rates of reproduction, regardless of size, were most likely to be snuffed out. Slow reproduction is the rule in large animals.

A more subtle possibility is that factors that caused humans to migrate contributed to the great Pleistocene extinctions. For example, we may have been driven to migrate by global climate changes that reduced our original food supplies. These factors would have weakened many other species, so they were more vulnerable to the shock of human invasion. Some archeologists contend that many of the species that disappeared were already in trouble due to climate change before humans arrived. This wasn't universally true, but it does raise the likelihood that humans had some help rendering some species extinct.

With the possible exception of dangerous predators, there is no reason to assume we deliberately brought about all these extinctions. Niles Eldredge (1991) makes the point that large animals in large numbers make their mark on the environment. The larger they are and the larger their numbers, the stronger their impact. We are large animals, and our numbers are larger than any other species our size today. Human populations in ancient times grew to substantial numbers, and their ability to take animals to the brink of extinction and beyond is well documented.

There were instances in which, after initial false starts, some immigrants managed to reach a balance with their environments. Examples such as New Zealand's Maori, Australian aborigines, and Amazon tribes come to mind. Some societies have retained their connection to nature like the Mbuti (Eldredge, 1991), who sing out "mother-forest" and "father-forest" as they move about the Ituri forest of Zaire. They understand that the forest gives them all they need, and that they must respect its needs. Nevertheless, most humans have reached the point where we feel that we aren't part of nature, and that it's there for us to exploit as we like.

This week's news:

Susan Wood, FDA director of the Office of Women's Health, has resigned her position. This is an excerpt from her announcement: "I can no longer serve as staff when scientific and clinical evidence, fully evaluated and recommended for approval by the professional staff here, has been overruled...The recent decision announced by the Commissioner about emergency contraception, which continues to limit women's access to a product that would reduce unintended pregnancies and reduce abortions, is contrary to my core commitment to improving and advancing women's health." (TomPaine.com)

NAIROBI: Increasing human encroachment on equatorial rainforests in Africa and Southeast Asia threatens imminent extinction of earth's dwindling population of great apes, the UN Environment Program (UNEP) said Thursday. It said habitat destruction posed as great a threat to great apes as poaching and disease. At least 90% of areas now inhabited by great apes—gorillas, chimpanzees, and bonobos (pygmy chimps) in Africa and orangutans in Southeast Asia—will be affected within 30 years unless urgent action is taken now, UNEP says. "If current trends continue, by 2032 99% of the orangutan range will suffer medium to high impact from human development, as will 90% of the gorilla range, 92% of the chimpanzee range and 92% of the bonobo range," UNEP's executive director says. "We have a duty to rescue our closest living relatives as part of our wider responsibility to conserve the ecosystem they inhabit." (terradaily.com)

Recommended reading:

Fagan, Brian M., *The Long Summer: How Climate Changed Civilization*. London: Granta Books, 2004.

79

GLOBAL COLLAPSE OF ECOSYSTEMS

(September 18, 2005)

Just as new species are constantly evolving, old ones are going extinct. Such *background level* extinctions are the consequence of natural selection, when species can no longer cope due to changes in their ecosystems to which they can't adequately adapt. One of those changes is the evolution of new species. The march of evolution is not, however, some sort of progress, in which species are becoming better and better. This is teleological thinking, like the Great Chain of Being, and has no scientific value.

There's also no evidence for built-in senescence or obsolescence of species, which makes sense: there's no way to select for such a trait, just as there's no way to select for a trait ahead of time that would be a useful response to a future, novel challenge. We can also rule out the idea that species can acquire traits through the life experience of individuals. There's no mechanism whereby your offspring can be made stronger through inheritance of a strength you've acquired by pumping iron, because there's no way to program your genes to specify the acquired property.

It is possible to make an ecosystem more vulnerable to collapse, by lessening its biological diversity. A healthy ecosystem rarely depends on any particular species for survival, because variations in the environment select in favor of naturally occurring diversification of the gene pool within and across species. An ecosystem is more fragile (subject to partial or total collapse through the extinction of many species) when this diversity is weakened, however. Within a species, decreased diversity limits the environmental challenges to which it can adapt. Across species, if all but one species serving a key role disappear (e.g., if there's only one kind of tree left in a forest), it becomes a *keystone* species, and without it the ecosystem collapses.

Widespread geographical distributions protect species. If a local community is demolished by the death of a reef or by a tsunami, the species that are widespread enough will survive. Habitat shrinkage to a small region of the globe can bring species to the brink of extinction.

What's different about a mass extinction?

- It has to be *global*, or nearly so, so that essentially *every member* of a species driven to extinction fails to propagate, wherever it may be.
- It has to be *broad-spectrum*, affecting enough species that it causes global ecological collapse.

- It has to be really, really *bad luck*, in that it isn't a routine challenge to species. other words, it has to be outside the realm of natural selection. We can't blame the victims' genes for their fate.
- It usually has to be *triggered*, so that once begun it's inexorable. There are two kinds of triggered processes: (1) those in which the system is semi-stable, like a house of cards (think mortally weakened ecosystem), or (2) those in which positive feedback (destabilizing) mechanisms overwhelm negative feedback (stabilizing) ones. In either case, the environmental challenge has to be big enough, widespread enough and long lasting enough to do the job.⟩

What kinds of challenges meet these requirements? There are two that have been documented, both of them more than once: global cooling and global decrease in oxygen.

Cooling has been brought about by supercontinent formation or drift of a continent over one of Earth's poles, and subsequent widespread glaciation, or by increased atmospheric albedo due to volcanism or meteorite impact. Note that glaciation also increases albedo, a positive feedback effect. Glaciation also causes a fall in sea level (oceanic regression).

Decreased oxygen has been associated with oceanic regression, in which sea levels dropped substantially. One proposed link is the exposure of large areas of continental shelf. When the sea organisms dwelling in these areas are exposed to air they die, and their subsequent decay consumes oxygen. Continental and oceanic organisms most dependent on oxygen die next as a result, initiating another positive-feedback cycle that remorselessly continues until the least oxygen-dependent or most-protected species are the only ones left.

The present mass extinction doesn't quite fit these descriptions. How's it different and what's driving it? Stay tuned.

This week's news:

The U.S. has asked the European Union and NATO for emergency assistance for the victims of Hurricane Katrina as salvage efforts in New Orleans and other cities begin to move from rescuing the living to recovering the dead. Britain will send 500,000 military ration packs, and Germany and Italy will make shipments of their own. Defense Secretary Reid promised as many flights as necessary to deliver British rations. Germany is sending its second Airbus with 15 tons of rations, and Italy is sending a cargo plane packed with enough emergency supplies for 15,000. Belgium and Spain sent teams of experts to assess needs on the ground. Pope Benedict asked the Vatican's central charity to co-ordinate Roman Catholic aid. Kuwait offered $500 million in oil products, impoverished Afghanistan pledged $100,000 and Iran offered to send supplies through the Red Crescent. The Bush administration is struggling to respond to national and international criticism over its response to the disaster. Survivors made no attempt to disguise their anger. An al-Qaeda group seized on the opportunity, in a website message: "God attacked America and the prayers of the oppressed were answered," it said. (*The Scotsman*)

Recommended reading:

Kennedy, Robert Francis, *Crimes against Nature: How George W. Bush and his Corporate Pals are Plundering the Country and High-Jacking our Democracy*. New York: Harper Collins, 2004.

80

IN MEMORIAM

(September 25, 2005)

In the current Age of Extinction, we're losing an estimated 17,000 tropical land animal species a year. That's not counting aquatic species, plants, and microbes. We only have around 10 million remaining species worldwide, down from 30 million.

The list below is just for the United States.

A PARTIAL LIST OF EXTINCT ANIMALS *OF THE U.S.* SINCE 1500
(from Wikipedia)

Fish

- Longjaw cisco, Great Lakes, 1970s
- Deepwater cisco, Great Lakes, 1950s
- Blackfin cisco, Great Lakes, 1960s
- Yellowfin cutthroat trout, Colorado, 1910
- Silver trout, New Hampshire, 1930s
- Thicktail chub, California, 1957
- Pahrangat spinedace, Nevada, 1940
- Phantom shiner, New Mexico, Texas, Mexico, 1975
- Bluntnose shiner, New Mexico, Texas, 1964
- Clear Lake splittail, California, 1970
- Las Vegas dace, Nevada, 1950s
- Snake River sucker, Wyoming, 1928
- Harelip sucker, Alabama, Arkansas, Georgia, Indiana, Kentucky, Ohio, Tennessee, Virginia, 1900
- Tecopa pupfish, California, 1942
- Shoshone pupfish, California, 1966
- Raycraft Ranch killifish, Nevada, 1960

- Pahrump Ranch killifish, Nevada, 1956
- Ash Meadows killifish, Nevada, 1957
- Whiteline topminnow, Alabama, 1900
- Amistad gambusia, Texas
- Blue pike, Great Lakes, 1971
- Utah Lake sculpin, Utah, 1928
- Lake Ontario kiyi, New York, 1967
- Alvord cutthroat, Nevada, Oregon, 1940
- Maravillas red shiner, Texas, 1960
- Independence Valley tui chub, Nevada, 1970
- Banff longnose dace, Alberta, 1982
- Grass Valley speckled dace, Nevada, 1950
- San Marcos gambusia, Texas, 1983
- Large-beaked shark

Amphibians
- Golden coqui, Puerto Rico, 1980s
- Web-footed coqui, Puerto Rico, 1980s

Reptiles
- St. Croix racer, St. Croix, U.S. Virgin Islands, 1900s

Birds
- Oahu Akepa, Hawaii, 1893
- Lanai Akialoa, Hawaii, 1894
- Oahu Akialoa, Hawaii, 1837
- Hawaii Akialoa, Hawaii, 1895
- Greater Amakihi, Hawaii, 1900
- Laysan Apapane, Hawaii, 1923
- Spectacled Cormorant, Aleutian Islands, 1850
- Kusaie Crake, Caroline Islands, 1828
- Lanai Creeper, Hawaii, 1937
- Eskimo Curlew, Texas, 1962
- Labrador Duck, New York, 1878

- Greater Kona Finch, Hawaii, 1896
- Kona Finch, Hawaii, 1894
- Lesser Kona Finch, Hawaii, 1891
- Heath Hen, Massachusetts, 1931
- Kioea, Hawaii, 1859
- Black Mamo, Hawaii, 1907
- Hawaiian Mamo, Hawaii, 1898
- Laysan Millerbird, Hawaii, 1923
- Oahu Nukupu'um, Hawaii, 1860
- Hawaii Oo, Hawaii, 1934
- Oahu Oo, Hawaii, 1837
- Molokai Oo, Hawaii, 1915
- Virgin Islands Screech Owl, Puerto Rico, Virgin Islands, 1980
- Mauge's Parakeet, Puerto Rico, 1892
- Carolina Parakeet, Missouri, 1905
- Louisiana Parakeet, South-central U.S., 1912
- Culebra Puerto Rican Parrot, Culebra Island, 1899
- Passenger Pigeon, Ohio, 1900
- Laysan Rail, Hawaii, 1944
- Hawaiian Brown Rail, Hawaii, 1964
- Hawaiian Spotted Rail, Hawaii, 1893
- Wake Island Rail, Wake Island, 1945
- Santa Barbara Song Sparrow, California, 1967
- Texas Henslow's Sparrow, Texas, 1983
- Amak Song Sparrow, Alaska, 1980
- Dusky Seaside Sparrow, Florida, 1987
- Kusaie Starling, Caroline Islands, 1828
- Lanai Thrush, Hawaii, 1931
- Oahu Thrush, Hawaii, 1825
- Edgington's Lesser Titmouse, continental U.S., 1900
- Ula-ai-hawane, Hawaii, 1892
- Bachman's Warbler, South Carolina, 1962
- San Clemente Bewick's Wren, California, 1927

Mammals
- Puerto Rican shrew, Puerto Rico, 1500
- Puerto Rican long-nosed bat, Puerto Rico, 1900?
- Puerto Rican long-tongued bat, Puerto Rico, 1900?
- Puerto Rican ground sloth, Puerto Rico, 1500
- Penasco chipmunk, New Mexico, 1980
- Tacoma pocket gopher, Washington, 1970
- Goff's pocket gopher, Florida, 1955
- Sherman's pocket gopher, Georgia, 1950
- Pallid beach mouse, Florida, 1946
- Giant deer mouse, Channel Islands, California, 1870
- Chadwick Beach cottonmouth, Florida, 1950?
- Gull Island vole, New York, 1898
- Louisiana vole, Louisiana, Texas, 1905
- Puerto Rican hutia, Puerto Rico, 1500
- Puerto Rican paca, Puerto Rico, 1500
- Lesser Puerto Rican agouti, Puerto Rico, 1500
- Greater Puerto Rican agouti, Puerto Rico, 1500
- Atlantic gray whale, Atlantic Coast, 1750
- Southern California kit fox, California, 1903
- Florida red wolf, Southeastern United States, 1925
- Texas red wolf, Oklahoma, Texas, 1970
- Kenai Peninsula wolf, Alaska, 1910
- Newfoundland wolf, Newfoundland, 1911
- Banks Island wolf, Banks and Victoria Islands, 1920
- Cascade Mountains wolf, British Columbia, Oregon, Washington, 1940
- Northern Rocky Mountain wolf, Idaho, Montana, Oregon, Wyoming, 1940
- Mongollon Mountains wolf, Arizona, New Mexico, 1942
- Texas gray wolf, New Mexico, Texas, 1942
- Great Plains wolf, Great Plains, 1926
- Southern Rocky Mountains wolf, West-central United States, 1935
- California grizzly bear, California, 1925
- Sea mink, New Brunswick, New England, 1890

- Wisconsin cougar, North-central United States, 1925
- Caribbean monk seal, Florida, 1960
- Steller's Sea Cow, Alaska, 1768
- Eastern elk, Central and eastern North America, 1880
- Merriam's elk, Southwestern United States, 1906
- Badlands bighorn, Montana, Nebraska, North Dakota, South Dakota, Wyoming, 1910
- Arizona jaguar, Arizona

This week's news:

Britain's soil releases 13 million tons of carbon/year above the amount released in 1978, which is more than the 12.7 million tons/year saved by cleaning up industrial pollution. Soil holds 300 times as much carbon as the CO_2 released from burning fossil fuels. Scientists thought soil could be a major buffer or "sink" for about a quarter of the industrial emissions of carbon dioxide, but these findings indicate that in a warming globe soil will actually become a new source of CO_2. "It's a feedback loop. The warmer it gets, the faster it is happening," says Guy Kirk of the National Soil Resources Institute at Cranfield University, who led the study. "Our findings suggest that the soil part of the equation is scarier than we thought. It means we've got twenty-five per cent more carbon to think about," Dr Kirk said. His analysis, published in *Nature*, revealed that carbon in the soil organic material is being lost at a rate of 0.6%/year. "We think it's largely coming off as carbon dioxide, but some may be leaching away as carbon. Industrial emissions in 1990 accounted for 150 million tons a year and this is about eight per cent of that, so it's really a very large amount. The consequence is that there is more urgency about doing something. The buffering that the vegetation sink was thought to give is probably smaller than accepted. Global warming will accelerate and the consequences of global warming will occur more rapidly." When asked if anything can be done to ameliorate the loss, Ian Bradley of Cranfield University, who took part in the original survey in 1978, replied: "If we were prepared to turn all of arable England back to trees, that would work—but there's no realistic possibility of that." Detlef Schulze and Annette Freibauer of the Max-Planck Institute for Biogeochemistry in Jena, Germany, said the scientific rationale behind the Kyoto treaty will have to be re-evaluated. (*The Independent*)

WASHINGTON: Warming in the Arctic is stimulating the growth of vegetation. A new report in the *Journal of Geophysical Research-Biogeosciences* indicates that as the number of dark-colored shrubs rises, the amount of solar energy absorbed will increase winter heating by up to 70%. The authors measured five adjacent sites in subarctic Alaska. They included areas covered by forest canopy, dotted with shrubs, and barren tundra. Mid-winter albedo was greatly reduced where shrubs were exposed and melting began several weeks earlier in the spring at these locations than in snow-covered terrain. The shrubs' branches also produced shade that slowed melting, so that the snowmelt finished at approximately the same time for all the sites they examined. Overall, warming in the region has stimulated shrub growth, which further warms the area and creates a feedback effect that can promote higher temperatures and yet more growth. This feedback could, in turn, accelerate increases in the shrubs' range and size over the 1.5 million square mile region. Satellite and photographic data show increasing plant growth across the Alaskan, Canadian, and Eurasian Arctic and continued warming will likely produce thicker stands of brush that protrude above the snow, replacing the smooth, white winter environment 8–10 months of the year. With almost 40% of the world's soil carbon in the Arctic, warming will significantly increase atmospheric CO_2. The combined effects of increasing shrubs on both energy and carbon could change the Arctic in a way that affects the rest of the world. (terradaily.com)

Recommended reading:

Stearns, Beverley Patterson, and Stearns, Stephen C., *Watching, from the Edge of Extinction*. New Haven, CT: Yale University Press, 1999.

81

THE END OF HUMANITY

(October 2, 2005)

There's a real possibility that we're killing ourselves, along with most other species on the planet. We're seeing how humans have driven species to extinction and how we're fueling the Sixth Extinction, the worst one ever, through overkill, farming and domestication, habitat fragmentation and destruction, pollution, and climate change. We're learning how these have all resulted from overpopulation and our inability to live sustainably. We're examining the options for averting our own extinction.

Before we go further, heed the words of four dedicated scientists, from 1991 to 2004, who take the end of humanity seriously:

Michael Boulter, paleobiologist, *Extinction: Evolution and the End of Man* (2002), quoted in Ellis, *No Turning Back: The Life and Death of Animal Species* (2004), writing about tales of the end of humankind: "'Now, it seems, the joke is that we are doing very well on our own, just with our use of fossil fuels. There is no need for nuclear weapons or the inventions of science fiction writers. It is our aggressive selfishness that has led to our lifestyle, and this has evolved its own political system to maintain the status quo. Now it's too late to change and we cannot organize ourselves to stop…' We have invented something that Boulter calls 'the self-organized mass extinction from within,' and with exquisite irony, that invention will prove the end of mankind."

Niles Eldredge, paleontologist and curator, American Museum of Natural history, *The Miner's Canary* (1991): "The great anthropologist Margaret Mead reportedly said to her companion just as she was dying: 'Nurse, I think I'm going.' The nurse replied, reassuringly: 'There, there, dear, we all have to die sometime.' Mead is said to have snapped back with her last display of verve and vigor: 'Yes, but this is different!' This *is* different. This is not just any ecosystem—it is *our* ecosystem, our own very existence as a species, that is at stake."

Richard Leakey, paleoanthropologist, and Roger Lewin, anthropologist, *The Sixth Extinction* (1995): "For each of the Big Five [mass extinctions] there are theories of what caused them, some of them compelling, but none proven.

"For the sixth extinction, however, we do know the culprit. We are."

Peter Ward, professor of geological sciences, adjunct professor of zoology, and curator of paleontology, *The End of Evolution* (1994): "I have a son. He is tall and gangly, with a face speckled by a galaxy of freckles. He is mischievous and playful, willful and happy, the normal mix of boyish hopes, dreams, and emotions. He is precious to me beyond belief.

"I keep having this vision, of living with him in the Amazon rain forest, where we exist in a small hovel no different from that inhabited by a fifth of humanity. And in this dream, my son is hungry. Behind our house sits one last patch of forest, and in that pristine copse is the nest of a beautiful bird, the last nest, it so happens, of that species. This vision is a nightmare to me, because even knowing that these birds are the last of their race, I don't have the slightest doubt what my actions would be: To feed my son, to keep him alive, I would do whatever I had to do, including destroying the last of another species.

"Anyone who thinks he or she might do otherwise is probably not a parent. There are a great number of parents currently on the earth, and many more on the way."

This week's news:

UNITED NATIONS—The United Nations Human Settlements Programme (UN-HABITAT) warns that countries will soon have to build almost 100,000 housing units *per day* to head off a massive crisis. Almost half of the world's six billion people already live in cities, a third of them in slums. "The housing crisis is already with us," the report notes. "The large-scale evictions from urban areas of Zimbabwe, Mumbai, India, or Malawi are all part of a larger problem of financing urban shelter." The number of city dwellers in developing countries will increase by more than two billion by 2030. Unless adequate urban shelters and services are developed, including clean water and sanitation, billions will be stuck with poverty, squalid housing, poor health, and low productivity. The report also points out the flaws of mortgage financing. While mortgages are increasingly available, only middle- and higher-income households have access to them. UN-HABITAT stresses that it's in the interests of all to extend mortgage markets down the income scale, as home ownership is beneficial economically, socially, and politically. Most urban poor can only build homes in stages, as money becomes available. Short-term small loans of one to eight years for $500–5,000 are more useful for incremental building than conventional mortgages. In response, shelter microfinance institutions and community-based funding initiatives have emerged in recent years. The number of lenders in the housing microfinance sector needs to be increased. Without massive investment by governments, adequate shelter would not emerge on its own. Governments need to subscribe to the main principle of the UN-HABITAT agenda, "adequate shelter for all." In particular, international support is needed for a global mechanism to assist the poor. (Inter Press Service)

DURHAM, NC: Humans have altered one third to one half of Earth's land surface. Duke University scientists have now analyzed many years' data with NASA Goddard Institute climate simulation programs and found that deforestation has widespread effects on rainfall. Deforestation in the Amazon Basin severely reduces rain in the Gulf of Mexico, Texas, and northern Mexico during the spring and summer seasons when water is needed for farming.

Deforestation of Central Africa causes a decrease of rain in the lower U.S. Midwest during spring and summer and in the upper Midwest in winter and spring. Deforestation in Southeast Asia alters rainfall in China and the Balkan Peninsula. Elimination of any of these tropical forests increases rainfall in Oman and Yemen. The combined effect of deforestation in all three regions causes the worst U.S. decline in California during the winter and further increases rainfall in the southern tip of the Arabian Peninsula. The Amazon Basin literally drives weather systems around the world. The tropics receive two thirds of the world's rain, and the evaporation-condensation cycles store and release heat, making the tropics the primary engine of heat redistribution. Tropical forest changes can have significant effects on water supplies, wildfires, and farming at remote locations. (terradaily.com)

Recommended reading:

Diamond, Jared M., *The Third Chimpanzee: the Evolution and Future of the Human Animal.* New York: Harper Collins, 1992.

Mythology and Religion

82

PALEOLITHIC MYTHOLOGY

(October 9, 2005)

All people and cultures are influenced by myths, stories that were not originally intended as accurate historical accounts but rather as allegories that provided guidance. To be effective, myths have to draw on powerful psychological forces. Psychologists in the early twentieth century drew upon the powers of myths to explore and describe such forces. Given their power and pervasiveness, myths are important keys to human behavior.

What we know of the origins of religion around 50,000 years ago is based on archeological studies of prehistoric artifacts and anthropological accounts of present and historical hunter-gatherer cultures. Paleolithic mythology reflected our newfound ability to ponder life and death, and to express our wonder and fear of an immense and implacable universe. To these early people, the world and life in it were sacred.

The hunter shared the same needs and hazards as other animals, and identified with his prey and with the predators that hunted him. Every hunt meant likely death for one or another of these participants. The horror of killing beings like themselves, who shed blood like their own, led hunters to perform rituals that were acts of propitiation, gratitude, and regeneration.

The myths that arose from the human condition were symbolic accounts that encouraged right behavior—an orientation in which practical, life-and-death policies and morality were inseparable. Since survival was at stake, the Paleolithic mythos was pragmatic, forcing people to face up to the realities of danger and death.

Initiation rites symbolized death and rebirth for the initiate. The symbolic death helped prepare the hunter to die. The symbolic rebirth promised an afterlife, and dead hunters were sent on their last journeys with their hunting tools.

The sacred nature of everything meant that behind every object in this world was another world with a deeper, truer reality. This was an early form of *idealism* that has persisted through the millennia. The seemingly deeper reality was experienced in dreams, hallucinations, and trances. Shamans were agents who "died," traveled to the other world, and were "reborn," to bring back accounts of this deeper reality. They told of archetypal beings that embodied ideal human abilities like hunting, finding water, and navigating, as well as all the living creatures of our world. They are timeless, whereas our world is a place of ceaseless flux, death, and birth.

It seems likely to me that idealism reflected a confusion about the cognitive process of *semantic association*, in which we recognize individual objects as members of classes, with associated properties, relationships to other classes, *affects* (emotional responses), and memories. There's historical evidence for other confusions about mental events, such as consciousness, mental disorders, and dreaming, which gave rise to a lot of misguided philosophy and psychology. These misinterpretations of mental events become embedded in a culture and profoundly influence peoples' worldviews.

Paleolithic people believed that in an earlier time there was less separation between the ideal world and our own. Their ancestors lived in a paradise of comfort and plenty, and a catastrophe brought that life to an end. Through a sacramental approach to the tasks of life, they could regain contact with that world.

The immense, distant sky, unreachable and unaffected by human actions, symbolized the other world. Gazing at it evoked dread, delight, and exaltation. It represented unattainable perfection, and became personified in the late Paleolithic as a creator god. Lesser gods such as Baal, Indra, and Kronos, with whom humans could communicate, deposed this unattainable god still later. Even today, heights are often associated with closer proximity to the other world. The metaphorical heights of heaven, however, were not really intended to be in the sky, but instead were a higher level of spiritual attainment.

Hunters were exclusively men, and they had mythical representation in the form of heroes, who embarked on dangerous quests. Women, with their special powers of procreation, had their representatives as well. For women to be fruitful they and their offspring must be fed; men must kill or be killed in the hunt. For this reason, female deities were usually both life giving and life destroying.

This worldview from 50 thousand years ago reflects the same fears, awe, solace, and promise that we experience today.

This week's news:

CURRENT STATE OF CLIMATE SCIENCE:
RECENT STUDIES FROM THE NATIONAL ACADEMIES
Statement of
Ralph J. Cicerone, Ph.D.
President, National Academy of Sciences
The National Academies
before the
Committee on Energy and Natural Resources
U.S. Senate
July 21, 2005

Good morning, Mr. Chairman and members of the Committee. My name is Ralph Cicerone, and I am President of the National Academy of Sciences. Prior to this position, I served as Chancellor of the University of California at Irvine, where I also held the Daniel G. Aldrich Chair in Earth System Science. In addition, in 2001 I chaired the National Academies committee that wrote the report, *Climate Change Science: An Analysis of Some Key Questions*, at the request of the White House.

This morning I will summarize briefly the current state of scientific understanding on climate change, based largely on the findings and recommendations in recent National Academies' reports. These reports are the products of a study process that brings together leading scientists, engineers, public health officials and other experts to provide consensus advice to the nation on specific scientific and technical questions.

The Earth is warming. Weather station records and ship-based observations indicate that global mean surface air temperature increased about 0.7°F (0.4°C) since the early 1970's (See figure). Although the magnitude of warming varies locally, the warming trend is spatially widespread and is consistent with an array of other evidence (including melting glaciers and ice caps, sea level rise, extended growing seasons, and changes in the geographical distributions of plant and animal species). The ocean, which represents the largest reservoir of heat in the climate system, has warmed by about 0.12°F (0.06°C) averaged over the layer extending from the surface down to 750 feet, since 1993. Recent studies have shown that the observed heat storage in the oceans is consistent with expected impacts of a human-enhanced greenhouse effect.

The observed warming has not proceeded at a uniform rate. Virtually all the 20th century warming in global surface air temperature occurred between the early 1900s and the 1940s and from the 1970s until today, with a slight cooling of the Northern Hemisphere during the interim decades. The causes of these irregularities and the disparities in the timing are not completely understood, but the warming trend in global-average surface temperature observations during the past 30 years is undoubtedly real and is substantially greater than the average rate of warming during the twentieth century.

Laboratory measurements of gases trapped in dated ice cores have shown that for hundreds of thousands of years, changes in temperature have closely tracked atmospheric carbon dioxide concentrations. Burning fossil fuel for energy, industrial processes, and transportation releases carbon dioxide to the atmosphere. Carbon dioxide in the atmosphere is now at its highest level in 400,000 years and continues to rise.

Nearly all climate scientists today believe that much of Earth's current warming has been caused by increases in the amount of greenhouse gases in the atmosphere, mostly from the burning of fossil fuels. The degree of confidence in this conclusion is higher today than it was 10, or even 5 years ago, but uncertainties remain. As stated in the Academies 2001 report, "the changes observed over the last several decades are likely mostly due to human activities, but we cannot rule out that some significant part of these changes is also a reflection of natural variability."

One area of debate has been the extent to which variations in the Sun might contribute to recent observed warming trends. The Sun's total brightness has been measured by a series of satellite-based instruments for more than two complete 11-year solar cycles. Recent analyses of these measurements argue against any detectable long-term trend in the observed brightness to date. Thus, it is difficult to conclude that the Sun has been responsible for the warming observed over the past 25 years.

Carbon dioxide can remain in the atmosphere for many decades and major parts of the climate system respond slowly to changes in greenhouse gas concentrations. The slow response of the climate system to increasing greenhouse gases also means that changes and impacts will continue during the twenty-first century and beyond, even if emissions were to be stabilized or reduced in the near future.

Simulations of future climate change project that, by 2100, global surface temperatures will be from 2.5 to 10.4°F (1.4 to 5.8°C) above 1990 levels. Similar projections of temperature increases, based on rough calculations and nascent theory, were made in the Academies' first report on climate change published in the late 1970s. Since then, significant advances in our knowledge of the climate system and our ability to model and observe it have yielded consistent estimates. Pinpointing the magnitude of future warming is hindered both by remaining gaps in understanding the science and by the fact that it is difficult to predict society's future actions, particularly in the areas of population growth, economic growth, and energy use practices.

Other scientific uncertainties about future climate change relate to the regional effects of climate change and how climate change will affect the frequency and severity of weather events. Although scientists are starting to forecast regional weather impacts, the level of confidence is less than it is for global climate projections. In general, temperature is easier to predict than changes such as rainfall, storm patterns, and ecosystem impacts.

It is important to recognize however, that while future climate change and its impacts are inherently uncertain, they are far from unknown. The combined effects of ice melting and sea water expansion from ocean warming will likely cause the global average sea-level to rise by between 0.1 and 0.9 meters between 1990 and 2100. In colder climates, such warming could bring longer growing seasons and less severe winters. Those in coastal communities, many in developing nations, will experience increased flooding due to sea level rise and are likely to experience more severe storms and surges. In the Arctic regions, where temperatures have risen more than the global average, the landscape and ecosystems are being altered rapidly.

The task of mitigating and preparing for the impacts of climate change will require worldwide collaborative inputs from a wide range of experts, including natural scientists, engineers, social scientists, medical scientists, those in government at all levels, business leaders and economists. Although the scientific understanding of climate change has advanced significantly in the last several decades, there are still many unanswered questions. Society faces increasing pressure to decide how best to respond to climate change and associated global changes, and applied research in direct support of decision making is needed.

My written testimony [below] describes the current state of scientific understanding of climate change in more detail, based largely on important findings and recommendations from a number of recent National Academies' reports.

The Earth is warming

The most striking evidence of a global warming trend are closely scrutinized data that show a relatively rapid increase in temperature, particularly over the past 30 years. Weather station records and ship-based observations indicate that global mean surface air temperature increased about 0.7°F (0.4°C) since the early 1970's (See figure). Although the magnitude of warming varies locally, the warming trend is spatially widespread and is consistent with an array of other evidence (e.g., melting glaciers and ice caps, sea level rise, extended growing seasons, and changes in the geographical distributions of plant and animal species).

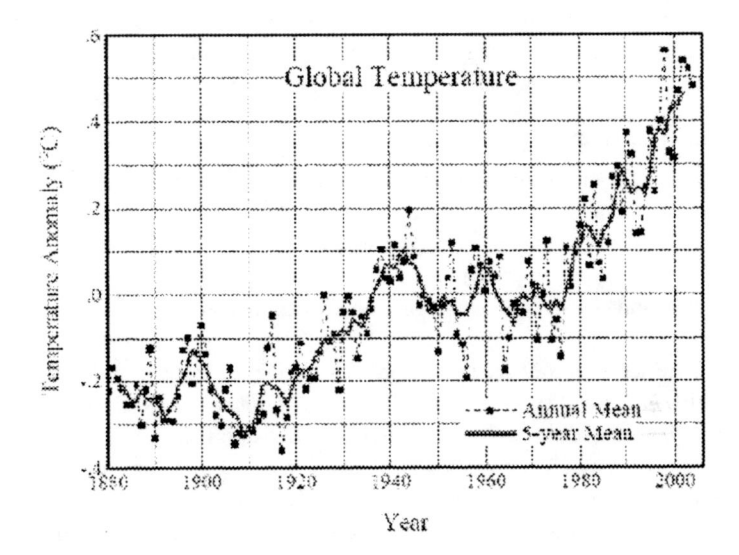

Global annual-mean surface air temperature change derived from the meteorological station network. Data and plots available from the Goddard Institute for Space Sciences (GISS) at http://data.giss.nasa.gov/gistemp/graphs/.

The ocean, which represents the largest reservoir of heat in the climate system, has warmed by about 0.12°F (0.06°C) averaged over the layer extending from the surface down to 750 feet, since 1993. Recent studies have shown that the observed heat storage in the oceans is what would be expected by a human-enhanced greenhouse effect. Indeed, increased ocean heat content accounts for most of the planetary energy imbalance (i.e., when the Earth absorbs more energy from the Sun than it emits back to space) simulated by climate models with mid-range climate sensitivity.

The observed warming has not proceeded at a uniform rate. Virtually all the 20th century warming in global surface air temperature occurred between the early 1900s and the 1940s and since the 1970s, with a slight cooling of the Northern Hemisphere during the interim decades. The troposphere warmed much more during the 1970s than during the two subsequent decades, whereas Earth's surface warmed more during the past two decades than during the 1970s. The causes of these irregularities and the disparities in the timing are not completely understood.

A National Academies report released in 2000, *Reconciling Observations of Global Temperature Change*, examined different types of temperature measurements collected from 1979 to 1999 and concluded that the warming trend in global-average surface temperature observations during the previous 20 years is undoubtedly real and is substantially greater than the average rate of warming during the twentieth century. The report concludes that the lower atmosphere actually may have warmed much less rapidly than the surface from 1979 into the late 1990s, due both to natural causes (e.g., the sequence of volcanic eruptions that occurred within this particular 20-year period) and human activities (e.g., the cooling of the upper part of the troposphere resulting from ozone depletion in the stratosphere). The report spurred many research groups to do similar analyses. Satellite observations of middle troposphere temperatures, after several revisions of the data, now compare reasonably with observations from surface stations and radiosondes, although some uncertainties remain.

Humans have had an impact on climate

Laboratory measurements of gases trapped in dated ice cores have shown that for hundreds of thousands of years, changes in temperature have closely tracked with atmospheric carbon dioxide concentrations. Burning fossil fuel for energy, industrial processes, and transportation releases carbon dioxide to the atmosphere. Carbon dioxide in the atmosphere is now at its highest level in 400,000 years and continues to rise. Nearly all climate scientists today believe that much of Earth's current warming has been caused by increases in the amount of greenhouse gases in the atmosphere. The degree of confidence in this conclusion is higher today than it was 10, or even 5 years ago, but uncertainties remain. As stated in the Academies 2001 report, "the changes observed over the last several decades are likely mostly due to human activities, but we cannot rule out that some significant part of these changes is also a reflection of natural variability."

Carbon dioxide can remain in the atmosphere for many decades and major parts of the climate system respond slowly to changes in greenhouse gas concentrations. The slow response of the climate system to increasing greenhouse gases also means that changes and impacts will continue during the twenty-first century and beyond, even if emissions were to be stabilized or reduced in the near future.

In order to compare the contributions of the various agents that affect surface temperature, scientists have devised the concept of "radiative forcing." Radiative forcing is the change in the balance between radiation (i.e., heat and energy) entering the atmosphere and radiation going back out. Positive radiative forcings (e.g., due to excess greenhouse gases) tend on average to warm the Earth, and negative radiative forcings (e.g., due to volcanic eruptions and many human-produced aerosols) on average tend to cool the Earth. The Academies recent report, *Radiative Forcing of Climate Change: Expanding the Concept and Addressing Uncertainties* (2005), takes a close look at how climate has been changed by a range of forcings. A key message from the report is that it is important to quantify how human and natural processes cause changes in climate variables other than temperature. For example, climate-driven changes in precipitation in certain regions could have significant impacts on water availability for agriculture, residential and industrial use, and recreation. Such regional impacts will be much more noticeable than projected changes in global average temperature of a degree or more.

One area of debate has been the extent to which variations in the Sun might contribute to recent observed warming trends. *Radiative Forcing of Climate Change: Expanding the Concept and Addressing Uncertainties* (2005) also summarizes current understanding about this issue. The Sun's brightness—its total irradiance—has been measured continuously by a series of satellite-based instruments for more than two complete 11-year solar cycles. These multiple solar irradiance datasets have been combined into a composite time series of daily total solar irradiance from 1979 to the present. Different assumptions about radiometer performance lead to different reconstructions for the past two decades. Recent analyses of these measurements, taking into account instrument calibration offsets and drifts, argue against any detectable long-term trend in the observed irradiance to date. Likewise, models of total solar irradiance variability that account for the influences of solar activity features—dark sunspots and bright faculae—do not predict a secular change in the past two decades. Thus, it is difficult to conclude from either measurements or models that the Sun has been responsible for the warming observed over the past 25 years.

Knowledge of solar irradiance variations is rudimentary prior to the commencement of continuous space-based irradiance observations in 1979. Models of sunspot and facular influences developed from the contemporary database have been used to extrapolate daily variations during the 11-year cycle back to about 1950 using contemporary sunspot and facular proxies, and with less certainty annually to 1610. Circumstantial evidence from cosmogenic isotope proxies of solar activity (^{14}C and ^{10}Be) and plausible variations in Sun-like stars motivated an assumption of long-term secular irradiance trends, but recent work questions the evidence from both. Very recent studies of the long-term evolution and transport of activity features using solar models suggest that secular solar irradiance variations may be limited in amplitude to about half the amplitude of the 11-year cycle.

Warming will continue, but its impacts are difficult to project

The Intergovernmental Panel on Climate Change (IPCC), which involves hundreds of scientists in assessing the state of climate change science, has estimated that, by 2100, global surface temperatures will be from 2.5 to 10.4°F (1.4 to 5.8°C) above 1990 levels. Similar projections of temperature increases, based on rough calculations and nascent theory, were made in the Academies first report on climate change published in the late 1970s. Since then, significant advances in our knowledge of the climate system and our ability to model and observe it have yielded consistent estimates. Pinpointing the magnitude of future warming is hindered both by remaining gaps in understanding the science and by the fact that it is difficult to predict society's future actions, particularly in the areas of population growth, economic growth, and energy use practices.

One of the major scientific uncertainties is how climate could be affected by what are known as "climate feedbacks." Feedbacks can either amplify or dampen the climate response to an initial radiative forcing. During a feedback process, a change in one variable, such as carbon dioxide concentration, causes a change in temperature, which then causes a change in a third variable, such as water vapor, which in turn causes a further change in temperature. *Understanding Climate Change Feedbacks* (2003) looks at what is known and not known about climate change feedbacks and identifies important research avenues for improving our understanding.

Other scientific uncertainties relate to the regional effects of climate change and how climate change will affect the frequency and severity of weather events. Although scientists are starting to forecast regional weather impacts, the level of confidence is less than it is for global climate projections. In general, temperature is easier to predict than changes such as rainfall, storm patterns, and ecosystem impacts. It is very likely that increasing global temperatures will lead to higher maximum temperatures and fewer cold days over most land areas. Some scientists believe that heat waves such as those experienced in Chicago and central Europe in recent years will continue and possibly worsen. The larger and faster the changes in climate, the more difficult it will be for human and natural systems to adapt without adverse effects.

There is evidence that the climate has sometimes changed abruptly in the past—within a decade—and could do so again. Abrupt changes, for example the Dust Bowl drought of the 1930's displaced hundreds of thousands of people in the American Great Plains, take place so rapidly that humans and ecosystems have difficulty adapting to it. *Abrupt Climate Change: Inevitable Surprises* (2002) outlines some of the evidence for and theories of abrupt change. One theory is that melting ice caps could "freshen" the water in the North Atlantic, shutting down the natural ocean circulation that brings warmer Gulf Stream waters to the north and cooler waters south again. This shutdown could make it much cooler in Northern Europe and warmer near the equator.

It is important to recognize that while future climate change and its impacts are inherently uncertain, they are far from unknown. The combined effects of ice melting and sea water expansion from ocean warming will likely cause the global average sea-level to rise by between 0.1 and 0.9 meters between 1990 and 2100. In colder climates, such warming could bring longer growing seasons and less severe winters. Those in coastal communities, many in developing nations, will experience increased flooding due to sea level rise and are likely to experience more severe storms and surges. In the Arctic regions, where temperatures have risen almost twice as much as the global average, the landscape and ecosystems are being altered rapidly.

Observations and data are the foundation of climate change science

There is nothing more valuable to scientists than the measurements and observations required to confirm or contradict hypotheses. In climate sciences, there is a peculiar relation between the scientist and the data. Whereas other scientific disciplines can run multiple, controlled experiments, climate scientists must rely on the one realization that nature provides. Climate change research requires observations of numerous characteristics of the Earth system over long periods of time on a global basis. Climate scientists must rely on data collected by a whole suite of observing systems—from satellites to surface stations to ocean buoys—operated by various government agencies and countries as well as climate records from ice cores, tree rings, corals, and sediments that help reconstruct past change.

Collecting and archiving data to meet the unique needs of climate change science

Most of the instrumentation and observing systems used to monitor climate today were established to provide data for other purposes, such as predicting daily weather; advising farmers; warning of hurricanes, tornadoes and floods; managing water resources; aiding ocean and air transportation; and understanding the ocean. However, collecting climate data is unique because higher precision is often needed in order to detect climate trends, the observing programs need to be sustained indefinitely and accommodate changes in observing technology, and observations are needed at both global scales and at local scales to serve a range of climate information users.

Every report on climate change produced by the National Academies in recent years has recommended improvements to climate observing capabilities. A central theme of the report *Adequacy of Climate Observing Systems* (1999) is the need to dramatically upgrade our climate observing capabilities. The report presents ten climate monitoring principles that continue to be the basis for designing climate observing systems, including management of network change, careful calibration, continuity of data collection, and documentation to ensure that meaningful trends can be derived.

Another key concept for climate change science is the ability to generate, analyze, and archive long-term climate data records (CDRs) for assessing the state of the environment in perpetuity. In *Climate Data Records from Environmental Satellites* (2004), a climate data record is defined as a time series of measurements of sufficient length, consistency, and continuity to determine climate variability and change. The report identifies several elements of successful climate data record generation programs, ranging from effective, expert leadership to long-term commitment to sustaining the observations and archives.

Integrating knowledge and data on climate change through models

An important concept that emerged from early climate science in the 1980s was that Earth's climate is not just a collection of long-term weather statistics, but rather the complex interactions or "couplings" of the atmosphere, the ocean, the land, and plant and animal life. Climate models are built using our best scientific knowledge, first modeling each process component separately and then linking them together to simulate these couplings.

Climate models are important tools for understanding how the climate operates today, how it may have functioned differently in the past, and how it may evolve in the future in response to forcings from both natural processes and human activities. Climate scientists can deal with uncertainty about future climate by running models with different assumptions of future population growth, economic development, energy use, and policy choices, such as those that affect air quality or influence how nations share technology. Models then offer a range of outcomes based on these different assumptions.

Modeling capability and accuracy

Since the first climate models were pioneered in the 1970s, the accuracy of models has improved as the number and quality of observations and data have increased, as computational abilities have multiplied, and as our theoretical understanding of the climate system has improved. Whereas early attempts at modeling used relatively crude representations of the climate, today's models have very sophisticated and carefully tested treatment of hundreds of climate processes.

The National Academies' report *Improving Effectiveness of U.S. Climate Modeling* (2001) offers several recommendations for strengthening climate modeling capabilities, some of which have already been adopted in the United States. At the time the report was published, U.S. modeling capabilities were lagging behind some other countries. The report identified a shortfall in computing facilities and highly skilled technical workers devoted to climate modeling. Federal agencies have begun to centralize their support for climate modeling efforts at the National Center for Atmospheric Research and the Geophysical Fluid Dynamics Laboratory. However, the U.S. could still improve the amount of resources it puts toward climate modeling as recommended in *Planning Climate and Global Change Research* (2003).

Climate change impacts will be uneven

There will be winners and losers from the impacts of climate change, even within a single region, but globally the losses are expected to outweigh the benefits. The regions that will be most severely affected are often the regions that are the least able to adapt. For example, Bangladesh, one of the poorest nations in the world, is projected to lose 17.5% of its land if sea level rises about 40 inches (1 m), displacing tens of thousands of people. Several islands throughout the South Pacific and Indian Oceans will be at similar risk of increased flooding and vulnerability to storm surges. Coastal flooding likely will threaten animals, plants, and fresh water supplies. Tourism and local agriculture could be severely challenged.

Wetland and coastal areas of many developed nations including United States are also threatened. For example, parts of New Orleans are as much as eight feet below sea level today. However, wealthy countries are much more able to adapt to sea level rise and threats to agriculture. Solutions could include building, limiting or changing construction codes in coastal zones, and developing new agricultural technologies.

The Arctic has warmed at a faster rate than the Northern Hemisphere over the past century. *A Vision for the International Polar Year 2007–2008* (2004) reports that this warming is associated with a number of impacts including: melting of sea ice, which has important impacts on biological systems such as polar bears, ice-dependent seals, and local people for whom these animals are a source of food; increased snow and rainfall, leading to changes in river discharge and tundra vegetation; and degradation of the permafrost.

Preparing for climate change

One way to begin preparing for climate change is to make the wealth of climate data and information already collected more accessible to a range of users who could apply it to inform their decisions. Such efforts, often called "climate services," are analogous to the efforts of the National Weather Service to provide useful weather information. Climate is becoming increasingly important to public and private decision making in various fields such as emergency management planning, water quality, insurance premiums, irrigation and power production decisions, and construction schedules. *A Climate Services Vision* (2001) outlines principles for improving climate services that include making climate data as user-friendly as weather services are today, and active and well-defined connections among the government agencies, businesses, and universities involved in climate change data collection and research.

Another avenue would be to develop practical strategies that could be used to reduce economic and ecological systems' vulnerabilities to change. Such "no-regrets" strategies, recommended in *Abrupt Climate Change: Inevitable Surprises* (2002), provide benefits whether a significant climate change ultimately occurs or not, potentially reducing vulnerability at little or no net cost. No-regrets measures could include low-cost steps to: improve climate forecasting; slow biodiversity loss; improve water, land, and air quality; and make institutions—such as the health care enterprise, financial markets, and transportation systems—more resilient to major disruptions.

Reducing the causes of climate change

The climate change statement issued in June 2005 by 11 science academies, including the National Academy of Sciences, stated that despite remaining unanswered questions, the scientific understanding of climate change is now sufficiently clear to justify nations taking cost-effective steps that will contribute to substantial and long-term reduction in net global greenhouse gas emissions. Because carbon dioxide and some other greenhouse gases can remain in the atmosphere for many decades and major parts of the climate system respond slowly to changes in greenhouse gas concentrations, climate change impacts will likely continue throughout the 21st century and beyond. Failure to implement significant reductions in net greenhouse gas emissions now will make the job much harder in the future—both in terms of stabilizing their atmospheric abundances and in terms of experiencing more significant impacts.

At the present time there is no single solution that can eliminate future warming. As early as 1992 *Policy Implications of Greenhouse Warming* found that there are many potentially cost-effective technological options that could contribute to stabilizing greenhouse gas concentrations.

Meeting energy needs is a major challenge to slowing climate change

Energy—either in the form of fuels used directly (i.e., gasoline) or as electricity produced using various fuels (fossil fuels as well as nuclear, solar, wind, and others)—is essential for all sectors of the economy, including industry, commerce, homes, and transportation. Energy use worldwide continues to grow with economic and population growth. Developing countries, China and India in particular, are rapidly increasing their use of energy, primarily from fossil fuels, and consequently their emissions of CO_2. Carbon emissions from energy can be reduced by using it more efficiently or by switching to alternative fuels. It also may be possible to capture carbon emissions from electric generating plants and then sequester them.

Energy efficiency in all sectors of the U.S. economy could be improved. The 2002 National Academies' report, *Effectiveness and Impact of Corporate Average Fuel Economy (CAFE) Standards*, evaluates car and light truck fuel use and analyzes how fuel economy could be improved. Steps range from improved engine lubrication to hybrid vehicles. The 2001 Academies report, *Energy Research at DOE, Was It Worth It?* addresses the benefits of increasing the energy efficiency of lighting, refrigerators and other appliances. Many of these improvements (e.g., high-efficiency refrigerators) are cost-effective means to significantly reducing energy use, but are being held back by market constraints such as consumer awareness, higher initial costs, or by the lack of effective policy.

Electricity can be produced without significant carbon emissions using nuclear power and renewable energy technologies (e.g., solar, wind, and biomass). In the United States, these technologies are too expensive or have environmental or other concerns that limit broad application, but that could change with technology development or if the costs of fossil fuels increase. Replacing coal-fired electric power plants with more efficient, modern natural-gas-fired turbines would reduce carbon emissions per unit of electricity produced.

Several technologies are being explored that would collect CO_2 that would otherwise be emitted to the atmosphere from fossil-fuel-fired power plants, and then sequester it in the ground or the ocean. Successful, cost-effective sequestration technologies would weaken the link between fossil fuels and greenhouse gas emissions. The 2003 National Academies' report, *Novel Approaches to Carbon Management: Separation, Capture, Sequestration, and Conversion to Useful Products*, discusses the development of this technology.

Capturing CO_2 emissions from the tailpipes of vehicles is essentially impossible, which is one factor that has led to considerable interest in hydrogen as a fuel. As with electricity, hydrogen must be manufactured from primary energy sources. Significantly reducing carbon emissions when producing hydrogen from fossil fuels (currently the least expensive method) would require carbon capture and sequestration. Substantial technological and economic barriers in all phases of the hydrogen fuel cycle must first be addressed through research and development. The 2004 National Academies' report, *The Hydrogen Economy: Opportunities, Costs, Barriers and R&D Needs*, presents a strategy that could lead eventually to production of hydrogen from a variety of domestic sources—such as coal (with carbon sequestration), nuclear power, wind, or photo-biological processes—and efficient use in fuel cell vehicles.

Continued scientific efforts to address a changing climate

The task of mitigating and preparing for the impacts of climate change will require worldwide collaborative inputs from a wide range of experts, including natural scientists, engineers, social scientists, medical scientists, those in government at all levels, business leaders, and economists. Although the scientific understanding of climate change has advanced significantly in the last several decades, there are still many unanswered questions. Society faces increasing pressure to decide how best to respond to climate change and associated global changes, and applied research in direct support of decision making is needed.

National Academies' Reports Cited in the Testimony

Radiative Forcing of Climate Change: Expanding the Concept and Addressing Uncertainties (2005)

Climate Data Records from Environmental Satellites (2004)

Implementing Climate and Global Change Research (2004)

A Vision for the International Polar Year 2007–2008 (2004)

The Hydrogen Economy: Opportunities, Costs, Barriers and R&D Needs (2004)

Understanding Climate Change Feedbacks (2003)

Planning Climate and Global Change Research (2003) *Novel Approaches to Carbon Management: Separation, Capture, Sequestration, and Conversion to Useful Products* (2003)

Abrupt Climate Change: Inevitable Surprises (2002)

Effectiveness and Impact of Corporate Average Fuel Economy (CAFE) Standards (2002)

Climate Change Science: An Analysis of Some Key Questions (2001)

Improving the Effectiveness of U.S. Climate Modeling (2001)

A Climate Services Vision: First Steps Towards the Future (2001)

Energy Research at DOE, Was It Worth It? (2001)

Reconciling Observations of Global Temperature Change (2000)

Adequacy of Climate Observing Systems (1999)

Policy Implications of Greenhouse Warming (1992)

Recommended reading:

Armstrong, Karen, *A Short History of Myth*. Edinburgh: Canongate, 2005.

83
NEOLITHIC MYTHOLOGY
(October 16, 2005)

The late Stone Age, or Neolithic period, roughly 9000—4000 BC, was the era of the first farmers. Just as the mythology of the Paleolithic period was shaped by the vicissitudes of hunter-gatherer life, Neolithic mythology centered on the very different trials of agricultural life.

The newfound capacity of the land to produce life was awesome, magical. The apparently inexhaustible fruitfulness of the land enabled people to multiply far faster than before, seemingly guaranteeing human survival. The commitment to agriculture as a means of survival was nevertheless a frightening leap of faith. Forests and savannas, the sources of hunter-gatherers' sustenance, had been cleared for firewood and farmlands. What if the gods withdrew the miracle of life springing from the ground? There was no going back to the old ways—the remaining forests and savannas couldn't support the swollen population. Agricultural life was as risky as hunter-gatherer life, although in different ways, with risks of drought, exhaustion of nutrients in the soil, poor harvests and famine, loss of seeds for the next year's planting. Agriculture was also much harder work, and nutrition was worse.

Religious emphasis shifted from male hero-deities to female earth goddesses. Some of these goddesses still retained their Paleolithic predecessors' powers to destroy as well as create life, as all living creatures must still return to dust when they die. Rituals centered on seasonal changes throughout the agricultural year, from planting in the spring to harvest in the fall. The analogies between life-giving earth and the human womb and between sowing of seed and human sexuality were obvious. The creation myths of some Neolithic societies depicted the first life, including humans, as sprouting from the ground. Ritual sex at planting time was common, and persisted in Israel well into the sixth century BC, even in the Jerusalem temple. Neolithic labyrinths were constructed as mystical paths back into the ground from which all life came.

Many myths were cyclical, in which male and female deities died at the harvest but through sexual union, often in the underworld during winter, they brought themselves back to life in the spring. Death was understood as a necessity for continued life, giving people some courage to face their own mortality. There is probably a connection here, to myths of reincarnation—although in some religions reincarnation is viewed more as a tribulation than a consolation.

Sacrifices were a way of sharing the bounty with the gods. They were an expression of gratitude as well as a bribe to the gods. Part of a meal would be sacrificed, and wine or blood would be spilled on the ground. Today's Santeria and Hindu sacrifices are probably traceable to Neolithic practices. In hard times, more elaborate sacrifices were performed, and when "necessary" humans were ritually killed. In some cases their flesh or blood was consumed. Ritual cannibalism is practiced by some societies even today.

Sacrifices are described in the Old Testament. Abraham, the Jewish patriarch, nearly killed his own son until Yahweh relented and let the son live. In a particularly grisly New Testament echo, the father-god allows his son-god to be tortured to death to save humanity, after the son instructs his followers at a last meal to eat his flesh and blood in his remembrance. This event is commemorated weekly, and given special emphasis every spring. Participants are urged to believe the ceremonial bread and wine are not symbolic, but are actually their savior's flesh and blood.

This week's news:

If CO_2 emissions continue as they are, the entire Southern Ocean and subarctic Pacific Ocean will be so acidic that the shells of marine organisms will dissolve, making them easier prey. Many organisms need shells for proper development and function. North Pacific salmon, mackerel, herring, cod, and baleen whales feed on sea butterflies, one of the imperiled species. "These are extremely important in the food chain and what happens if they start to disappear is a great unknown," says Jim Orr, lead scientist on a study at the Laboratory for Science of the Climate and Environment in Gif-sur-Yvette, near Paris. "Within decades, there may be serious trouble brewing in these polar oceans. Unlike climate predictions, the uncertainties here are small." (*The Guardian*)

Polar bears are disappearing fast. Thinning ice and longer summers are wrecking the bears' habitat, and as they starve, they go into human settlements looking for food and are killed. Stranded polar bears try to swim hundreds of miles to find ice floes, and drown. Climate change threatens the survival of thousands of species. The vast majority are migratory animals—sperm whales, polar bears, gazelles, garden birds, and turtles whose survival depends on the habitats, food supplies, and weather conditions that, for some species, can stretch for 6,500 miles. A report being presented this week reports faster and more frequent freeze-thaw cycles make it harder for caribou to dig out lichens to eat. Wetter, warmer winters are reducing calving success, and increasing insects and disease. Migratory wading birds, the northern seal, and the spoon-billed sandpiper face extinction. Species "cannot shift further north as their climates become warmer. They have nowhere left to go…We can see, very clearly, that most migratory species are drifting towards the poles…The habitats of migratory species most vulnerable to climate change were found to be tundra, cloud forest, sea ice, and low-lying coastal areas. Increased droughts and lowered water tables, particularly in key areas used as 'staging posts' on migration, were also identified as key threats stemming from climate change." Four out of five migratory birds listed by the UN face problems from lower water tables, increased droughts, spreading deserts, and shifting food supplies in their crucial "fuelling stations" as they migrate. One-third of turtle nesting sites in the Caribbean would be swamped by a sea level rise of 20 inches. Shallow waters used by the endangered Mediterranean monk seal, dolphins, dugongs, and manatees will slowly disappear. Whales, salmon, cod, penguins, and kittiwakes are affected by shifts in distribution and abundance of krill and plankton, which has "declined in places to a fraction of their former numbers because of warmer sea-surface temperatures." Increased dam building in response to water shortages and growing demand is affecting the natural migration patterns of South American river dolphins, "with potentially damaging results." Fewer chiffchaffs, blackbirds, robins, and song thrushes are migrating from the UK due to warmer winters. Egg laying is two to three weeks earlier than thirty years ago. "Are we fighting a losing battle? Yes, we probably are," one naturalist says. The U.K. is still unable to persuade the American, Japanese and Australian governments to admit that humankind's gas emissions are the biggest threat. These three continue to insist there is no proof that climate change is largely human-made. (*The Guardian*)

Recommended reading:

Eliade, Mircea, *Patterns in Comparative Religion*. Lincoln, NE: University of Nebraska Press, 1958.

84

ZARATHUSTRA AND THE MAGI

(October 23, 2005)

Three Kings Day, commemorating the visit to the manger by the three Magi, is celebrated on January sixth in Spanish-speaking countries. The Magi actually go much further back. The earliest accounts place the Magi in the ninth millennium BC, in Persia (today's Iran). They initially continued the Paleolithic tradition of worshipping the elements of nature, personified as gods.

Zarathustra, or Zoroaster, was a Magus who converted his native religion to monotheism. His god was a god of sun and fire. Zarathustra may be the earliest individual remembered from prehistory. He renounced the old nomadic ways and exhorted his followers to settle down and farm the land. In the oldest Zoroastrian stories, handed down by oral tradition, his cult was opposed and possibly oppressed by the nomadic followers of Paleolithic gods. He was a prophet of the new age, guiding his followers through the Paleo-Neolithic transformation. His emissaries spread his religion throughout northern Africa, the Middle East, the Aegean, and India. In more recent, but still ancient, texts he became transformed into a well-established demigod, among the first of a new class of divine humans.

Upon the advent of farming, the accumulation of possessions and the establishment of chiefdoms, the first wars were fought, judging from the sudden appearance of battle scenes in cave art and a plethora of arrowheads throughout Western and central Europe. Undoubtedly, crime was a problem in the new non-egalitarian agrarian societies. The new life was harsh, with backbreaking work and new diseases, a true fall from the relatively Edenic existence of hunter-gatherers in the times of plenty. The old gods had become anthropomorphized, and people sought solace through them. A new form of spirituality was needed for this new way of life.

Although it's not certain, many scholars believe Zarathustra's reforms included a shift from orgiastic rites to the worship of order and the right way of life, similar to the *tao* of China. The Magi were an ascetic sect. They were widely respected for their learning and wisdom. Their moral teachings undoubtedly helped propagate new codes of behavior that promoted harmony in this new world of scarcity, inequality, unrelenting labor, and crowding.

The sequence of events isn't certain, and there may have been more than one Zarathustra. In fact, the name may have been a generic term for Zoroastrian priests. This would be consistent with the fact that the reforms attributed to Zarathustra probably took generations to bring about. In any case, Zarathustra's religion was a precursor to pre-Classical (Egyptian, Mycenaean), Classical (Greek and Roman) and Mosaic (Judaic, Christian, Muslim) traditions. Zoroastrianism and offshoot religions are still practiced today.

Zarathustra promoted farming, building, and a respect for the environment, and he sought to replace the warrior and priestly classes with cultivators of the earth. He emphasized personal responsibility for the fate of the world and one's soul. His followers were fervent missionary activists.

Rather than believing that history was perpetually cyclic, Zarathustra preached a theory of progress in which the world evolved from a beginning to a final goal. At the end of time, a savior would lead the forces of light through a universal conflagration that would destroy all traces of evil.

This week's news:

Rep. Richard Pombo (R-CA) ushered his emasculation of the Endangered Species Act through the House of Representatives last week. If it passes, the Senate Bush will sign it. "I can't remember a time when any major environmental statute was under greater threat," said John Kostyack, senior counsel at the National Wildlife Federation. Pombo's bill would obliterate critical-habitat protections, making it impossible to prohibit projects on lands necessary for recovery of endangered species. The bill would also force the government to reimburse landowners if the presence of an endangered species limits their development options. The Fish and Wildlife Service has no money for this, so they couldn't restrict commercial development. "It gives developers the right to say to the government, 'Either grant me a permit to destroy sensitive habitat, or pay me market value not to,'" said Patrick Parenteau, director of the Environmental Law Clinic at Vermont Law School. The measure could actually stimulate more development plans for sensitive habitat, says Erich Zimmermann of Taxpayers for Common Sense. "Pombo's bill creates a perverse incentive for landowners to come up with a plan to develop on the most biologically sensitive areas of their property, simply so that they can cash in on a government rebate." The bill is so rife with loopholes, said Zimmermann, that "there's nothing in it that would stop landowners from collecting multiple times on proposals for species-threatening projects on the same piece of property." The bill would also eliminate pesticide limits for 5 years, grant political appointees more decision-making powers, and limit evaluation of land-use proposals to 180 days, with landowners allowed to pursue development plans if they don't receive a response by then. Pombo advertises the bill as improving species recovery. He says, "We protect private property owners. That is what leads to recovery." Environmentalists prefer a rival bill by Reps. George Miller (D-CA) and Sherwood Boehlert (R-NY) that proposes incentive-based programs, but omits some of the more controversial provisions in Pombo's version, maintains key components of the critical-habitat provision, and keeps the pesticide restrictions intact. It lost narrowly in a vote of 206 to 216, compared with Pombo's bill's vote of 229 to 193. The Boehlert-Miller bill contains provisions that Sen. Lincoln Chafee (R-RI), chair of the Senate Fisheries, Wildlife, and Water Subcommittee, is considering for an ESA-update bill. But Chafee's good intentions don't assuage the fears of environmentalists. "Even if the Senate passed the best bill imaginable, it's essentially dead on arrival in conference," says Jamie Rappaport Clark, executive VP of Defenders of Wildlife. She believes that negotiations would produce something more like Pombo's bill than Chafee's, because the lead negotiator for the Senate would be James Inhofe (R-OK), chair of the Senate Environment Committee and no friend of the environment. Sen. Mike Crapo (R-ID), a member of the Agriculture Committee, plans to unveil his own bill. "I think the House bill is a very good bill," he says. "My objective here is to make sure that we get a bill that has as much of those reforms that the House has and maybe even some more that we can get consensus on through the Senate." (*Grist Magazine*)

Recommended reading:

Clark, Peter, *Zoroastrianism: An Introduction to an Ancient Faith*. Brighton, England: Sussex Academic Press, 1998.

85
MITHRA
(October 30, 2005)

Cultural evolution during the Neolithic (late Stone Age) Period brought about a profoundly dysfunctional psychological change, which is seldom acknowledged. Call it *substitution of faith for reality*. As life became more desperate for the majority of people, some of whom were slaves, and as chiefly despots became more powerful, some of the unfortunate subjects began to confuse parable and symbol with reality—just as people today confuse fiction with truth, or propaganda with news. They really began to believe in divine powers upon which they could call for help, supernatural magic that could make their world a better place. It was a natural consequence of desperately wishful thinking, in which people yearned for the heavenly existence before the fall, which their oral history recounted in idealized form.

With the promise of succor, however, came two corollaries: the belief that their now-powerful gods could harm as well as help them, and the notion that to avoid angry retribution and earn godly good will one must appease the gods. This yearning and fear could be taken advantage of by kings and queens who professed to be gods or their representatives on earth (e.g., Egyptian pharaohs), and by the priestly caste, and by their successors to this day. By promising divine (and worldly) rewards for the faithful, and fire and brimstone (or real-world punishments) for dissenters, they could control their subjects.

Worse yet, the subjects began to confuse the codes of behavior imposed by their religious leaders with morality. Horrific crimes were justified as moral acts. Even today, many believe morality cannot exist without heavenly guidance, mediated through authoritative priests. This has led, for example, to virtually continuous wars of aggression and proxy wars launched by our own nation since its founding, with the endorsement of many (but not all) religious leaders. Brutal racism is still justified with theological arguments.

The late Neolithic, Bronze, and Iron Ages continued the process of humanizing the gods, heroes and demigods of Europe, Africa, and Asia to the point where they descended to petty squabbles and frivolous interference in human affairs. Their only remaining differences from humans were their immortality and their comic book powers. Their moral standards were, one hopes, inferior to those of their worshippers. We know much more about these imaginary creatures than earlier gods because of the invention of writing.

Somewhat less frivolous were the monotheistic gods who were created during these times. They nevertheless were prone to criminal capriciousness.

The Persian cult of Mithra was a particularly successful example of these trends. Variously described as a hero and a god, he may have actually preceded Zarathustra. His popularity was eclipsed during the Zoroastrian age, only to rise again in the fourth century BC. The cult was widespread throughout the Roman Empire, Persia, and India. Mithraism didn't completely disappear from the Roman Empire until the fourth century AD.

Mithra was born on December 25, of either an incestuous union between the sun god and his mother, or a mortal virgin, or a rock fertilized by lightning. Shepherds and gift-bearing Magi witnessed his birth. He raised the dead, healed the infirm, and cast out devils. After a last supper with his twelve disciples, he adjourned to a cave and ascended to heaven at the spring equinox (Easter). His followers commemorated his ascension with a meal of bread marked with a cross. Only celibate males could join his priesthood, and women were barred from the cult.

Mithra's followers believed the world had earlier been drowned in a flood, when a man saved himself and his cattle in an ark. The faithful were baptized to prepare for ascension to heaven. They were puritanical, and believed sinners would go to hell. Their religion was a confrontational, militant one, popular in the Roman military, and in 307 AD the emperor declared Mithra the "Protector of the Empire." They celebrated the same seven sacraments as the Christians who followed them, and they believed there would be a great battle between the forces of light and darkness during the Last Days, which would destroy the earth. In 376 AD, Christian bishops of Rome seized the Mithraic temple on Vatican Hill and adopted the Mithraic high priest's title of *Pater Patrum*, which became *Papa*, or Pope. In 813 AD, the Christian church appropriated the Mithraic festival of the Epiphany, in which the Magi arrived at the savior's birthplace.

Christians blamed the remarkable parallels between their religion and Mithraism on the devil.

This week's news:

Last week Malawi was declared a disaster area. Nearly half the country's twelve million people could starve in six months without massive and immediate food aid. So far, no food has been donated. Local people blame the food shortages on drought and a bad harvest. "We had erratic rainfall but it was by no means a disaster situation," said Rafiq Hajat of the Institute for Policy Interaction. "People are now lethargic and listless because of starvation. They don't have seeds or fertiliser. How on earth are they going to plant for next year?" Others blame Bingu wa Mutharika, the president, whose recent spending included £264,000 for a limousine. MPs are debating impeachment. Aid agencies fear the political crisis will put off donors. One of the president's campaign promises was subsidized fertilizer. It never materialized, cutting yields for small farmers dependent on chemical fertilizers. "Malawi should be the bread-basket of southern Africa," said Mr Hajat. "We have fertile soil, we have plentiful water. There is no reason why we should be starving." But 14% of adult Malawians have HIV/Aids. The pandemic has killed or incapacitated many parents of working age. Some villagers survive by diving for nyika, black and bitter-tasting water-lily tubers at the bottom of the Shire River. Alice, a local woman, says, "One person in my village died because of a crocodile but I haven't any choice. Sometimes when I am in the water my body feels weak because I've got no food but I still have to do it." In another region, it has been reported that women and children have resorted to frying and eating termites. (*The Guardian*)

MAKUHARI, JAPAN: Honda announced a new hydrogen-powered fuel-cell concept car Wednesday, which uses a refueling system that generates hydrogen from natural gas supplied to households. It reforms natural gas to make hydrogen, supplies electricity to the home and recovers heat during power generation for water heating. It will reduce carbon dioxide emissions by 40% and cut in half the total cost of household electricity, gas and vehicle fuel, Honda said. "Ultimately, what we should aim for is to circulate zero-pollution energy through solar panels," Honda president Takeo Fukui told reporters. "There are many methods (to achieve the goal). En route to that, we will use natural gas, which is conventional infrastructure, and gradually add use of solar-panel energy," he said. (terradaily.com)

NEW YORK: A CBS News telephone poll indicates most respondents don't accept the theory of evolution. Fifty-one percent believe God created humans in their present form. Thirty percent believe that while humans evolved, God guided the process, and only 15% thought humans evolved independently. Sixty-seven percent felt it's possible to believe in both God and evolution, and 29% disagreed. Although most demographic groups say it's possible to believe in both God and evolution, more than half of white evangelical Christians say that is not possible. (terradaily.com)

Recommended reading:

Campbell, Joseph, et al., *The Power of Myth*. New York: Doubleday, 1988.

86

THE OLD TESTAMENT

(November 6, 2005)

The Old Testament, like most religious scriptures, provides an account of the divine creation of humans and their universe. It's also ostensibly a history of the Jewish people and their god YHWH.

Textual analysis reveals the Old Testament was assembled from multiple sources, and heavily edited over the centuries by several authors. It wasn't begun, as it suggests, in the tenth century BC in David's or Solomon's Judah, because Judah wasn't literate then. It contains anachronisms, such as references to camels as beasts of burden, which date some of its authors to the seventh century BC, in the time of King Josiah. The book took its final form between the fifth and second centuries BC, in the Persian and Hellenistic periods.

There are two creation myths in Genesis, written by two authors. They're mutually inconsistent, meaning they can't both be right. There are inconsistent versions of other stories as well. There's no evidence that the Patriarchs immigrated around 2000 BC to Canaan from Mesopotamia as suggested by the Bible. Instead, archeological digs show that the first Jews were Canaanite hill people who vacillated between pastoralism and sedentism in Israel and Judah. According to the Bible, God promised Abraham and his descendants all the land between the Nile and the Euphrates, in return for worshipping him as the only god.

Egyptian records of the period described in Exodus are rich with detail, but they don't mention any enslaved Hebrews. There's no archeological evidence of an exodus by 600,000 escaped Hebrews returning to their homeland, and it's virtually impossible that they could have left Egypt without any Egyptian account of it or of the plagues that were supposed to have preceded it. There's no archeological confirmation of this large group wandering in the Sinai for forty years. There were no kings of Edom for them to encounter. Many of the populated sites mentioned in Exodus didn't exist until around King Josiah's time—six centuries later.

There was no conquest of Canaan by the returning Israelites under the leadership of Joshua. Correspondence among the rulers of other states in the region chronicles other conflicts at that time, but never mentions this one. Canaan was a thinly peopled Egyptian province of small, unfortified towns. Archeological digs confirm the existence of Egyptian military strongholds throughout the region. There were no walls of Jericho, no signs of destruction from that period. The description of towns in the region fits the time of King Josiah, not the supposed time of Joshua.

The kingdom of David in Judah did not exist at the time recorded in the Bible. Instead, the evidence points to his kingdom being much later. His kingdom and Solomon's were minor

Middle Eastern chiefdoms, not the opulent and powerful nations depicted in the Bible. In fact, rather than Israel being a pariah state repeatedly punished by YHWH for its idolatrous ways and Judah being a significant regional force, Israel attained the status of city-state centuries before Judah, and Judah only rose to power after the conquest of Israel by the Assyrians.

Priestly editors who lived in the time of King Josiah, when manuscripts were "discovered" behind a wall in the Temple of Jerusalem in 622 BC probably introduced many of the distortions of the Old Testament. The history of the Jews was revised to solidify the myth that they were a single family descended from Abraham. The myth that they were chosen by JHWH to worship him and cast out false religions provides them with a shared burden and shared reward (the promised land), suited to Josiah's ambition of unification, religious reform and dominion.

JHWH was as capricious and cruel as gods who came before him, and his vengeance for straying from his path was swift and terrible. What better way for the chief Josiah to consolidate power than through a theocratic chiefdom with a long history of God's rule through hegemonic rewards and devastating punishments?

This remains the central myth of Israel to this day, and the theological rationale for much of Israeli policy.

This week's news:

Biologists and wildlife-watchers in Scotland say the record heat wave is fooling many species into behaving as though winter has already passed. Animals and plants may perish in large numbers if the UK has its coldest winter for years, as predicted by some experts. Normally hibernating animals may be in trouble after stopping preparations for the winter because of the unseasonable warmth. Plants are blooming too early, with no insects to pollinate them. Migrating geese are arriving in Scotland after flying south from Iceland for the winter. Pink-footed geese in SE Scotland are searching for food at night. Rare and tropical species are appearing in Scotland. Basking sharks are swimming in the previously cold North Sea. The Wall Butterfly died off in Scotland after cold summers in the 1860s, but it has reappeared in East Lothian. Experts are concerned that deer, hedgehog, and frog young won't survive the cold. A spokeswoman for the Scottish Society for the Prevention of Cruelty to Animals, said: "Deer feed on the young shoots of plants and in cold conditions they can starve to death as the shoots don't grow. Hedgehogs produce two sets of young and the latest set will not have had time to put on enough weight to hibernate. If people see hedgehogs running around during daylight hours then they are looking for food. People can help by putting out non-fish-based cat and dog food to help them." Experts at Scottish Natural Heritage claim that while the cold weather could help control pest species such as midges and caterpillars that have boomed following recent mild winters, rare bird species could be hit hard. (*The Scotsman*)

Recommended reading:

Finkelstein, Israel, and Silberman, Neil Asher, *The Bible Unearthed: Archaeology's New Vision of Ancient Israel and the Origin of its Sacred Texts*. New York: Free Press, 2001.

87

THE GREAT CHAIN OF BEING AND TELEOLOGY

(November 13, 2005)

There's a vision of the world, going way back into prehistory, called "the Great Chain of Being." It varies from one culture to another, but the main idea is that there is an unbroken chain of beings which extends from the perfect (god) to the utterly imperfect (formless matter). All creatures fall in between, with humans at the top of the earthly domain. In many religions, there is also a hierarchy of beings in the heavenly domain and another in the infernal domain. On Earth, humans are the pinnacle of life, so all other earthly life forms are inferior. They are there to meet our needs, and we "own" and "oversee" them. This is *exceptionalism*, the notion that humans are an exception to the rules that govern nature.

Such influential thinkers as Aristotle, Aquinas, Dante, Goethe, Shakespeare, Descartes, Leibnitz, and Spinoza took all this very seriously. It was accepted by all levels of European society from the Classical Period through the Middle Ages, and still influences many people's thinking today. And exceptionalism is what got us into our ecological disaster.

The Great Chain was not only a theoretically useful theological and philosophical construct; it provided the rationale for social hierarchy. Everyone had his or her place in life. Princes of the church and secular kings were viewed as having natural dominion over their subjects. The peasant was inferior to the nobleman, and it was his duty to obey. The lord of the manor had a responsibility to care for his serfs and a duty to serve his king. Women were inferior and subservient to men. While this provided a degree of social stability it also led to abuses, such as the *droit du Seigneur*, the right of a feudal lord to deflower a vassal's bride on her wedding night. Implicit in the Great Chain was a supposed hierarchy of races and nationalities, useful in justifying slavery and the caste systems of many societies. Although the Enlightenment loosened the bonds of the Great Chain, there are many people today who still feel that membership in one group confers superiority over members of another.

In the natural sciences, previously called natural philosophy, the Great Chain shaped thinking about biology. It was thought that humans were so different from the closest life forms, the apes, that there must be an intermediate form—one origin of myths of the abominable snowman, sasquach, and yeti.

The evolutionists of the nineteenth century were not immune to this concept. Many, like natural philosophers before them, were members of the clergy. They considered evolution to be the work of God, and humankind to be its culmination. This line of thought still appeals to some today, although the concept is no longer part of scientific thinking.

The idea that natural phenomena are guided by some purpose is *teleological thinking*. Scientists have developed very useful explanations of natural phenomena assuming causation goes from an antecedent cause to a subsequent effect. Teleology reverses the temporal order of causation. Instead of saying A causes B, teleological thinking asserts that A happens in order for B to happen: the subsequent B causes the antecedent A. No one has developed useful explanations of natural phenomena with teleology.

This is not to deny that animals, including humans, lack purposive behavior. It's perfectly natural, but wrong, for people to anthropomorphize their world, giving it purpose. For many even today, a world without purpose would be unbearable, hence, it is unthinkable. Nevertheless, we must avoid teleology and the Great Chain of Being if we hope to understand and address the problems we face.

This week's news:

A state of emergency has been declared in the Amazon River basin, undergoing the worst drought in four decades. More than one thousand towns and hamlets that use the river for transportation have been cut off and are running out of drinking water, medical supplies, and provisions. Wildfires are occurring more often. Greenpeace blames deforestation and climate change for the drought: "The Amazon is caught between these two destructive forces, and their combined effects threaten to flip its ecosystems from forest to savannah." (*The Sydney Morning Herald*)

UNITED NATIONS: "Climate Change Futures," by the Center for Health and the Global Environment at Harvard Medical School, the United Nations Development Program (UNDP) and Swiss Re, a private health insurance company, indicates that severe health impacts of global warming will probably cause severe economic problems. "Global climate change and the ripples of that change will affect every aspect of life, from municipal budgets for snowplowing to the spread of disease," the report says. For example, the effects of hurricanes "can extend far beyond coastal properties to the heartland through their impact on offshore drilling and oil prices." Paul Epstein, associate director of the Center for Health and the Global Environment, says, "Health is the final common pathway of all that we see around us." Warming and extreme weather trigger the breeding of mosquitoes, which carry the malaria parasite. "For human health, malaria is clearly danger number one, killing 3,000 African children every day. This is a dramatic increase from the 1950s and '60s, when we thought we could control and contain this disease and the numbers were really dipping." Malaria is spreading in the mountains of Africa, as warmer temperatures allow mosquitoes to migrate higher. West Nile, Lyme disease, and asthma are more prevalent. "Carbon dioxide is affecting the plants, spores, and fungi. (It has) an impact on public health that we had not even thought of several years ago," says Epstein. "In the U.S. alone, asthma prevalence has quadrupled since 1980, which costs the American taxpayer up to 18 billion dollars a year." Although developed countries produce the most greenhouse gases, developing countries suffer the worst consequences.

"Poor countries and their people are most vulnerable to the increased risks from rising water levels, more frequent and intense extreme weather events, the spread of infectious diseases including malaria, intensified water scarcity and failing agricultural crops, and the extinction of species", said Brian Dawson, a senior climate change policy advisor with UNDP. (*Inter Press Service*)

NEW YORK: A study sponsored by the Northeast states shows the plan by nine governors to cut greenhouse emissions would raise electricity rates only 0.3–6.9%, not the 23% predicted by a business group calling themselves the New England Council. The states hope to pass a Regional Greenhouse Gas Initiative to cut greenhouse gas emissions 10% by 2010. Governor George Pataki, (R-NY), initiated the plan in a break with George Bush, who dropped out of the Kyoto Protocol. RGGI would reduce greenhouse gases by capping emissions and setting up an emissions trading market similar to the European Union's. Mark Breslow, director of the Climate Action Network in Massachusetts, one of the RGGI states, said the plan would not raise electricity rates by much because its ambitions to cut emissions are "reasonably modest" compared to Europe's. He also said RGGI allows a wide variety of ways to reduce levels of greenhouse gases including planting trees and broadening the use of alternative energies such as wind and solar. (Reuters)

Recommended reading:

Lovejoy, Arthur O., *The Great Chain of Being: A Study of the History of an Idea. The William James Lectures Delivered at Harvard University, 1933*. Cambridge, MA: Harvard University Press, 1936.

88

CAN RELIGION FIND MORE OIL?
(November 20, 2005)

This is not a facetious question. Recently, in Dover, Pennsylvania, a court case that could only occur in the US took place. The school board was sued because it ordered schools to read a statement to biology classes that the theory of evolution is not established fact and that gaps exist in it. The statement mentions intelligent design as an alternative theory and recommends a book on the subject.

At issue was the question whether schools should be required to teach intelligent design as a plausible alternative to the theory of evolution. That decision *should* hinge on whether the concept of intelligent design is a scientific one, supported by the same weight of evidence as evolution. It rests on the assertion that the complexity of biological organisms could not arise by evolution.

Intelligent design is not a theory. It is an unfounded objection to a theory, claiming that such complex structures as the mammalian eye could not arise through the processes of small genetic variations and natural selection. There is no evidence for this contention, and plenty of evidence against it. A recently published book, *The Plausibility of Life*, by Kirschner and Gerhart, summarizes that evidence. This evidence alone should be enough to ban intelligent design from science curricula.

Intelligent design is a thinly disguised version of creationism. In some biology textbooks "intelligent design" simply replaces the phrase "creation science" found in earlier editions. Creation science is based on the stories of Genesis. They imply the universe is only a few thousand years old, in spite of geological and cosmological evidence that it is billions of years old. This myth also indicates that all species were created at the same time, contrary to geological and paleontological evidence. If true, it would imply that present species aren't the result of descent with modifications, in defiance of evidence from biochemistry and molecular biology.

"Creation science" bears none of the hallmarks of a scientific theory, such as the ability to generate useful predictions. People who could not have been elected or appointed without the support of creationists now dominate all three branches of our government. Cynical leaders who seek power and wealth have duped the faithful. Those leaders justify their actions with appeals to faith rather than reason.

Overpopulation, global warming, and the current mass extinction are dismissed by our current government with lies and appeals to faith. We now face declining oil supply and increasing demand. Fact-based science cannot find us much more oil. Can faith-based creationism?

As England's dependence on fossil fuels skyrocketed during the machine age, a pressing need arose for ways to find oil and coal. Peter Ward, in *The End of Evolution*, describes how William Smith observed that fossils, rather than rock types, could be used to judge the ages of geological strata. By looking at fossil types, geologists could judge the likelihood of finding coal or oil. The reason this worked, of course, was that different species existed at different times. They were not all created at once; they evolved over very long periods. During one geological era, fossil fuel-producing materials were deposited, contemporaneously with the organisms whose fossils geologists now associate with fossil fuels. But creation "science" tells us this can't be so.

How ironic that our president's and vice president's fossil-fuel cronies depend on evolutionary phenomena to find energy resources. Evolutionary phenomena that the faithful vehemently deny.

There are many other religions around the world. Their adherents have no problems with evolution, and they're intellectually equipped to deal with overpopulation, climate change and mass extinctions. Only some branches of one religion are intellectually incapable of dealing with the problems we face. They are the swing voters who empowered the current US government. Soon our only hope may be for the rest of the world to save us all from the consequences of Christian fundamentalist ignorance.

This week's news:

Although industry officials testifying before the Senate Energy and Commerce committees denied it last week, a White House document proves big oil executives met with Cheney's energy task force in 2001. Officials from ExxonMobil, Conoco, Shell, and BP met with Cheney aides preparing the national energy policy. Although not mentioned in the document, the Government Accountability Office says Chevron was one of the companies that "gave detailed energy policy recommendations" to the task force. Cheney met separately with BP's chief executive, according to an insider. Environmentalists were not allowed to attend. The meetings were held in secret and the White House refused to release a list of participants. Judicial Watch and the Sierra Club sued unsuccessfully for records from the meetings. Sen. Frank Lautenberg (D-N.J.), who posed the question about the task force, said he will ask the Justice Department today to investigate. "The White House went to great lengths to keep these meetings secret, and now oil executives may be lying to Congress about their role in the Cheney task force," Lautenberg said. A spokeswoman for Cheney said the courts have upheld "the constitutional right of the president and vice president to obtain information in confidentiality." Committee Democrats criticized Commerce Chair Ted Stevens (R-Alaska) for not swearing in the executives, but making "any materially false, fictitious or fraudulent statement or representation" to Congress is punishable by fine or imprisonment for up to five years. A Conoco manager confirmed a meeting: "We met in the Executive Office Building, if I remember correctly." A spokesman for ConocoPhillips said James J. Mulva, the chief executive, had been unaware of this when he testified at the hearing. In a brief phone interview, former Exxon vice president James Rouse, named in the White House document, said, "That must be inaccurate and I don't have any comment beyond that." A spokesman for BP declined to comment. A spokeswoman for Shell said she didn't know if Shell officials met with the task force, but that they often meet members of the administration. Chevron said its executives did not meet with the task force but confirmed that it sent President Bush recommendations in a letter. The person familiar with the task force's work, who requested anonymity out of fear of retribution, said the document was based on records kept by the Secret Service of people admitted to the White House complex. (*Washington Post*)

Recommended reading:

Kirschner, Mark W., and Gerhart, John C., *The Plausibility of Life: Great Leaps of Evolution*. New Haven, CT: Yale University Press, 2005.

What to Do?

89

GETTING OUR PRIORITIES STRAIGHT

(November 27, 2005)

Even more than New Year's, election time should be a time for reflection, for stepping back, looking at the world, and asking "what's right and what's wrong with this picture?" Often it seems that one's vote counts for little. That's true in a sense. But the course of world events can be drastically different, depending on the outcomes of even the closest elections. No one should doubt that the future of our planet has been and will be irreversibly affected by the outcomes of the 2000, 2002, and 2004 elections, at all levels of government. News stories like those at the end of these chapters, all too common today, would have been unthinkable in 2000. Three more years of these trends will bring even worse horrors by 2008. Momentous issues, all interrelated, depend on the wisdom of American voters. Let me share some of them with you in the form of questions about outcomes.

Which is more promising: more foreign fossil fuel dependence, more oil wars, more expensive and destructive drilling and mining, apocalyptic global warming, and mass extinctions—or a crash program to replace fossil fuels with independent, secure, and sustainable wind, sun, wave and hydrogen power? Which would you prefer, more nuclear power and armaments, more nuclear waste and pollution, and more nuclear proliferation? Or an international program to ban all nukes and clean up and secure nuclear materials?

How do you like globalization, which transfers wealth from people of all nations to the coffers of corporations and privileged individuals? Or would you like economic fairness better?

If you could choose, would you rather have stronger corporate control of "democracies"—or genuine democracy? High-quality, low-cost health care for the privileged; mediocre, expensive corporate-controlled "managed medicine" for the middle class; and no health insurance at all for millions of Americans—or genuine health care for all at reasonable cost? Quality education and excellent job prospects for the privileged and tragically poor education and prospects for the rest of us—or genuine equality of opportunity?

Do you want a world with continued coerced childbirth and the misery that attends it, dictated by medieval religions and social policies? Or a world with marital and reproductive choice, free family planning, pre- and postnatal care, contraception, abortion, sterilization and adoption? More overpopulation and degradation of all life? Or genuine respect for life and one child per couple until we reach a sustainable population?

How about one too many superpower (US)? More military bullying? More resource wars? At a cost that will destroy US as it destroyed the Soviets? More unemployment and underemployment, so the only way for many to get educations or jobs is to enlist? Or would you prefer high employment driven by a transition to a sustainable economy, and a small peacetime military?

More corporate-run prisons? More victimless crimes classed as felonies? More denial of ex-felons' voting rights? More illegal purging of people from voting rolls because they have names like ex-felons? A rogue Supreme Court? An illegitimate administration with criminal disregard for national and international laws and treaties? Or equal justice for all under fair and decent laws?

Are we a country of cowards who lash out blindly, destroying nations whenever our government and media use fear to manipulate us? Or are we courageous enough to make a society free of fear, with our liberties restored? Are we doomed to be spoiled, childish slaves to sleazy consumerism, or can we maturely invest our labor and our wealth in better health care and education, meaningful employment, more leisure time, earlier retirement, a long and secure old age, and enjoyment of the finer things in life?

Do we want lives of ignorance shaped by corporate-controlled media? Leisure time filled with pointless diversions? Or truly informative journalism and a sense of purpose and community in our lives?

Can we loose the chains of faith-driven ideology and take up the mantle of evidence-driven realism? Can we combat deceit, manipulation, and secrecy with honesty, integrity, and openness? Can we finally abandon greedy materialism and self-defeating opportunism to dedicate ourselves and our society to the good, the true, and the beautiful?

If you're smart enough to read this book, I'm confident you're smart enough to be part of the answers to our problems, not part of the problems.

This week's news:

In Kuwait, the world's second-largest oil field is starting to decline. The output from the Burgan field will be about 1.7 million barrels a day, rather than the 2 million a day engineers had predicted. The emirate may invite companies such as Exxon Mobil Corp., Royal Dutch Shell Plc, and BP Plc to invest about $8.5 billion to almost double output at the emirate's northern fields by 2025. The project would be the first time since the 1970s that foreign companies operate Kuwaiti oilfields. (Bloomberg)

MADISON, WI: The World Health Organization (WHO) reports in the journal *Nature* that human-induced climate changes cause at least 5 million cases of illness and more than 150,000 deaths per year. Temperature fluctuations influence the spread of infectious diseases and cause illness-producing heat waves and floods. The regions that contribute the least to global warming are the most vulnerable. These include coastlines along the Pacific and Indian Oceans, and sub-Saharan Africa. Megacities act as "heat islands," causing temperature-related health problems. Co-author Diarmid Campbell-Lendrum of the WHO says, "Many of the most important diseases in poor countries, from malaria to diarrhea and malnutrition, are highly sensitive to climate. The health sector is already struggling to control these diseases and climate change threatens to undermine these efforts." The report underscores the moral obligation of countries with high per-capita emissions, such as the US and European nations, to reduce the health threats of global warming. It also points to the need for rapidly growing economies like China and India to develop sustainable energy policies. Computer models predict that: • Climate-related health problems will more than double by 2030. • Flooding will affect up to 200 million people by the 2080s. • Heat-related deaths in California could more than double by 2100. • Hazardous ozone pollution days in the Eastern US could increase 60 percent by 2050. Individuals can make a difference: "Our consumptive lifestyles are having lethal impacts on other people around the world, especially the poor. There are options now for leading more energy-efficient lives that should enable people to make better personal choices." (terradaily.com)

Several studies in rodents show that what they eat can switch certain genes on or off. An article in *New Scientist* indicates foods may do the same in humans. While some disorders in humans are caused by changes of DNA, others such as some cancers are caused by genes switching on or off. There are thousands of genes in the body, but not all of them are active. Scientists have been looking at what factors might control gene activity and have found some evidence to suggest that diet is important. One illustration is the result of a study in which adult rats were injected with the amino acid L-methionine. The animals were "less confident" when exploring new environments and produced higher levels of stress hormones. L-methionine altered a gene for glucocorticoid, a hormone involved in stress responses. The researchers are now investigating whether trichostatin A (TSA) can cause the opposite effect. TSA does produce an effect on genes that is the opposite of L-methionine's effect. Professor Ian Johnson at the Institute of Food Research is investigating whether colon cancer in humans might be triggered by diet through control of gene expression. He's studying healthy people before cancer starts. He says "It's quite a strong possibility that nutrients might cause DNA changes. We think diet may have a role to play as a regulator in genes. Ultimately, one would want to choose diets that would give you the most beneficial pattern of DNA methylation in the gut. But it is too early to say that we know the dietary strategy to do that. We need much more research. Genes regulate all the processes in the body and things that change gene expression, therefore, may be linked to a number of health issues other than cancer too." He said one nutrient that might influence gene expression is folate. Folate deficiency has been linked to increased risk of breast and colon cancers. (BBC News)

Recommended reading:

Ward, Peter D., *The End of Evolution: on Mass Extinctions and the Preservation of Biodiversity.* New York: Bantam Books, 1994.

90

REDEFINING SUSTAINABILITY

(December 4, 2005)

What if our future depended on everyone accepting some obvious math? It does.

It's now widely reported that climate change will be our undoing if we don't reverse it soon. Less widely reported, the current mass extinction may soon cause ecological collapse (not the end of all life, but loss of a large fraction of species, probably including us). Climate change is one of the major causes of the current mass extinction, the others being habitat destruction, overkill and pollution.

The one crucial process affecting climate change and mass extinction that is virtually never reported in the mass media is overpopulation. This is partly because the subject is taboo to many, and partly because our formulation of overpopulation is imprecise. We need to understand overpopulation in terms of sustainability. Sustainability is the use of our planet's resources at or below their replacement rate. This applies to all resources: clean air, water, and land, and materials we extract from them like metals, fossil fuels, and food. Among our resources are the absence of excesses, like too much carbon dioxide, too many prescription drugs in our water supply, and too much ultraviolet penetrating the ozone layer. Our resources also include essential processes, such as atmospheric and ocean currents, chemical cycles, and the negative feedback cycles that maintain ecospheric stability.

Overconsumption of any essential resource is unsustainable—it can't last forever. If we have sufficient resources to start with, sustainability requires that the rate of consumption C of any resource x is less than or equal to the rate of its restoration R:

$$C_x <= R_x \tag{1}$$

("$<=$" means "less than or equal").

If this holds for all resources x, we're living sustainably.

Many practices we consider unsustainable, like carbon dioxide production, would be sustainable if practiced by a small enough number of people. In this case, think of production of CO_2 as consumption of anti-CO_2, so we can stick with simple terminology. Total consumption C_x is per capita consumption c_x times population size p:

$$C_x = c_x p. \tag{2}$$

Restoration of x, R_x, is determined by both natural and human restoration processes, N_x and H_x respectively:

$$R_x = N_x + H_x. \tag{3}$$

Substituting for R_x in equation (1),

$$c_x p <= N_x + H_x. \tag{4}$$

Rearranging equation (4), a sustainable population size p_s is one that satisfies this condition:

$$p_s <= (N_x + H_x)/c_x$$

for all resources x.

We can increase the permissible size of p_s by increasing n_x or h_x, or decreasing c_x. In other words, we can increase the sustainable population size by increasing the natural or human processes of restoration of natural resources or by decreasing per capita consumption (increasing the efficiency of use of a resource, or lowering the "standard of living" in order to consume less).

This gives us a new, *functional definition of overpopulation* p_0:

$$p_0 > (N_x + H_x)/c_x \tag{5}$$

for any resource x (">" means "greater than").

This probably describes the situation for some resources anywhere in the world, even in countries we consider sparsely populated. The reason this has come about is simple.

At first, we ignore the subtle decline of our resources as we increase our consumption and expand our population. Eventually we have to increase consumption just to maintain the status quo, e.g., we're forced to expend more energy to obtain enough oil, food, and water. We've been doing this ever since we began to drive species to extinction and were eventually forced to take up agriculture. To make matters worse, most countries' economies depend on growth, the accumulation of wealth based on resource consumption, in search of economic security.

Usually we can't tell from our local patterns of consumption whether they're unsustainable because of global trade. We blame shortages on "inappropriate distribution." The prices of goods are manipulated of much that prices aren't reliable indicators. Ecological imbalances go unnoticed more often than not because of the complexities of ecosystems, inadequate monitoring and regulation, and the corruption, secrecy, and narrow self-interest of consumers, regulators and legislatures.

We are the frogs slowly being boiled to death. Most of us frogs are heedlessly turning up the heat by living unsustainably: using too many resources too fast and breeding like—well, like frogs. And we're starting to die, like the frogs all around the world that are going extinct from global warming, the thinning ozone layer, pollution, and habitat destruction.

A recent conversation illustrates the absurdity of our behavior. I was on the local solid waste authority task force, and we were discussing a technological fix for our fossil fuel crisis. The option under consideration wasn't renewable energy, it was conversion of solid waste to fuel. A participant remarked, "We're going to have a world population of 12 billion by mid-century and we have to be ready to provide enough energy for them." This was a university professor with a track record of many clever and important inventions. There was no suggestion that, before we reached such an obscene population density, we would have condemned ourselves to extinction. There was no suggestion that our primary efforts should be shifting to renewable energy and population reduction.

Growth can't continue indefinitely in a world of finite resources, whether it be population growth or growth of consumption. To make matters worse, we've interfered with natural restorative processes, to the point where natural restoration processes N for many resources is declining fast in a positive feedback loop.

We have to switch to a recovery mode of life, in which

$$p << (N + H)/c \tag{6}$$

("<<" means "much less than"), until N can return to a livable level. This means a rapid decline in human population worldwide coupled with rapid decline in consumption and greatly increased human restoration H of resources.

Conservation and renewable energy can't restore our planet without population reduction. In order to reduce our population fast enough, we all have to understand the math and commit to a maximum of one child per woman and per man worldwide. Even then, we'll need a crash program to convert to a non-growth economy, lower consumption, and turn to fully renewable energy.

This week's news:

WASHINGTON:—Deep-sea sediment and ice core studies in the journal *Science* show CO_2 levels are rising faster than they have for thousands of years. The ice core project examined air bubbles trapped in 10,000 feet of ice from Antarctica and found today's atmosphere has more CO_2 than any time in 650,000 years. The ocean sediment study showed sea level has risen twice as fast in the past 150 as in the previous 5,000 years. In the ice core study, European researchers found CO_2 levels were stable until 200 years ago. The current rate of increase is 200 times faster than any earlier rise found in the samples. CO_2 levels were well correlated with temperatures. Geosciences specialist Edward Brook of Oregon State University says "There's no natural condition that we know about in a really long time where the greenhouse gas levels were anywhere near what they are now. And these studies tell us that there's a strong relationship between temperature and greenhouse gases. Which logically leads you to the conclusion that maybe we should worry about temperature change in the future." (MSNBC)

Recommended reading:

Schor, Juliet B., and Taylor, Betsy (eds.), *Sustainable Planet: Solutions for the Twenty-First Century*. Boston: Beacon Press, 2002.

91

HOW MUCH TIME DO WE HAVE?

(December 11, 2005)

Well, what do we need to accomplish? What's the latest we can reach our goal of sustainability before it becomes impossible? What's the earliest we can expect results if we start right now?

Old adages say that to end war or crime we must first have justice. Population control and a sustainable civilization will require justice, which can only come from security and the satisfaction of a life worthwhile. Before everyone can commit to major changes of lifestyle, we'll need some basic guarantees. Everyone will demand food, water, shelter, and decent jobs. We'll expect good health care, a decent environment, education, and security in our old age. We'll want a sense of purpose in life, and the happiness that comes from the enjoyment of beauty, civility, friendship, and love. We'll want the right of self-determination and protection from domination, foreign or domestic.

There are some corollaries to these needs. The widening gap between wealthy and poor has to be greatly reduced. The developed world's economy, currently based on perpetual growth and uncontrolled depletion of resources, will have to be replaced with an economy based on sustainability.

We'll have to stabilize our environment, which involves more than not polluting it. Many pollutants are long-lived and will continue exerting their damaging effects for decades or even centuries unless we take positive steps to remove them. Even removing them may not suffice, if natural restorative processes are not sufficiently robust. Further research is urgently needed.

We'll need to bring our net consumption of resources to zero, or even negative levels. In other words, we'll have to re-use non-renewable resources such as metals and plastics, and consume renewable ones such as water, forests, and arable land at rates that allow their recovery. The higher our per capita rate of consumption of resources, the smaller our population must be. We'll have to achieve not just zero population growth but negative growth for some time. Even if we were to stop reproducing entirely tomorrow, which is not a good idea, it would take decades for enough people to die of natural and human causes to bring world population down to a sustainable level.

Already, the world's largest cities have over 15 million people, many of whom squat in vast, crowded slums where they fight for food, water, and shelter, and live in a state of anarchy. In some slums, children are hunted for sport like rats in a garbage dump.

The long lag period between action and recovery of the planet means starting now, when many people still feel no serious effects. Like some diseases, once everyone has become uncomfortable, it's too late. We also need enough slack to make mistakes.

The period of adjustment will take about 100 to 250 years, *if we start now*, depending on how successful we are and the standard of living we want. The faster we can adapt, the more likely we are to succeed and the better the world will be in the end, because less damage will have occurred.

That's our timetable for population reduction: start serious action this year, continue lowering population for 100–250 years. We have especially urgent deadlines with regard to eliminating fossil fuel use and saving our ecosystem.

This week's news:

LOS ANGELES: When Tina Roggenkamp and her husband Mark decided not to have children, they considered their shared desire for greater freedom, something that enabled her to get a graduate degree and start a small consulting business. There was also their enjoyment of what she calls "smaller things." But there were also environmental and overpopulation concerns. "We worry about global warming. We worry about what the world will be like in the future. There's so much uncertainty and I can't see bringing a life into such a world." The latest Census Bureau data show 18% of women between 40 and 44 haven't conceived. In 1976, it was 10%. Not everyone is comfortable with this trend. Jennifer Shawne, author of "Baby Not on Board: A Celebration of Life without Kids," has been accused of being un-American for deciding not to have children. "There is this assumption that all women have a biological clock that one day is going to start ringing, and we're going to become baby maniacs who have to give birth no matter what," she says. "But that's just not true." Philip Morgan, professor of sociology at Duke University, says, "Childlessness is not new." During the Depression, many Americans decided not to have children because they couldn't afford them. "Childlessness levels now are not higher than those in the 1930s," he said. Albert Mohler, president of the Southern Baptist Theological Seminary in Louisville considers a married couple deciding not to have children to be a violation of God's will. "I am trying to look at this from a perspective that begins with God's creation," Mohler said in a phone interview. "God's purpose in creation is being trumped by modern practices. I would argue that it (not having children) ought to be falling short of the glory of God. Deliberate childlessness defies God's will," he said. Mohler, rather than being concerned about over-population, is concerned about under-population. "We are barely replenishing ourselves. That is going to cause huge social problems in the future." Not all Christians, even Evangelicals, agree with Mohler. (Chicago Tribune)

LONDON: Royal Society President Robert May of Oxford is urging scientists to oppose the climate change "denial lobby." He warns that core scientific values are "under serious threat from resurgent fundamentalism, West and East." May applies the term fundamentalism not only to organized religions, but also to lobby groups on both sides of the climate change debate. He maintains that both the climate change "denial lobby" and organizations opposed to nuclear power are capable of denial and misrepresentation of scientific facts. May also criticized US religious groups for campaigning for creationism to be taught in science courses. "By their own writings, this group has a much wider agenda, which is to replace scientific materialism by something more based on faith." May says scientists must speak out. (terradaily.com)

Recommended reading:

United Nations Human Settlements Programme, *The Challenge of Slums: Global Report on Human Settlements*. Sterling, VA: Earthscan Publications, 2003.

92

REMEDIES FOR OVERPOPULATION

(December 18, 2005)

We know some of the factors associated with success and failure of population control. For example, population growth can be greatly alleviated by making affordable good reproductive health education, contraception, sterilization (tubal ligation and vasectomy) and abortion services. The cost of such services is far less than the costs of overpopulation. It's in the interests of developed countries to foot the bill, both for themselves and for undeveloped and developing countries. It's very important that developed nations take the lead in simultaneous programs to reduce their own populations, switch to renewable energy, and rescue endangered species.

In traditional male-dominated societies undergoing population explosions, the women often wish to lower their fertility. But the men consider abundant children to be testimony to their masculinity. Large extended families have more social influence. Arranged marriages help consolidate economic and political power. Women are powerless, so childbearing is coerced. Men block family planning. Social pressure is strong. Childless marriages can lead to persecution, divorce, and even murder of the wife.

Population control can only succeed if these deterrents are eliminated, by popular consensus, so that couples—especially women—are genuinely free to have fewer or no children. Women must have the right and the means, if need be, to walk away from husbands who attempt to coerce procreation.

Religion can impede population control. In many religions, an abundance of children is considered an obligation or a sign of God's blessing. In many societies, freedom of religion is as unthinkable as women's rights. This is true for a large segment of our own religious right. We have to stand up to them.

In societies where women have adequate nutrition, health care, education, and career opportunities, they are more likely to have fewer babies. A healthy economy with modern industry can provide better economic opportunities for women, from which may follow women's rights. This process can be accelerated by education about the population explosion and its consequences, and about family planning. Such education has been directed primarily at women in the past, but it needs to be provided for men as well because they must be persuaded to relinquish control of women.

Unfortunately, prosperity isn't a guarantee of women's rights, including reproductive rights, as illustrated by the U.S. In the United States, where women nominally have equal rights by law, there is still sex discrimination with regard to family support for higher education, employment opportunities, pay levels, property ownership, credit, and social roles. Contraception and abortion are (only recently) legal, yet couples still face strong pressure to have children, and women who seek abortions are often subjected to harassment and sometimes violence. Although the majority of Americans believe in the freedom to decide how many children to have and the right to contraception and abortion, it still takes courage for a political candidate to speak up in public for a woman's right to choose. The traditional alliance of authoritarian religion, strongly patriarchal extended families, and pride in high fertility breeds intense opposition to family planning.

This alliance undemocratically dominates U.S. policy on population control. Our government's opposition to international family planning efforts is based largely on specious claims that funds are spent to support abortion (including coerced) and infanticide, in spite of the facts that investigations have failed to confirm these contentions and that voluntary abortion is legal in this country. Blocking affordable family planning leads to more abortions, not fewer. If we don't fight back, there's no chance of population reduction at home, let alone abroad.

To the best of my knowledge, we have never had a national leader who has spoken out about the need for population control. This will have to change by the 2008 elections.

This week's news:

Scientists at the University of Wisconsin-Madison have combined satellite images with agricultural census data to make detailed maps of land use. "In the act of making these maps we are asking: where is the human footprint on the Earth?" says Amato Evan, a member of the team. This map shows land use for 2000, but the scientists also have data going back to 1700. "The maps show, very strikingly, that a large part of our planet (roughly 40%) is being used for either growing crops or grazing cattle," says Dr. Navin Ramankutty, another member of the team. By comparison, only 7% of the world's land was being used for agriculture in 1700. "One of the major changes we see is the fast expansion of soybeans in Brazil and Argentina, grown for export to China and the EU," at the expense of tropical forests in both countries, says Ramankutty. Intensive farming in the U.S. and Europe has decreased cropland areas and urbanization is spreading. "Except for Latin America and Africa, all the places in the world where we could grow crops are already being cultivated. The remaining places are either too cold or too dry to grow crops," says Ramankutty. "The real question is how can we continue to produce food from the land while preventing negative environmental consequences such as deforestation, water pollution, and soil erosion?" (*The Guardian*)

Recommended reading:

Feldt, Gloria, *The War on Choice: The Right-Wing Attack on Women's Rights and How to Fight Back*. New York: Bantam Books, 2004.

Flanders, Laura, (ed.), *The W effect: Bush's War on Women*. New York: Feminist Press at City University of New York, 2004.

93

HUMANE POPULATION POLICY

(December 25, 2005)

Scientists have recently reported how the die-off of a huge algae bloom in the Gulf of Mexico has created a 7,000-square-mile dead zone without oxygen to sustain sea life. Much of the biosphere is under a similar death sentence unless environmental degradation is quickly reversed. For this we need to achieve a better-than-sustainable life style, restoring our environment to a pre-industrial level. This is possible only with significant population decline.

In free societies restrictions on human activities must be agreed upon by the majority, and then only with respect for human rights. One of the many rights assumed over the ages has been the right to have children, sometimes when the circumstances are totally inappropriate. Everyone, not just the majority, needs to be far better educated about sustainability before they will agree to population reduction.

The first step in population decrease is to provide the human rights and social security everyone needs.

At the local level: Family planning education and services in every municipality are essential. Good nutrition and health care will ensure those of working age will be healthy enough to work and care for their children and the elderly; women will have affordable access to day care, child and maternal health care, contraception, sterilization, abortion, and adoption services, and men will have to understand the need for them.

At the national level: Population control is meaningless unless women are free of pressure to reproduce. They need the financial and social independence to walk away from harmful circumstances. This requires equality of education and employment opportunities as well as the right to vote, own property and make their own decisions about marriage and family life. This will require increased legal reform to protect women against discrimination.

Such steps require a strong economy, with little unemployment. The few unemployed need to be supported adequately, and government services have to be adequate to meet universal human needs such as education, health care, infrastructure maintenance, and social security. Local food and goods production are crucial in order to minimize debt and maximize employment. Each nation will have to produce most of its own needs in an era of energy scarcity, plus an excess that can be traded for locally unavailable goods and materials, maintaining a zero balance of trade.

These goals can be met only if businesses are locally owned and controlled, all revenues are locally used, and all utilities are nationalized. Local oligarchs will need to be persuaded that they face the same doom as everyone else. They must reform: implement democracy, land and economic reform, and root out corruption.

At the global level: Many of these reforms can be assisted by the same international organizations that have forced undeveloped countries to globalize: the International Monetary Fund, World Bank, and World Trade Organization. The United Nations and many NGOs are well positioned to take central roles. The leadership of these organizations will need equal representation of all nations, not just developed ones.

The developed nations: Developed nations need to embrace the same better-than-sustainable policies required in undeveloped countries, including zero balance of trade. The oligarchs in developed countries, like those in undeveloped countries, will have to be persuaded it's in their best interests. Since the developed nations are in the best position economically to implement changes, they consume the most resources, and they generate the most CO_2 and other pollutants, they must take the lead and help undeveloped nations to follow. Not bloody likely, you say? I agree.

It's not at all certain that the people of the world will voluntarily limit themselves to one child per couple for one generation, let alone several. It is certain, however, that attempts to coerce such a policy would meet fierce resistance.

This week's news:

A research team that includes UC Irvine's Susan Trumbore has used radiocarbon dating to determine the ages of trees in Amazonia. Their article in the *Proceedings of the National Academy of Sciences* reports that up to half of all trees more than 10 cm in diameter are over 300 years old. Some are 1,000 years old. These results have implications for Amazonia's contribution to global CO_2 levels. The Amazon forests contain about one third of the carbon in land vegetation, but they have less capacity to absorb atmospheric carbon because of their ages and slow growth rates. Most Amazon forest trees grow so slowly because the soil is poor and they are in the deep shade of larger trees. This means that the forests will recover exceptionally slowly, if at all, from the massive ongoing deforestation. (*ScienceDaily*)

WASHINGTON: The first large offshore US wind farm proposed for Nantucket Sound would be aborted by an amendment to a Coast Guard budget bill. The amendment, offered by pro-oil Rep. Don Young (R-AK), would outlaw new offshore wind facilities within 1.5 nautical miles of a shipping lane or ferry route. The installation would consist of 130 turbines in a grid that would occupy 24 square miles in the sound. Each tower, with its turbine and blades, would rise 420 feet above the water. Cape Wind Associates says the wind farm could produce three quarters of the electricity for Cape Cod, Martha's Vineyard, and Nantucket combined. Christine Real de Azua of the American Wind Energy Association says if Young "were really worried about safety issues, he would be worried about oil rigs that are allowed within 500 feet of a shipping channel." In Denmark, which has the most experience with offshore wind farms, there are turbines only one-quarter mile from shipping channels into Copenhagen Harbor and a mile from another in the Baltic. They handle much more traffic than Nantucket Sound. (*New York Times*)

Recommended reading:

Mazur, Laurie Ann (ed.), *Beyond the Numbers: a Reader on Population, Consumption, and the Environment.* Washington, DC: Island Press, 1994.

94

THE NEW LEADERS

(January 1, 2006)

Improving the standard of living and education are not effective in themselves for achieving population control. Contraception, abortion, and sterilization are not enough in themselves to reduce family size. Remarkably, even the privations brought about by overpopulation may not lead to population reduction, short of famine, drought, or disease. Yet some societies have managed to live sustainably for long periods, until they encounter some challenge beyond their ability to meet.

What is a world to do? Our clue comes from the finding that the major determinant of family size is desired family size, and the strongest factor influencing desired family size is the *perceived affordable size*. It's likely that the greatest source of miscalculation is an inability to recognize that a society is living beyond its means, due to inadequate understanding of environmental effects, the belief that trends are better than they are, or an over-reliance on a way of life that has become unsustainable. These miscalculations are partly rational judgment based on wrong information and partly simple irrationality. Unethical corporate, religious, and government leaders bombard us with misinformation and foster an anti-rational climate.

Our most immediate need, then, is for accurate information and a rededication to rational discourse, policy, and action. Certainly, education, better standards of living, and family planning services are more humane than their alternatives. Governments and non-governmental organizations have essential roles to play. But we have to wean ourselves from dependency on government and profiteers for accurate information or sound plans of action. People like you and me have to face the facts of overpopulation, resource depletion, climate change, and loss of biodiversity. We must take the lead, with or without the support of our governments and religious groups.

We, the new leaders, face multiple challenges. We must first fend for ourselves and those who will join us, at the community level. We have to take individual action as leaders in other communities are already doing, and demonstrate the benefits of conservation, recycling, smaller families, sustainable energy, and the reclamation of our environment. It's too late to address anything less than the full spectrum of problems we face. We cannot wait the 100–250 years it will take to reduce world population to sustainable levels before decreasing our deadly impact on the planet. To do this we have to teach everyone, young and old, to recognize the damage all around us. We have to feel in our bones that the largest affordable number of children per woman or man is one, and that that number will drop like a rock as long as we take no action.

Our new citizen-leaders have already been working for decades to raise people's awareness of our problems. Our environmental problems are becoming so obvious that everyone is aware of them. Our examples of news items from around the world show that rational, informed people are taking charge in and out of government. There's been a remarkable transformation even in the two years covered by this book. These are grounds for hope.

Here in the U.S., our "post-industrial" (non-productive), "globalizing" (mercantile) "consumer" (borrowing) society is in decline. Our country can no longer dominate even small nations economically or militarily and other nations realize we're no longer a stabilizing but a destabilizing force in the world. It remains to be seen whether the efforts of an informed, rational, and moral U.S. majority can prevail over the ones in control. If not, it's we who will be the losers, with no one to blame but ourselves.

This week's news:

NEW YORK: As of Jan. 1, 2006, buyers of some hybrid vehicles can get sizeable tax credits. The credits for some vehicles almost cancel the cost of the hybrid technology. Until now, some hybrid vehicles qualified for tax deductions. Tax credits are much more valuable than deductions. (CNN)

Recommended reading:

Arendt, Hannah, *On Revolution*. Westport, CT: Greenwood Press, 1982.

95

DO THESE THINGS NOW

(January 8, 2006)

Overpopulation, global climate change, and mass extinctions have reached the point where it would be prudent to start changing our lifestyles to improve our individual and collective well-being. Here are some suggestions.

Food. There's no slack in the global food system to accommodate drops in production or increases in demand. Almost all arable land is in production (except in countries where subsidized food imports have driven native farmers out of business) and crop rotation and multiculture are largely abandoned. Much production depends on petroleum-derived fertilizers and pesticides, and the genetic diversity needed to adapt to changing conditions has largely been eliminated. Genetically modified foods and patented food genomes will further reduce that diversity. As population increases, farmland is depleted, the cost of oil climbs, the climate changes, and wild species that support the biosphere go extinct, there will be less food available, its quality will decrease, and it will cost a lot more.

Get used to eating less meat, poultry and seafood to save money and decrease demand for items at the top of the food chain. Far more nutrition can be obtained per acre of farmland by consuming lower on the food chain. Toxic substances are less concentrated in food that's lower on the food chain, and plants don't transmit zoonoses (diseases we can catch from animals). Most of us would also benefit from eating less. As much as possible, eat locally produced foods and support local farmers. Eat "green" foods: organically grown, free-range, additive-free. Grow some of your own food. Demand local inspection of food and farmland for toxic substances. Get yourself tested for toxics. Contact the Mormon Church (Latter-Day Saints) for advice on food storage and stock up a year's worth of food (including bulk dried, canned, and dehydrated) for each member of your family. Don't forget salt, spices, and a can opener. Be prepared to feed people less fortunate than you.

Water. In many places, there isn't enough potable water. Oppose waste of local water in unwise industrial processes and the production of bottled water or soft drinks. Press for better water quality monitoring and treatment. Equip yourself with a small water purification system for your drinking water. Prepare to store enough drinking water for a month. Plan to share it.

Shelter. Leave places where local food or water supply will be inadequate. Leave places where the climate will be too harsh or where flooding, earthquakes, mudslides, or avalanches are becoming more common or where pollution will be worse. Make improvements in your home to adapt to changing conditions. Use local contractors and materials, encourage recycling of demolition materials. Be prepared to take in people less fortunate than you.

Energy. Be prepared for electric and other energy shortages. You should be able to keep warm and cook food. Kerosene space heaters and lanterns, and primus stove with good supplies of fuels will all come in handy. Have several flashlights, batteries, and light bulbs available. You'll need matches. Buy sleeping bags. Be prepared to cook and preserve everything in your refrigerator and freezer when the power fails.

Clothing. As clothing becomes more expensive, begin to supply yourself with several years' supply of basics that don't require special care like dry cleaning. Buy larger sizes for growing children. Plan on a cold house in winter, a hot one in summer.

Health. Be prepared to stock up on medications and get prescriptions for antibiotics and antivirals. There won't be enough to go around if a rapidly moving epidemic strikes. Get surgical masks and hoods, and sturdy rubber gloves, boots, jackets and pants. Stock up on chlorine bleach to use as a disinfectant. And have lots of packages of rehydration salts on hand for cases of diarrhea.

Transport. Get used to walking, bicycling and taking public transit. Sell your gas-guzzler and get the best-mileage car you can afford for those rare occasions when you really need one. Forget about going more than one or two hundred miles.

Energy. If you live somewhere solar or wind power will repay their purchase cost in less than 20 years, have enough power capacity installed to meet 100% of your needs. Reduce your energy consumption sharply.

Jobs. Learn several useful skills that will always be needed: farming, food preparation, green construction, solar and wind power, making clothing, repairing things.

Money. Pay your debts and stop wasting money on things you don't need. Save your money in the form of tangible, useful goods like storage food, rental property, tools. Invest in local businesses and work to help them succeed. Get out of the stock market if you can (some pension plans don't allow you to). If you can't get out, switch to "green" investments like the new energy economy. Bonds and even Treasury notes may become worthless. Don't count on Social Security or profit sharing plans, or even pension plans that are based on paper, for your old age. Don't retire, until you just can't work any longer.

Family and Friends. Cherish them. Support them. Convince them to take the same precautions as you. Have an emergency plan for reestablishing contact after periods of disruption. Declare a ten-year personal moratorium on having children and after that have no more than one per mother or father. Encourage family planning and negative population growth. Don't be seduced by arguments to the contrary. Express strong disapproval of excessive breeders.

Government. What can I say? Think globally, act locally. Become involved, don't be afraid to be assertive. Show up at community meetings and forums with legislators and demand action. Ask political candidates what they're going to do about climate change, mass extinctions and (gasp) overpopulation. Make it clear these are the issues that will determine your votes.

Information. Cherish and support fact-based reality. Learn all you can about the topics central to this book. Teach. Oppose misinformation.

Civic Action. Find the areas of life in your community where you can have the fastest and greatest impact. Do it. Set an example.

This week's news:

Since 2003, Lake Victoria has been drying up. A report entitled *Study on Water Management of Lake Victoria states,* "The dropping water levels will lead to extinction of marine life forms and affect breeding grounds of fish," Fish spawning areas could be cut off from the lake as the water level continues to drop. "We could be harvesting the last stocks of fish from the lake because the breeding should have been reduced by lowering of the water," said a source. The lake, which is shared by the three countries, sustains more than 30 million people. The report blames the water loss on its use to generate hydropower. Other causes that were evaluated include drought, which may have reduced the water level by another 1/2 meter. Another possible contributor is increased evaporation due to global warming. A study by the East African Community found that flower growing in the three countries uses a lot of water from the lake. In Entebbe, around Ngamba, the water level has dropped by over a meter in depth, but the shoreline has retreated by more than 40 meters. This has led the chimpanzee sanctuary and Wildlife Conservation Trust to build a new electric fence, at a cost of 10m Ugandan shillings ($5200). (*The New Vision*)

The Natural Environment Research Council reports that plankton in the North Sea is at half its previous levels. The North Atlantic current has changed course due to global warming, and is carrying the plankton farther north. Newly hatched cod need adequate plankton levels to survive. Dr Martin Angel, who directed the study, says, "No amount of fishery regulation is going to bring the cod stocks back. The fate of cod is commercial extinction. Probably the main reason for the lack of success of cod to recover after overfishing is that the ecosystem has changed and the [plankton] is no longer in the right place at the right time. Normally this plankton is carried to cod spawning grounds, but the current has changed and they are now drifting to the southeast of Greenland. At this critical period, when cod require to feed on high plankton densities, the plankton they favor is no longer there. The engine of the Atlantic current—the cold, salty water that used to drive it—has been switched off. This will have a major impact on global climate because the oceans distribute energy over the surface of the globe. The seas off Britain appear to be getting warmer and the animals are responding to that." Dr Richard Dixon, director of WWF Scotland, remarks, "These findings do look very significant and it may already be too late for some species. While we have seen interesting changes with birds moving further north, butterflies changing their distribution and trees flowering at the wrong time of year, this is clear evidence of a fundamental change in the marine ecosystem, which will make a really big difference. Half of Scotland's plants and animals live in the sea and a large part of our diet and economy are based on there being a diversity of life in the oceans." Michael Park, chair of the Scottish White Fish Producer's Association, says Angel's findings were supported by the experience of Scottish fishermen. (*The Times*)

GONAÏVES, HAITI: In September of 2004, Tropical Storm Jeanne struck Haiti, flooding Gonaïves and killing more than 1,900 people. Deforestation is causing desertification, which in turn is causing more flash floods and mudslides. Most Haitians use charcoal for cooking, and tree cutting for fuel and lumber—at the rate of 50 million trees/year—is one of Haiti's few sources of livelihood. Forests have declined by 98%. Without tree roots, topsoil is eroded and the ground can't absorb water. Rainwater pours down the bare hillsides, bringing mudslides and floods. Criss-crossed with rivers and ringed on three sides by stripped mountains, this western coastal city didn't stand a chance. "There are many more Gonaïveses in Haiti just waiting to happen. We're sitting on a time bomb," says one of Haiti's environmental leaders, Jean Andre Victor. Eighty percent of Haitians have no other source of energy for cooking. There's little electricity, and most Haitians can't afford gas or kerosene. Haiti received pledges of $1.3 billion in foreign aid, of which less than 2 percent is for the environment. (*Newsday*)

BOB WHITE, WV: Mountaintop removal involves blowing the top off a mountain to expose coal seams. Ninety-five percent of surface mining in southern West Virginia uses this method. It devastates the ecosystem. More than 1,200 miles of streams have been buried and 350 square miles of mountain land destroyed. The Environmental Protection Agency, Army Corps of Engineers, Fish and Wildlife Service, and Office of Surface Mining have issued an Environmental Impact Statement on the practice. Its original purpose was to minimize harm from mountaintop mining, but in 2001, Deputy Interior Secretary Steven Griles narrowed it to "centralizing and streamlining coal mine permitting"—which it does. The government is weakening environmental regulations to help the mining industry. "A lot of what has renewed the growth of mountaintop removal mining in the last four years has to be attributed to Bush administration policies that have removed any obstacle, including local citizens, standing in between industry and the mountains," says Joan Mulhern, legislative counsel at EarthJustice. She cites a 2002 rule change that redefines the rubble as "fill" rather than as waste, and will allow mining closer to streams. Luke Popovich, spokesman for the National Mining Association, says, "You have to look at what you're disturbing versus what you're offering your country." Coal technology is cleaner than ever, and becoming more essential for US energy needs, he says. He mentions the many mining jobs and related work in West Virginia "These are the kinds of jobs that can hold a community together year after year and sustain it through ups and downs." The mining industry touts its reclamation of mined mountaintops. "Level land is a rare commodity in the steep slopes of southern West Virginia," says Bill Raney, president of the West Virginia Coal Association. Popovich lists developments like golf courses. Scientists point out that it's impossible to replace forest ecosystems when the headwaters of streams are buried and the steep slopes are gone. "They're living in this world of denial," says Ben Stout, an ecologist at Wheeling Jesuit University, of the reclamation claims. Long-time resident Maria Gunnoe calls reclamation "putting lipstick on a corpse." She points out other problems: a school where most of the children have respiratory problems, a spot where a mine dumps wastewater into abandoned mine shafts, and which sometimes erupts and causes landslides. Gunnoe has become an advocate for solar and wind power instillations on leveled mountaintops. "If they want to call me an environmentalist, that's fine," Gunnoe says. "But they need to realize the issues I'm talking about are human issues. Anybody who enjoys clean water and clean air is an environmentalist.... I don't fight just to save a mountain; I fight to save the people at the bottom of the mountain." (*Christian Science Monitor*)

Recommended reading:

Stiglitz, Joseph E., *Globalization and its Discontents*. New York: W. W. Norton, 2002.

96

RISING ENERGY COSTS

(January 15, 2006)

The next few years will be a turning point for energy costs, and it will bring about big changes in our lifestyles. We've all been hit by the increased cost of gasoline. This winter most of us will be hit by the higher cost of heating our homes. The cost of electricity will continue its climb next year. As temperatures go up, so will air conditioning costs.

There will be serious effects on our economy, because everything in our lives takes energy. Food is grown with artificial fertilizer and harvested with machines, which are energy-intensive. Most of our food is trucked long distances instead of grown locally the way it should be. The same is true of most of our other necessities and luxuries. With the enormous cost of intercontinental shipping, globalization will eventually wither.

This is the optimal time to change your economic behavior. Trade in your cars for one hybrid: I think the Toyota Prius is the best buy. You can recoup the cost of the technology with federal, and in many states, state tax breaks. The net cost of my new Prius, after tax savings, was $16,000. Cut your annual mileage by half, by doing all your errands once a week, car pooling, and using public transportation.

Stop buying things you don't need, and buy local products as much as possible. Pay off your debts and stop using credit, with four exceptions: your home (the rent you pay could be earning you equity), your hybrid car, passive solar heating, and solar electricity. I just used a home equity loan to buy a 4kW photovoltaic system for $30,000. Now that I've cut my electricity use to an average of 1 kW, my electricity savings will pay for it in 17–20 years assuming an annual inflation rate for electricity of 20%, and my electricity will be free from then on.

I've designed a passive solar heating renovation for my home that will cost about $12,000 and will pay for itself in about 5 years (my natural gas costs increased almost 50% last year). In many states, tax breaks allow you to do even better.

Don't fear that this will be a losing investment or that your timing may be off. This isn't a scheme for enrichment; it's an investment in security. If you can't do it all at once, do it in stages over the next three years. You may have more time than that, who knows? But in terms of security, the sooner the better.

Most of our population can afford these costs, if they adjust their priorities. The required lifestyle changes will be forced on us anyway and those of us who wait won't be able to lock in current savings: rising energy costs will price hybrids, photovoltaics, and even passive solar out of reach. With the coming inflation many people will lose real income. That includes retirees. If your community becomes sufficiently self-reliant you should be able to afford the necessities.

Unfortunately, a growing segment of our population can't afford these expenses. There will be little or no help for them. Be prepared to help them. Some will be family, friends, or neighbors.

Don't bring any more children into this world, for your sake and for everyone's. Support Planned Parenthood and population reduction. Speak out against those who favor coerced procreation.

The greatest threat to our survival is our government, which is controlled by international corporations, the rich, and the super-rich. We now live in a fascist nation, with little to choose from among candidates for national office; rigged elections; widespread corruption in all three branches of government including the Supremes; and the replacement of competent career civil servants with incompetent cronies. The unholy alliance of corporations, the military-industrial complex, and the religious right control our national agenda, condemning us to war and loss of civil liberties. Dissent, and even scholarly work, are more and more problematic. Our now-privatized prison system is overflowing with more of us than ever. Our government spies on us and can "disappear" us and torture us with impunity. Corporations have more rights than people do: they can loot our money (by defaulting on our pensions, which are our earned money), while we often can't even declare bankruptcy.

By preparing yourself and your community, there's a chance you'll make it. Start by securing the essentials: shelter, food, water, and a stable local economy.

This week's news:

Lawyers with the EC warn that a draft concordat between Slovakia and the Vatican would be illegal. The pact would permit doctors in Catholic hospital to refuse to perform abortions or fertility treatment on "conscience" grounds. The EU Network of Independent Experts on Fundamental Rights said "certain religious organizations" should have the right not to perform "certain activities where this would conflict with [their] ethos or belief." But it added, "It is important the exercise of this right does not conflict with the rights of others, including the right of all women to receive certain medical services or counseling without any discrimination. There is a risk that the recognition of a right to exercise objection of conscience in the field of reproductive healthcare will make it in practice impossible or very difficult for women to receive advice or treatment...especially in rural areas." Keith Porteous Wood, of Britain's National Secular Society, said, "We welcome this opinion, which shows conscience clauses in EU member countries cannot be taken advantage of regardless of the consequences for others. This concordat would enable those Slovaks wishing to enforce Catholic doctrine, for example, on abortion and contraception in the performance of their duties regardless of the adverse implications on the patients, which could be severe. The draft also discriminated in favour of Christians in certain areas to the detriment of those of other faiths or none." (*The Guardian*)

Recommended reading:

Shuman, *Going Local: Creating Self-Reliant Communities in a Global Age*. New York: Routledge, 2000.

97

MAKING YOUR BUILDING ENERGY-EFFICIENT

(January 22, 2006)

As the cost of energy goes up, it will drive up the cost of everything else. You should put a high priority on minimizing energy costs in your home, place of business, or other buildings.

The first thing you should do is assess your building location. Is it in a flood plain or an earthquake zone? If it is, move. Is it in an exposed location, where strong, cold wind or hot sun increases your heating or cooling costs? You may want to move, or plant some trees to provide a windbreak or some shade, or add a finished basement where you can spend most of your time. Many people in Newfoundland spend most of their home time in basements blasted from solid rock, in order the save on heating bills. The same thing works for cooling bills in hot climates. Once you get around 5–10 feet below ground level, the temperature holds steady at the average annual temperature. That's around 60°F, and going up, in my area.

Next, assess your space needs. Do you really need all the space in your building? If not, move to a smaller one or rent out part of the space.

An energy audit will usually save you lots of money. Find the places where your house leaks heat (in or out), and install better storm windows, storm doors, weather stripping, and insulation as needed. Fifteen years ago, I glazed in sheets of glass in the window frames of my house because I couldn't afford storm windows. Caulk cracks and crevices in the sides of your building. Insulated drapes or cellular blinds on your windows can make a big difference.

Replace old, energy-inefficient appliances like refrigerators, washing machines, driers (clothes and hair), furnaces, air conditioners, and hot water heaters with newer, energy-efficient ("Energy Star") ones. When replacing your hot-water heater, get an on-demand one that only heats the water when it's flowing. Set it for the lowest temperature that will meet your needs. Take shorter showers (no baths) and wash clothes with cold water in maximum-size loads. You can often sell your old appliances by placing, and buy almost-new ones by searching, classified ads.

In winter, divert the hot-air exhaust (**NOT the flue**) into your living space with a diverter-with-lint-trap you can pick up at an appliance store. When it's warm and dry enough, dry your clothes on a clothes line, unless particulate pollution or allergens make this unworkable. Remember you can humidify your living space by drying clothes indoors in the winter.

Replace incandescent lighting with full-spectrum fluorescents, including screw-in replacements. They consume about one-third the electricity. LED (light-emitting diode) lights use negligible amounts of electricity but are too expensive for now. Switch from an electric to a natural gas stove and use a pressure cooker to cut cooking times to a third and energy costs to even less. Cook batches large enough for two or three meals and reheat leftovers with a microwave. Minimize use of an exhaust fan when cooking and showering if you're heating the house. Put your remote-controlled entertainment center on a switched outlet so everything's really turned off when not in use.

Keep room doors closed to prevent drafts except when cross-ventilating with outside air. Replace your thermostat with a programmable one that allows different settings for times when you're in the building and when you're absent. Switch from electric or fuel oil to natural gas heat, with one exception: you can heat specific rooms in your house, like the bathroom, kitchen, or bedroom independently of other rooms, with timer-controlled baseboard heaters. Zone control can also be achieved by opening duct vents and turning on radiators in occupied rooms, and doing the opposite when leaving a room for more than an hour (don't shut heat off entirely if there's a risk of frozen pipes). Forced-air heat is more efficient than radiators, and you can install zone-control dampers on the air ducts. Install a humidifier on your forced-air furnace: cooler winter temperature settings are more tolerable when the air isn't dry, and your respiratory system will be a lot healthier.

In winter, lower your thermostat, put rugs on the floors, and dress warmly. Lower your bedroom temperature setting, wear pajamas, and use more blankets. In spring and fall, open the windows and doors (screens will be a must, due to increased mosquito-borne diseases brought to us by global warming), and use fans for cross ventilation when the outside temperature is comfortable; close windows and doors when it's too hot or cold out. In summer, raise the thermostat, take up those rugs, wear light, loose clothing and sandals, use fans, drink cool liquids, and sleep naked without sheets. An attic fan will lower cooling bills.

You'll be surprised: these measures can save half your energy bills. And don't forget, you can live more responsibly and reduce energy costs by having no children.

This week's news:

SYDNEY: Ministers from the United States, Australia, Japan, China, India and South Korea and more than 100 top executives from big business have pledged new approaches to climate change, but refused to stop using fossil fuels. The six countries account for nearly 50% of the world's gross domestic product, population, energy consumption, and greenhouse gas emissions. "We recognised that fossil fuels underpin our economies, and will be an enduring reality for our lifetimes and beyond," the six nations' representatives said in a statement. "It is therefore critical that we work together to develop, demonstrate and implement cleaner and lower emissions technologies that allow for the continued economic use of fossil fuels while addressing air pollution and greenhouse gas emissions." Greenpeace representative Catherine Fitzpatrick says the agreement is more of a trade pact than a solution. "The short-term interests of the fossil fuel sector have been put ahead of the long-term health and welfare of ordinary people," she said. The World Wildlife Fund said the plan would cause global temperature to increase eight degrees Fahrenheit by 2050. (terradaily.com)

Recommended reading:

Editors of E/The Environmental Magazine, *Green Living: The E Magazine Handbook for Living Lightly on the Earth*. New York: Plume, 2005.

98

HOME SOLAR ELECTRICITY

(January 29, 2006)

Solar electricity generation depends on the *photoelectric effect*. When photons (light particles) strike certain types of atoms, they dislodge electrons, which can provide an electric current. *Photovoltaic (PV) panels* produce *direct current (dc)* electricity when exposed to light. *Inverters* can be used to convert dc to *ac, alternating current*, which is compatible with the electric power provided by your electric utility company. In fact, your system can be attached to the utility's electric meter so that you have power 24 hrs/day and lower power bills.

A 21" × 48" PV panel can produce 75 watts (W), 0.075 kilowatts (kW) of power: enough for one or two fluorescent light bulbs. A solar power installation costs about $7–8 per watt including installation by a qualified contractor. The panels should be installed facing due south, in an area devoid of shade all day, tilted at an angle equal to the location's latitude. If you have a roughly south-facing roof, these angles can be achieved with rooftop installation. For greater power output, they can be installed on mounts that track the sun, as long as none of them shadows others at any angles—you'll have to space them apart or put them on a common platform. One kW rated capacity will require 1/0.075 = 13 1/3 solar panels, which will take up roughly 8' × 12'.

You can estimate your average power consumption U, in kW, by adding up the number of kWh from your electric bills for one year and dividing by 8760, the number of hours in a year. The average hourly cost C is the sum of the amounts you were billed for divided by 8760. The cost per kWh is C/U; I pay about $0.07/kWh.

Let's say you've followed my guidelines to maximize energy efficiency in your house, and your average power consumption U is about 0.8 kW. A good rule of thumb to determine how large a system you need is to assume that you have the equivalent of full sunlight 8 hours a day, times the percent of days that the sun is not blocked by clouds (in my town that percentage is probably about 60%) to get the average number of hours of full sunlight each day. Divide this by total hours per day to get the efficiency e, for your system:

e = (8 hours/day × 60%)/24 hours/day = 0.2,

meaning in my location I'll only average, day and night, about 1/5 of the rated power of my system.

If you'd like to break even (zero net cost per year), the annual kWh produced by your system should equal the annual kWh you use. For our assumptions, P, the necessary power rating of your system, is estimated as:

$P = U/e = 0.8/0.2 = 4$ kW.

At $7.50/W, a 4kW system costs $30,000 and will take up about 12' × 40'.

You can figure out how long it will take to pay for your investment in electric bill savings. Assuming your future electric bills will be the same as they are now, say $0.07/kWh, you'll save 0.8 × 8760 kWh/yr × $0.07 = $491 per year, and it will take you a little over 60 years to pay off your $30,000 investment (longer if you financed it). Pretty poor, for a system that's good for forty years. But if you assume that energy costs will increase by 10% per year, a conservative estimate, amortization time is reduced to a little over 26 years, which is a lot better. At 20% annual inflation, which may come sooner than you think, it will take a bit less than 21 years, which is great for a system that will last 40 years. You'll have many years of essentially free electricity (actually, inverters have to be replaced occasionally), leaving more utility power for others and cutting down on power plant emissions.

Starting in 2006, there's a federal tax incentive for solar power installations, and many states also have tax incentives. One factor essential for this rate of payback is *net metering*, where the power you put on the grid during excess supply from your system (e.g., at noon) is deducted from the power you take from the grid during excess demand for your building (e.g., at night). Forty states have net metering, and I believe the federal energy act of 2005 mandates net metering by 2007.

This week's news:

Vultures in India, Pakistan, and Nepal are going extinct from consuming carcasses of cattle treated with diclofenac. The anti-inflammatory drug causes kidney failure in the vultures. Debbie Pain, a scientist with the Royal Society for the Protection of Birds in England, says, "Populations of three vulture species affected by diclofenac in South Asia have declined by more than 97 percent since the early 1990s." Pain explains that in India and Pakistan it's customary to leave the carcasses where they die. Vultures quickly reduce them to piles of bones. With vultures almost extinct, the dead animals rot, attracting wild animals, which are a danger for people. The rotten flesh also can host diseases like anthrax. The only way to save the birds may be to protect them in captivity, since the governments have refused to ban the drug. (ABC News)

OSLO: UN University reports 800–1,000 people have died each year in landslides over the last 2 decades. About 100 experts will meet in Tokyo to discuss prevention and alleviation of landslides as global warming may make them worse through heavier rainfalls. "If climate change predictions are accurate you will expect…more intense and extreme rainfalls," said Srikantha Herath, senior academic officer. Herath said the experts will probably recommend better monitoring and early warning, protection of cultural sites, and better aid for victims. (Environmental News Network)

Recommended reading:

Pieper, Adi, *The Easy Guide to Solar Electric, Part II: Installation Manual.* Santa Fe, NM: ADI Solar, 1999.

99

PASSIVE SOLAR HEATING

(February 5, 2006)

Much of the energy to heat our buildings can be obtained from contemporary, rather than ancient, sunlight. The use of sunlight without any energy-consuming technology to heat buildings is called *passive solar heating*. It's implemented by using a few simple design principles.

The first principle is *site placement*. Pick a site where hills and buildings to the south don't block the sunlight. It also pays to have some deciduous trees for summer shade on the south side. Once they've lost their leaves, they pass sunlight in the winter. On the other sides of the building it's worth having evergreen trees or other buildings for shade and wind blocks.

The second principle is *conservation*. The building should be no larger than needed and it must be well insulated, with good windows and storm doors. For buildings with substantial traffic in and out, airlocks and revolving doors are advisable.

The third design principle is proper *design and placement of windows* on the south side of the building. They should pass visible light well, and infrared poorly. Sunlight is absorbed indoors as heat, but not lost as infrared radiation: the greenhouse effect. Also, less heat is admitted as infrared into the building in the summer. Ordinary glass exhibits these properties, but there is specially treated glass that is even better. It's more expensive than conventional window glass, but worth the money in savings. We can enhance this difference with heat barriers: good insulating drapes or blinds, or outdoor shutters or panels on the south windows that are kept closed at night in the winter to prevent heat loss, and all the time in summer to prevent heat gain.

The fourth principle involves use of *overhangs* for maximum heat capture in winter and minimum heat capture in summer, by using southern roof overhangs of the proper angles and depths. A properly designed overhang will shadow the south windows in the summer when the sun is higher in the sky, but will allow sunlight to pass under it and through the windows in the winter when the sun is lower in the sky.

The fifth principle is use of *heat mass* for heat storage. In south-facing rooms, it's necessary to have a heat mass that will easily absorb light and radiate heat. It can consist of dark tiles on a concrete floor, or a stone insert in a wood floor. Alternatively, it can be an inner wall a few inches inside the window, painted a dark color to absorb light and radiate heat. It should be built of heavy, heat-conducting material such as concrete, as opposed to light, poorly heat-conducting material like drywall on studs, in order for it to absorb sufficient heat. By having an air space under and over it, air can circulate from the colder floor region, heating and rising as it goes past the warm heat mass, and spreading throughout the room, the way air circulates in a room heated by a radiator. Rooms will be slightly warmer in the daytime and cooler at night.

The warmed air in south-facing rooms can be used to heat north-facing rooms if there is adequate air exchange, with perhaps an assist from small register fans. Inside walls that don't come up to the ceiling or down to the floor are best for passive air circulation. If the sunlight reaches the inner wall of the south-facing room, the wall itself can serve as a heat mass for both north and south rooms. Another solution is to make the building only one room deep, with a windowless northern corridor connecting rooms through doors on their north sides.

The figures illustrate a retrofitted passive solar heating design I devised for my 1908 foursquare home using the Web resources listed at the end of this chapter.

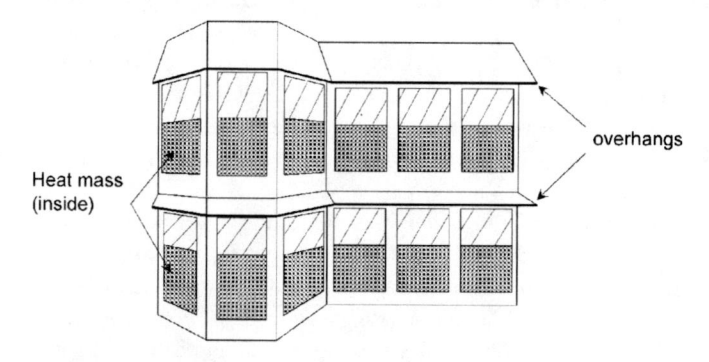

A passive solar building should provide solar lighting as well, as independent as possible of temperature. A simple method is to vary the available sunlight hitting the heat mass with shutters, drapes, or blinds, to control the temperature. An additional drape or blind can be used to block excess sunlight coming around the heat mass on the sides or the top. With appropriate design, this should provide reasonably independent control of the heat and light. Light from the other sides of the structure can also be admitted through the few windows on those sides.

Remember: sunlight can bleach paint, fabrics, rugs, wood, paintings, and prints—even book covers. Keep this in mind when furnishing and decorating passively heated rooms.

Cellular curtains

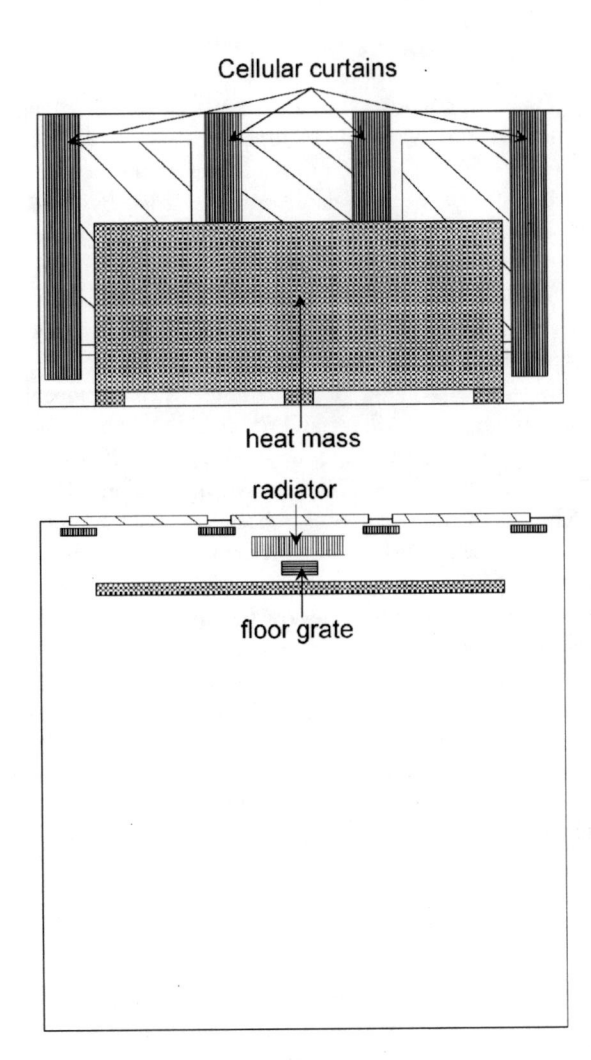

heat mass

radiator

floor grate

    There are computer programs to calculate parameters for the overhang, amount of solar window area, and necessary heat mass to maintain comfortable temperatures on winter days. This will depend, like solar electric systems, on latitude and the amount of winter sunlight available in your area. In most locales, you'll need conventional heating as backup for overcast days.

    It's generally easier and more economical to construct these buildings from scratch instead of retrofitting older ones. They cost a little more than well-insulated non-solar buildings, but the savings in heating bills will more than pay for the difference in a few years, especially as heating costs from fossil fuel use continue to rise.

This week's news:

The current debate among scientists now is whether climate is changing so fast that soon we'll be unable to stop it. Many say it's urgent that we cut CO_2 emissions faster or risk irreversible damage. There are three principal areas of concern: coral bleaching that could damage marine life within three decades; sea level rise by 2100 that would take tens of thousands of years to reverse; and a shutdown of the ocean current that moderates temperatures in northern Europe by 2200. James E. Hansen, director of NASA's Goddard Institute of Space Studies, notes that global temperature has risen almost 1°F since the mid-70s, and another increase of 4° by 2100 would "imply changes that constitute practically a different planet. It's not something you can adapt to. We can't let it go on another ten years like this. We've got to do something." Michael Oppenheimer, professor of geosciences and international affairs at Princeton, says the Greenland and West Antarctic ice sheets hold about 20% of the world's fresh water. If either one disintegrates, sea level would rise almost 20 feet over 200 years, swamping south Florida and downtown Manhattan. The Greenland and Antarctic ice sheets are melting faster than anticipated. "Once you lost one of these ice sheets, there's really no putting it back for thousands of years, if ever." A report from a scientific symposium last year says disintegration of the two ice sheets becomes likely if temperatures rise by more than 5°, "well within the range of climate change projections for this century." A mere 1.8° increase "is likely to lead to extensive coral bleaching," wiping out essential fish nurseries. This fall temperature increases of 2° killed corals from Texas to Trinidad. The Atlantic thermohaline circulation is already slower than it was thirty years ago. According to simulations, odds are 50:50 that the current will collapse within 200 years. Peter B. deMenocal, an associate professor at Columbia, says that about 8,200 years ago, sudden cooling switched off the Atlantic conveyor belt. Land temperature in Greenland dropped more than 9° in only 10–20 years. "It's not this abstract notion that happens over millions of years," deMenocal said. "The magnitude of what we're talking about greatly, greatly exceeds anything we've withstood in human history." (*Washington Post*)

"It's an ugly, gut-wrenching thing to watch," says University of Washington researcher Julia Parrish, who watched nesting murres starve last summer. Winds and currents essential to the marine food web weren't on schedule last year. The breeding failures were preceded by tens of thousands of dead sea birds washing ashore in Washington, Oregon and California. In Washington where 8,000 glaucous-winged gull chicks normally fledge, only 88 did last year. Scientists also noted low catches of juvenile salmon and rockfish, and sightings of emaciated gray whales. They were preceded for the first time in Washington of squid never found earlier north of San Francisco. Plankton typical of San Diego bloomed off the Northwest coast. Off Southern California, zooplankton are down 70%, fish larvae 50%, and there have been massive die-offs of kelp. John McGowan of the Scripps Institution of Oceanography says we can expect these kinds of phenomena as the planet warms. The annual Aleutian Low normally triggers strong winds from north to south. The Earth's rotation deflects the winds, which push the Pacific's surface waters to the southwest, leaving a gap near shore. Cold water loaded with nutrients wells up: food for plankton, the base of the food chain. In 2005, the winds were 2 months late. (Seattle Post-Intelligencer)

Recommended Web resources:

http://www.susdesign.com/overhang/index.php
http://www.nesea.org/buildings/passive.html
http://www.greenbuilder.com/sourcebook/PassSolGuide1-2.html

100
IT'S THE CARBON, STUPID
(February 12, 2006)

Have you noticed the flurry of efforts to reduce dependence on foreign oil? Here's a sampling from a week ago.

English drivers are mixing vegetable oil with their diesel, but the government is trying to stop them.

A Maryland chicken farm has devised a way to burn chicken manure to make steam for converting chicken parts into animal feed.

A bill before the Colorado legislature would require use of biodiesel to power state vehicles. A second bill calls for a study of the use of biomass to heat state buildings.

Bills in Washington State would give loans and tax breaks to the biofuel industry and require oil companies to add Washington-produced vegetable fuels.

A bill in the Kentucky General Assembly would require development of a strategy for production of liquified coal (artificial gasoline), ethanol, and biodiesel.

Ford is investing millions on a car that would run on soybeans.

A coalition of interests wants to show there's a demand for plug-in hybrid autos to reduce U.S. oil consumption.

The federal government is planning to award large grants for development of clean coal power. This is not an alternative, it's a plan to (among other things) bury CO_2 underground—about as useful as dumping nuclear waste in the ocean (been there, done that).

With the current high CO_2 level in the atmosphere, global air, water, and land temperatures are rising. We have now entered a positive feedback phase, where decreased albedo and release of greenhouse gases from land and water are causing global warming to accelerate. Soon this feedback process will reach the runaway phase where nothing we do can stop it. For all we know we may have reached this stage. We can expect extinctions to accelerate. For example, in the last few years fish and amphibian populations have plummeted.

What can be done to stop global warming, given all the constraints that apply, such as peak oil, population growth, tropical forest destruction, pollution, and species extinctions?

Since we don't know how much time is left, we must assume that immediate emergency action is imperative. What if I'm wrong? In that case, we will have averted catastrophe at an earlier stage, at less cost to us and the planet, and we can move to stable sustainability more quickly.

We must stop polluting and clean up existing pollution—this will slow extinctions, but it will take decades. We must restore habitats, especially forests. This will slow extinctions and provide a higher CO_2 elimination rate, but it will take several decades at best, centuries at worst.

We must stop overfishing. Cutting back on overfarming will conserve petroleum and permit more extensive reforestation. Instead of subsidizing farming, we can subsidize reforestation and truck farming. We can do this quickly.

We must bring the global human fertility rate down to less than 1. This will reduce global population by half roughly every 30–50 years after an initial lag of about half that time. Our goal should be to continue shrinking our population until recovery has advanced enough for us to calculate a sustainable world population size. I would guess that will fall between one-tenth (650 million) and one one-hundredth (65 million) of the current world population. If those estimates seem ridiculously low, keep in mind that our species began living unsustainably and degrading the ecosystem when our world population was even smaller. In any event, a program of rapid population reduction is badly needed, for the foreseeable future.

We must find means to raise albedo. I don't know how—paint the melting tundra and deserts white? Orbit giant reflective parasols?

What's wrong with all the initiatives in the news articles mentioned above is that too few will reduce carbon consumption by too little, too slowly, too expensively. We have to stop human CO_2 production entirely. We must stop fooling ourselves that biomass will help, because it still involves burning carbon. We can't use nuclear energy because we already have more nuclear waste than we know what to do with. Electric vehicles powered by carbon-derived electricity are no help. We must *stop burning carbon.* Everywhere. In ten years, starting with reduction by half right now through rationing.

We could do it if all the industrialized nations worked together with the developing nations. By cooperating and sharing sacrifices, we can eliminate the mindless competition among national economies currently driven by multinational corporations. ***Every dollar (and we're talking about trillions of dollars in the next decade or so) wasted on biomass, CO_2 sequestration, and nuclear, let alone fossil fuels, is a dollar that will not be spent on wind and solar power for years to come.*** We don't have years. We won't have a chance to correct the mistakes we're embarking upon.

The world still has the technology, the work force, and the wealth necessary to provide non-carbon power and fuel adequate for everyone, at a modest standard of living, if we abandon consumerism and conspicuous consumption, and redirect half our carbon and nuclear industries to fabrication and installation of solar and wind power today. We must end globalization and empire building, and switch from a growth economy to a recovering, shrinking one.

There's no hope that the greedy psychopaths running the world, especially the US, will voluntarily relinquish their control until too late for all of us. They must be deposed. No, I don't know how to do it, especially here, where our veneer of democracy has been stripped away.

This week's news:

A report by the U.S. Fish and Wildlife Service says, "There is insufficient habitat in South Florida to sustain a viable panther population. The prospects for population expansion into south-central Florida are questionable at this time." The report suggests moving some Florida panthers to wilderness areas elsewhere. Panthers once roamed throughout several southern states, but officials estimate there are about 80 panthers left in the wild, mostly in an area with some of the state's fastest development. State and federal agencies designated 600,000 acres of privately owned land as prime habitat for the panther in the 1990s. However, the same agencies have subsequently granted permits for a university, churches, roads, golf courses, and subdivisions within the habitat. The proposed plan is unworkable because it requires tripling the population in the current habitat, which is too small. The Fish and Wildlife Service agency repeatedly ignored its own biologists' advice, permitting development that wiped out panther habitat, more than once. Panther expert Andy Eller says his bosses feared angering political contributors who might need development permits. In 2004 he filed a complaint that the Fish and Wildlife Service was using flawed science and jeopardizing the panther. He was fired. Last year the agency admitted Eller was right and reinstated him—in a different state. They didn't overturn any permits they had approved over his objections. The original panther recovery plan described the inadequacy of current rules to protect panther habitat, but the wildlife service removed that part. (*St. Petersburg Times*)

Recommended reading:

Field, John G., et al. (eds.), *Oceans 2020: Science, Trends, and the Challenge of Sustainability*. Washington: Island Press, 2002.

101
A RATIONAL ENERGY ECONOMY
(February 19, 2006)

We've developed the logic for a rational energy economy that could be implemented in a decade if our government weren't in the pockets of the fossil and nuclear energy industries. Humans can virtually eliminate the greenhouse gas pollution that's killing our planet. No jobs would be lost, and the same energy fat cats could even remain very wealthy, but most of them see no need to switch. The current oil shortage is pouring our dollars into their pockets at an unprecedented rate. Fossil fuel and power companies' stocks are doing very well, thank you. They've made massive investments in fossil and nuclear energy, and they don't want those investments to go to waste. Only a few energy companies like BP and Shell are getting into renewable energy. However, several automobile manufacturers are working on hydrogen vehicles. Instead of devoting all their efforts to fuel cells, they should be required to use current technology to put internal combustion hydrogen vehicles on the road next year, and oil companies should be required to use current technology to provide hydrogen electrolysis (no, not hair removal) and fuelling stations.

Two ingredients are needed to break the power these corporations have over energy. The first is a means of storing energy to compensate for natural fluctuations in supply of sunlight and wind, the two largest potential sources of renewable energy. The second, more difficult, is the fortitude to fight the powerful corporations and our own government when they obstruct the changeover to rational energy.

Power can be stored in many ways. During periods of excess supply, when the sun is shining, the wind is blowing, and demand is low, excess electrical energy can be stored in batteries, as it is in off-grid solar homes. During periods of excess demand, battery power can be tapped and fed into the grid. But batteries are expensive. They wear out and our recycling system is pitifully inadequate to the task of keeping their toxic materials out of our food, water, and air.

We can also use excess electrical energy to power electric motors to pump water uphill, storing potential energy. During excess demand we can convert the potential to kinetic energy by letting the water flow back downhill, driving the electric motors backwards as generators, putting power back into the grid. This can be a relatively benign process if implemented properly.

A third method of storing energy is to use an element that will be integral to the rational energy economy anyway, namely hydrogen. During excess electrical energy supply, we can produce hydrogen by hydrolysis, releasing oxygen to the atmosphere. The stored hydrogen can be recombined with oxygen to produce electricity during periods of excess demand. Although stored hydrogen tends to leak, it's easy to recapture leaked hydrogen and generate electricity from it. Fuel cells are not essential for now. Coal and (especially) natural gas-fired power plants can be converted to hydrogen quickly, easily and inexpensively.

The transition to renewable energy, alternative-fuel vehicles, and a hydrogen economy is already in motion. Such changes usually occur only when economically advantageous, i.e. in Europe where fossil fuel energy is cost-prohibitive, and where people and governments are relatively enlightened. They aren't occurring in fossil-fuel rich nations, i.e., in the Middle East. In the US, where our tax dollars and military bullying subsidize fossil fuel costs, the people don't recognize the truly horrible hidden costs of fossil fuels and the government doesn't have enough integrity to quit propping up the fossil and nuclear energy business. The US is far behind Eurasia in switching to renewables, and the resulting economic disadvantage will really, really hurt soon.

Nevertheless, economic forces are making renewable energy more attractive. New technology is driving costs of renewable energy down while the cost of fossil and nuke energy climbs. People are catching on to the dangers of greenhouse gases and nuclear waste.

The changeover has begun, even in the US. Maybe economic forces will swing it, and the fossils and the nukes will just die off. Maybe the American people won't have to grow a backbone for a little longer. But don't count on it.

The notion that massively expensive societal endeavors only occur in response to economic pressures is largely a myth anyway. Nations have been galvanized into heroic endeavors by the First and Second World Wars, Vietnam, and the Iraq War. These were enormously expensive projects, each one costing more than the efforts recommended in this book. Yet many fortunes were made. Whereas the aggressions that sparked them were uncalled for and the conflicts were very destructive, the measures proposed here are both necessary and constructive.

This week's news:

Dr A.K.M. Mosharraf Hossain, an associate professor at Bangabandhu Sheikh Mujib Medical University, says asthma patients now comprise about 10% of residents of Dhaka city. "Dust covers the whole city due to unplanned road digging activities. Smoke emitted by automobiles and from factories mixes with fog and creates poisonous smog. Therefore responsible for causing asthma," Dr Mosharraf said. In a 1999 study on 5,642 Bangladeshi people the prevalence of asthma was 6.9%. The prevalence of asthma in children (5–14 years) was higher than in adults (15–44). (*The New Nation*, Bangladesh)

Recommended reading:

Markowitz, Gerald, and Rosner, David, *Deceit and Denial: The Deadly Politics of Industrial Pollution*. Berkeley, CA: University of California Press, 2002.

102

AN EMERGENCY NATIONAL POLICY

(February 26, 2006)

Let's review a few basics, which will be enough to shape an emergency energy policy:

- The population explosion requires that all nations take emergency measures to reduce their numbers.
- We reached peak domestic oil (after which domestic oil production has fallen behind domestic demand) around 1970
- We are near peak imported oil.
- The global supply of *relatively* clean-burning coal has almost peaked. However, even this coal is killing people and wildlife at unacceptable rates with mercury and acid rain.
- We have reached, or soon will reach, runaway global warming. We may have passed the time limit for remedial action.
- The capacity of the planet to store and eliminate greenhouse gases, especially CO_2, is declining and will decline faster due to deforestation and killing of aquatic ecosystems worldwide.
- The accelerating mass extinction is approaching global ecosystem collapse.
- We don't know how much time we need, and we don't know how much time we have.

This is sufficient rationale for immediate, emergency action. The US has to act unilaterally, because we are causing worse harm than any other nation. Other nations will have to follow suit quickly, and many will need our help.

To do otherwise would amount to the greatest crime against humanity the world has ever seen. In fact, our delay already amounts to such a crime.

A recovery mode of life will take generations to heal our planet, and it will involve much more than an energy policy because the global changes we need involve much more than energy. They will involve restructuring (not abandonment) of our economic system, the end of globalization, the end of corporate control of government, the end of consumerism, and the beginning of a truly kinder, gentler way of life.

If we survive to look back on this period, we may come to realize that the simultaneous population explosion, mass extinction, energy crisis, global warming, and the development of renewable energy technology were fortunate, because combined they provided a strong enough impetus and a means for change.

First, we have to stop the non-essential use of fossil fuels, in carefully planned stages, over a period of no more than ten years, starting with 50% reduction right now, by means of rationing. We need an immediate moratorium on nuclear energy, a tragic mistake that never should have happened. We've reached the end of our century-long love affair with cars, the end of long-distance trucking, and the end, for a while, of all air and most ocean travel. It will involve rationing fuels, cutting off imports and exports (ending globalization), and rapidly phasing out fossil-fuel power plants. Many Americans will have to move closer to their jobs, and many jobs will have to move as well, to cut down on travel and transportation costs.

We'll have to rebuild our railroads. We'll have to use the best available technology to retrofit all vehicles for hydrogen. It will require diverting a large portion of our capital, labor, and domestic oil and natural gas to generate electricity to run the factories that will make such retrofits possible, and to manufacture the solar cells, wind turbines, and infrastructure for our new renewable-electricity power grid. Before that new electricity supply is completed, we may have to go to rolling blackouts in order to stay on schedule for the fossil-fuel shutdown. At the same time, we will have to devote much of our scientific and technical expertise to energy conservation and research on the ecological effects of solar, wind and hydrogen technologies.

We will have to share our technology with other nations, giving them license-free access to patented and proprietary processes, and helping them develop the infrastructure to rebuild their societies. The health of our nation will depend on the health of theirs, because their peoples will not agree to sacrifices without a payoff, just as ours won't.

In order to cut national travel to the bone, we'll have to go local for the essentials of life: food, shelter, water, clothing, health care, education. An emphasis on locally-owned businesses will provide a much-needed multiplier effect for local economies and will break the grip of giant corporations on our democracy. Recycling and remanufacturing will be important components of local economies, lowering the energy cost of essential materials.

These aren't hippie-dippy pie-in-the-sky proposals. They are our only viable options. They will require a metamorphosis of our society unlike any other in world history.

This week's news:

MAPUTO: "Since the start of this year, we have received in our hospitals 1,506 cholera patients of whom three died," deputy director for health Martinho Dgedge says. Mozambique's ambassador to Portugal on Monday said the southern African nation was bracing for a cholera epidemic. "We expect that at any moment cholera will spread to the entire country, especially the areas worst-affected by the heavy rains," Ambassador Miguel Mkaima told a Lisbon news conference. At least 21 people have died since December because of heavy rains that have swept the entire country, where the rising water has affected tens of thousands of families.

Recommended reading:

Flannery, Tim F., *The Weather Makers: How Man is Changing the Climate and What it Means for Life on Earth*. New York: Atlantic Monthly Press, 2005.

103

WHAT KATRINA TAUGHT ME

(March 5, 2006)

Although I'm proposing the use of incompletely tested approaches and technologies, remember that we have to choose those that appear most likely to serve our needs (saving our necks). No method is ever "proved," any more than a scientific theory is ever proved, because new evidence can come along that invalidates it. This is clearly the case with fossil fuels and nuclear energy, both of which seemed reasonable once upon a time. The level of certainty that our fossil fuel use is causing catastrophic global warming is high enough to require that we find alternatives, test them as thoroughly as we can, and implement the least noxious. The longer we wait, the less time we'll have to test them.

That said, what can we learn from Katrina? First, a consensus is emerging that more extreme weather patterns are one consequence of the current global warming. So are rising sea levels. Central America, Florida and the Gulf Coast are particularly vulnerable to both, since they are so low-lying and in Hurricane Alley. Bangladesh is the only other big, highly populated continental area I can think of which is as obviously vulnerable to both. Much, maybe all, of the Gulf Coast and Florida probably will become too risky to inhabit in the next decade or two. To make matters worse, some of our oil drilling and most of our refining capacity are in that area, and population is exploding there faster than most other places in the U.S.

We should just get out. When a city like New Orleans or Biloxi is devastated, we shouldn't waste our rapidly dwindling resources rebuilding it, because we would have to rebuild it over and over, eventually so often that it will become clear that we never should have rebuilt in the first place. It's less expensive to move. Clean up the toxics and abandon the region to wildlife. Our wildlife deserves better, but it's a start.

How can we throw away the billions of dollars invested in the area? Where will the money come from to relocate a coastline full of cities, especially now that we're up to our nostrils in debt? The communities to which people are resettled will be foreign to them, with a different mix of needed skills and available jobs. What about the oil and the refineries? What about other vulnerable areas, like the eroding east coast and the earthquake-prone coast of California? A very large portion of our population, like everywhere else in the world, lives on our coasts.

When does anyone abandon a losing investment rather than continue to pour money into it? *When it becomes clear that there's no hope of recouping losses.*

For the moment, many of the demolished refineries and oil rigs in the Gulf will have to be restored. Our tax dollars recently handed to the oil companies can pay for it, but they shouldn't get a dollar more. Meanwhile, U.S. oil consumption has to be curtailed with rationing. This is preferable to raising gas taxes because it doesn't discriminate according to financial status. We do need to raise taxes, though: taxes on the profiteers who have looted us since we abandoned a balanced Federal budget and mounted our preemptive attack on those who challenged our oil supremacy.

The next catastrophe after Katrina should be the first opportunity to try disaster relocation on a small scale. Starting with the next disaster in a federally designated vulnerable zone (areas vulnerable to natural disasters), each affected resident can be given the choice of the usual federal aid to stay (call them "sessiles"), or the same amount plus 100% of real estate value, to leave (call them "mobiles"). The abandoned properties should be consolidated, cleaned up and used for wildlife reserves, and mobiles' portions of federal aid to the community (e.g., for infrastructure reconstruction) should go to the cities to which the mobiles move. If a mobile moves to a vulnerable zone, they get no aid, nor does the community to which they move.

The ratio of aid provided mobiles relative to sessiles should go up on an annual basis, so that it goes from 1:1 this year to 1:0 in five or ten years, after which relocation of disaster victims should be mandatory, eventually leaving only a skeleton population to manage cleanup and wildlife management. We will want to start closing our most vulnerable ports, as we shift from our decreasingly feasible import/export economy to a local one. As our fishing industry shrinks to a sustainable size, we'll have less need for fishing towns.

We need to pull our troops back from around the world, leaving abandoned bases minus military facilities for the host countries. Yes, this will have geopolitical repercussions. They won't be any worse than the present ones. We'll need to reverse the funding priorities for the Departments of State and Defense. Some of our forces should be assigned to UN peacekeeping forces, under UN supervision. The rest will be needed at home, as a civilian emergency corps to respond to emergencies and prepare for coming emergencies. We also need to release all prisoners who are not a danger to society, and put them, underpaid service workers and able welfare recipients to work, under appropriate supervision, with appropriate retraining, in good environments, building our new energy infrastructure. We can start by converting our prisons to solar and wind power equipment factories. Subsidize renewable energy, not fossil or nuclear energy. Subsidize transportation using electricity and hydrogen, not fossil fuels. Assist startup companies in essential new industries. Let existing energy and transportation companies make the transition or not, as they choose. Start recycling parts from the old-technology power plants, starting with the ones that are most damaging to the environment. Start recycling gas guzzling vehicles to make new transportation systems.

Yes, we will have to pursue many parallel transformations, including a drastic drop in transportation, and negative population growth. Sound crazy? Not as crazy as what we're doing now.

Will we do these things? Only if we all understand the compelling necessity and put shared interests over individual interests.

This week's news:

Graeme Pearman, a former CSIRO senior scientist and internationally recognized expert on climate change, claims he was encouraged to resign after he spoke out on global warming. He believes there's increasing pressure on researchers in Australia whose work or professional opinions aren't in accord with the Federal Government. A growing number of scientists are concerned about government interference with "intensely political" areas of research. Dr Pearman had joined the Australian Climate Group, which hopes to encourage Australian political leaders to adopt a target of reducing greenhouse gases 60% by 2050. The Federal Government recognizes the reality of global warming but doesn't support mandatory targets. Dr. Pearman said he was reprimanded for "making public expressions of what I believed were scientific views, on the basis that they were deemed to be political views. In 33 years (with CSIRO), I don't think I had ever felt I was political in that sense. I've worked with ministers and prime ministers from both parties over a long period of time, and in all cases I think I've tried to draw a line between fearless scientific advice about issues and actual policy development, which I think is in the realm of government," he said. Dr Pearman is one of three leading climate experts quoted on ABC's *Four Corners* who have been repeatedly censored. Dr. Pearman is one of a dozen senior climate change experts who have left in the last three years. He was also at odds with the CSIRO's emphasis on wealth-generating research, arguing "public good" science was being lost. "I don't think it is something that has been specific to (Australia). It's a sign of the times that governments seem to want to get on with the job of making decisions based on the ideology they have presented in their elections, and they are more reluctant to seek open and fearless advice from scientists, from economists, from the judiciary, from groups...(who) might not agree with their position." His views are similar to those of James Hansen, top climate change expert at NASA, who says the Bush Administration tried to censor him after he called for rapid reductions of greenhouse emissions. (*The Age*, Australia)

Recommended reading:

Foreman, Dave, *Rewilding North America: A Vision for Conservation in the 21st Century*. Washington: Island Press, 2004.

104

NATIONAL LEGISLATORS

(March 12, 2006)

Climate change, mass extinctions, and overpopulation are the most ignored political issues of our times. The national legislature is rotten with corrupt lawmakers who sacrifice the future for the large corporations who own them. We need to make the most sweeping changes in the history of our nation. It's time to become single-issue voters.

I can't review all the members of Congress. Due to the tight party discipline that has so polarized Congress and the people, I only need to consider the exceptions to a simple stereotype. In general, Republicans are anti-environment because they're pro-industry, and opposed to population control, the reproductive rights of women, sex education, and family planning—predominantly for religious reasons. The Republican majority in Congress means that all committee chairs are Republicans. A favorite tactic in committee hearings is to stack the witnesses so that unqualified "experts" paid by fossil-fuel industries to dismiss scientific evidence on global warming, CO_2 emissions, and endangered species as liberal propaganda outnumber legitimate scientists publishing relevant research in peer-reviewed journals. Peer review of a scientist's work is scoffed at as elitism or politics. In actuality, it's the mechanism that maintains high scientific standards. It's not perfect, but it reduces quackery.

Democrats usually oppose Republicans on all these issues, so you should usually vote Democrat if you agree with me that we can't afford to lose more precious time before reforming our energy economy and encouraging single-child families. I urge you to consult the Web site for the League of Conservation Voters (http://lcv.org/), which is where I get my information. Unfortunately, I can't find a rundown of voting records on population-related issues.

Here are the men and women who are *exceptions* to the rules above. This list is slightly out of date, as the column was written shortly before the 2004 election. The percentages represent environmental voting records (0 = no votes pro-environment, 100 = all votes pro-environment). My criterion for finding exceptions was records above 50% for Republicans or below 50% for Democrats, plus all Independents. Sorted by state (2nd column):

Senators:

Blanche Lincoln (DEM)	AR	32%
Mark Pryor (DEM)	AR	42%
John McCain (REP)	AZ	53%
Joseph Lieberman (DEM)	CT	42%
Zell Miller (DEM)	GA	0%
John Breaux (DEM)	LA	11%
Mary Landrieu (DEM)	LA	21%
Susan Collins (REP)	ME	68%
Olympia Snowe (REP)	ME	74%
Max Baucus (DEM)	MT	42%
John Edwards (DEM)	NC	37%
Byron Dorgan (DEM)	ND	47%
Ben Nelson (DEM)	NE	21%
Judd Gregg (REP)	NH	53%
Lincoln Chafee (REP)	RI	79%
James Jeffords (IND)	VT	89%

Representatives:

Robert Cramer (DEM)	AL-5	35%
Marion Berry (DEM)	AR-1	40%
Mike Ross (DEM)	AR-4	45%
Robert Simmons (REP)	CT-2	70%
Christopher Shays (REP)	CT-4	90%
Nancy Johnson (REP)	CT-5	70%
Michael Castle (REP)	DE-AL	70%
Sanford Bishop (DEM)	GA-2	35%
Jim Leach (REP)	IA-2	90%
Mark Kirk (REP)	IL-10	80%

Timothy Johnson (REP)	IL-15	75%
Ken Lucas (DEM)	KY-4	30%
William Jefferson (DEM)	LA-2	45%
Rodney Alexander (DEM)	LA-5	20%
Chris John (DEM)	LA-7	5%
Wayne Gilchrest (REP)	MD-1	55%
Vernon Ehlers (REP)	MI-3	55%
Jim Ramstad (REP)	MN-3	75%
Collin Peterson (DEM)	MN-7	20%
Richard Gephardt (DEM)	MO-3	5%
Rodney Frelinghuysen (REP)	NJ-11	55%
Frank LoBiondo (REP)	NJ-2	85%
Jim Saxton (REP)	NJ-3	75%
Christopher Smith (REP)	NJ-4	85%
Sue Kelly (REP)	NY-19	70%
Sherwood Boehlert (REP)	NY-24	65%
Brad Carson (DEM)	OK-2	45%
John Murtha (DEM)	PA-12	45%
John Tanner (DEM)	TN-8	45%
Max Sandlin (DEM)	TX-1	35%
Charles Stenholm (DEM)	TX-17	15%
Jim Turner (DEM)	TX-2	25%
Solomon Ortiz (DEM)	TX-27	35%
Ralph Hall (DEM)	TX-4	15%
Bernard Sanders (IND)	VT-AL	90%
Alan Mollohan (DEM)	WV-1	30%

Look to the Congressional leaders as the worst villains, followed closely by the chairs of committees charged with legislation affecting such vital areas as energy, the environment, health, and international relations. Vote the criminals out of office, and hold their successors' feet to the fire. You can have a significant impact if you become an activist. You can do several things, including volunteering for organizations that protect the environment, lobby lawmakers, campaign for the few good men and women who are nominated for public office, and get out the news that is so frequently buried by the corporate-run media. Or you can contribute money to these groups to help them get their job done.

This week's news:

MANILA: An avalanche of mud and boulders buried Guinsaugon in the eastern Philippines, once a 100-acre farming village of 2,500 people. "There are no signs of life, no rooftops, no nothing," Southern Leyte province Gov. Rosette Lerias said. 375 homes and a school were buried under mud up to 30 feet deep. Two other villages were affected, and about 3,000 evacuees were at a municipal hall. Lerias said many residents evacuated the week before, but returned home when the rains appeared to be abating. In 1991 6,000 were killed in floods and landslides triggered by a tropical storm. 133 died in 2003. (*L.A. Times*)

Bottled water is contributing to climate change. Britons consume more than two billion litres a year. The industry emits as much greenhouse gas as electricity for 20,000 homes, mostly for transport. Most water bottles are made from plastic, a crude-oil product. Most go to landfills, where they take about 450 years to break down, or are incinerated. Next year one company will use a biodegradable bottle made from corn, which will compost in 10 weeks. Environmental groups have urged consumers to return to tap water, which is 10,000 times cheaper, just as healthy and much less environmentally costly. A standard carbon tap filter costs about £35. The water-extraction facilities for Coca Cola's Dasani line in India have caused water shortages in more than 50 villages. (*The Times*)

Recommended reading:

Gelbspan, Ross, *Boiling Point: How Politicians, Big Oil and Coal, Journalists, and Activists are Fuelling the Climate Crisis—and What We Can Do to Avert Disaster*. New York: Basic Books, 2004.

CONCLUSION

What have we learned from our two-year journey together? We've learned some of the currentscientific thinking about questions that have preoccupied thinkers since thinking began:

- The origin of the universe, of life and of ourselves, and our place in this world.
- The evolution of species and our planet, and the intricate web of interdependencies among them.
- Extinctions and their causes, including the current mass extinction.
- Our cultural evolution and the roles of language, nomadism, sedentism, and agriculture in shaping our behavior.
- The evolution of mythology and religion to fit our changing circumstances.
- The evolution and organization of human societies, urbanization, writing, and civilization.
- Some of the relationships among all of these to the three horsemen of overpopulation, global warming, and the current mass extinction.

We've also lived through two years of history in a crucial period for human existence. We've witnessed the advance of the three horsemen, the growth of scientific understanding about their mechanisms and their effects, and the response of people and governments to them.

We've witnessed the death and ruin wrought by the three horsemen, and we've watched the efforts of intelligent, honest, and brave men and women to alert the human race to their ravages. Like a Greek tragedy, the course of events seems more and more to lead to an inexorable fate, in which we are damned for our hubris by forces far superior to our own. We're left to ponder the moral of this tragedy and to reflect on what the future will hold. We can't help asking some very fundamental questions about right and wrong, and the basis for our values.

We've also learned a good deal about what we're doing wrong and what we could be doing right. And we've learned to our dismay that the obstacles to our doing what's right may be insurmountable. What's worse, those obstacles are all in the realm of human behavior. *They should be avoidable.*

We see how foolish are many of the simplistic remedies put forth by our erstwhile leaders, such as burning biomass, depending on fuel cells, nuclear energy, the "green revolution" in agriculture, and reversing population decline. Successful plans for recovery can only come from an appreciation of the interacting physical, biological, psychological, and social phenomena. We're rapidly developing the knowledge we need on all fronts, but we're only beginning to discern the outlines of an integrated plan that will work for the world and the people in it. We do know, however, what many of the constraints of such a plan must be, including little-discussed issues of economics and human justice, which are outside the scope of this book. We'll begin to deal with those in the next volume.

We may be too late, but the only moral path is for us to do everything we can to minimize the disaster. We don't know how much time we have before global warming and mass extinctions hit the runaway stage. Therefore, we must take quick, strong action. It must be our top priority. Here are some ideas to chew on, some of which are undoubtedly all wet. They involve considerable sacrifice of many things we don't need, and a commitment to many changes that we need but that don't immediately bring perceptible rewards. My proposed timeline is based on my own perception of the fastest, strongest response of which we're capable. I don't know if it's too little, too late. If this turns out to be too much, too soon, we will have averted disasters earlier, more easily, with less lasting damage (e.g., fewer extinct species, less desertification).

Here's our action timeline:

- 10 years for elimination of carbon combustion.
- 20 years for atmospheric CO_2 reduction—as opposed to 150–200 years naturally—if we can develop the technology.
- 40 years for habitat restoration.
- 50 years for pollution cleanup.
- 100–250 years for population reduction to 1/10 (ca. 1700 AD level)–1/100 (ca. 7000 BC level) of the current 6.5 billion. We'll need to undershoot our eventual sustainable level, in order for the planet to heal.

With regard to global implementation:

- In keeping with rights of self-determination, each country must devise its own methods, with international cooperation and assistance.
- Globalization will wither as transportation costs rise. Put it out of its misery now, and benefit everyone.
- Each country will need to withdraw its global presence and provide incentives for nationals to return home. This will discourage large families, encourage autonomous economies, and reduce transportation energy expenditure.
- The UN, World Bank, International Monetary Fund, etc., must shift away from globalization toward international cooperation and assistance. The World Trade Organization is dying. Good riddance.

In order to eliminate carbon combustion:

- Rationing is more equitable than taxes. Ration all carbon fuels, cutting rations by half each year, go to zero after 10 years. Ration daily use of electricity accordingly.
- Replace carbon with distributed, renewable sources (e.g., solar, wind) for electricity.
- Generate hydrogen from water (not carbon compounds) using electricity.
- Upgrade the power grid.
- Use hydrogen for energy storage and as fuel for transport. Use existing internal combustion engines and carbon-burning power plants to burn hydrogen until fuel cells or batteries are competitive.

Transportation must be drastically curtailed:

- Move most people away from suburbs, rural areas, and areas prone to natural hazards such as floods, droughts, high winds, volcanoes, and earthquakes, into towns and cities. Gradually (5 years) outlaw non-business, private vehicles.
- Move farms and factories into and around the cities to minimize transportation.
- Go to local production of essentials (e.g., food, water, clothing, shelter, energy), local repair and conversion of equipment.
- Ration continental personal travel, use public trains, and boats.
- Eliminate intercontinental travel except for implementation of plan.

Renewable use of resources:

- Reduce fossil fuel extraction and mining to zero in 10 years.
- Reduce logging and fishing by half each year, undershooting until sustainable.
- Ban pollution, non-recyclable products.
- Practice:
    - ecoagriculture: no genetically modified organisms, fertilizers, herbicides, pesticides, or drugs; use multiculture, crop rotation, recycling of crop and animal wastes.
    - green architecture: solar space and water heating, solar and wind microgeneration, use of locally available materials, no concrete (manufacture of which produces CO_2).
    - green cities: non-polluting public transport, bicycle and pedestrian routes, urban reforestation, corridors for animals, zero waste to landfill.
    - green chemistry: use of manufacturing processes that minimize toxic wastes, recycle waste.

- Strengthen local autonomy, ban new development (by renovating, replacing existing structures, recycling building materials)

Rural land use:

- Habitat restoration: captive breeding and reintroduction of endangered species, regeneration of extinct species, reforestation, wildlife preserves, migration corridors, transplantation zones. Move whole ecosystems together to prevent their breakup.
- Electricity production: ecologically appropriate sites for wind, solar.
- Hydrogen production locations:
  - for energy storage: near converted fossil-fuel power plants.
  - for transport: at fueling stations.

Population reduction:

- *Universal public education*, including the subjects of this book.
- Sex education.
- National health service.
- Equal rights for women, including freedom from coerced procreation.
- Real social security: government pensions.
- Maximum one child per person, followed by vasectomy, tubal ligation.
- No immigration or emigration.

None of these programs can succeed without major changes of behavior at all levels of society, across the globe. Universal agreement and support (not unanimous, but large majority) will be necessary for success. It's the principal topic of the next two volumes of *Notes from a Dying Planet*, for 2006–2010. To whet your appetite, here are some preliminary thoughts:

- Structural changes:
  - one person, one vote:
    - no lobbying.
    - public funding and administration of campaigns; no private contributions or support, including candidates' funds.
    - recall petitions and votes at all levels.
  - corporations:
    - eliminate fiction of corporate "personhood."
    - eliminate all corporate privileges with regard to taxes, eminent domain, government subsidies, and legislative and regulatory special treatment.

- similarly eliminate all privileges for the rich and special interest groups.
- eliminate all out-of-country corporate activities.
- require corporate charters to include responsibility of the corporation, its officers, workers, and shareholders to all stakeholders, not just shareholders, with criminal penalties for all parties responsible in cases of criminal activity—including full financial reimbursement of any victims.
- in cases of corporate bankruptcy, require fulfillment of all legal obligations such as pensions, legal judgments, and settlements, with liability extending to corporation, current and past officers, workers, and shareholders.
- restore antitrust laws and government regulation of industry.
- taxes:
  - restore true graduated income tax, with no loopholes.
  - tax all unearned income more heavily than earned income.
- government (specifically U.S., although other governments will need to take many similar steps):
  - establish a cabinet-level agency to coordinate implementation of the plan.
  - implement full transparency, with no exceptions for national security.
  - restore a government agency to review all matters of fact and science presented to legislators and regulatory agencies for accuracy, and to review "expert's" credentials. This agency should adhere to peer review practices of relevant disciplines, relying on accepted expertise of entities in those disciplines.
  - include in all government agencies participation at all levels (including regulation and enforcement) by all stakeholders.
  - de-privatize all major public functions, including defense, prisons, and health insurance.
  - phase out all U.S. physical presence in other countries, and all foreign government physical presence in the US, except as needed to implement this plan.
  - phase out the military and its armaments, and the intelligence establishment, down to a level appropriate for defense of U.S. soil and for U.N. activities.
  - implement whistle blowing in all government bodies.

- separate the Department of Justice as a fourth branch of government.
- implement a national disaster response and relocation program.
- release all prisoners that pose no public risk, shift sentences to restitution and public works.
- provide a public works program for released prisoners, homeless and unemployed, with competitive salaries, to implement the plan.
- Nationalize utilities, including infrastructure for national power generation and transmission, communications, water, sewage, drainage, transportation, waterways, and public resource management.
- end the sale of public resources, including land, and the exploitation of those resources by private businesses.
- keep national patent protection and proprietary secrets to encourage research and development, but eliminate them internationally in order to disseminate technology globally.
- eliminate patents of natural products such as natural genomes and products.
- eliminate terminator seed technology and other restraints of trade and obstacles to use of heritage stocks.
- enact accuracy of expression laws throughout society, including (in alphabetical order) industry, news, politics, religion, and science. Statements that do not reflect established fact or current theory to be identified as unsupported or contrary opinion.

- Behavioral changes:
  - individuals:
    - concentrate on things best done by individuals, e.g., set an example, work for change.
    - keep informed, spread the word about the urgency of the dangers we face.
    - support individuals and groups, including our public servants, working to meet these challenges. Join them in their efforts.
    - take personal responsibility: reduce your "ecological footprint" by conserving, eliminating debt, buying only essentials, using public transportation, generating your own electricity (microgeneration), space heating, and hot water from solar and wind power, and recycling.
    - make no more than one child, adopt as many as you like and can afford.
    - participate in civic affairs, helping shape them toward sustainability.

- press others to do the same.
- don't be afraid of disapproval, be patient but firm with those who disagree.
- seek support groups or counseling if you have difficulty doing these things. They're hard.
- groups—set an example, work for change:
  - concentrate on things best done by groups, e.g., those that require more than one person.
  - shape the goals and behavior of your group to fit the plan.
  - endorse other groups that support the plan, pressure those that don't.
- polities:
  - concentrate on appropriate things.
  - outlaw pollutants, non-recyclables.
  - recycle, go for zero waste to landfill.
  - set policies that reinforce local vs. outside businesses, recycling, use of public transport, conservation, microgeneration.

A special note on mental illness: our society is fraught with horrific mental disorders. Indeed, it engenders them. Many street people are dysfunctional or marginally functional individuals who need our help. The American people are manipulated by government, Madison Avenue, and religious organizations to the point where most of us have become fearful, depressed, materialistic, criminally selfish, ignorant, exceptionalist, prejudiced, sociopathic, paranoid, or hopelessly dependent on authority. Examples:

- we're obsessed with terrorism out of all proportion (compare with other countries where terrorist acts are continual occurrences).
- antidepressants are among our most-prescribed medications.
- we go into major debt for things we don't need, such as SUVs and exorbitant homes.
- we willfully deprive others of necessities, for our own gain (inequitable taxes, no national health coverage).
- we know more about sports and entertainment than we do about science or current events, and we place our trust in people as ignorant as we are (George W. Bush).
- we apply different standards of international relations, justice, and morality to others than to ourselves (military and economic imperialism; denying human rights and due process to non-citizens, poor and minorities; sexual and reproductive double standards).

- prejudice against women, minorities, and other nationalities and religions suffuse every level of our lives.
- our government, corporate, and religious leaders are notoriously corrupt, and routinely defy the law and social standards. Power attracts sociopaths.
- we're prone to conspiracy theories, including those used by our leaders to maintain their hold on power. We distrust our government, our employers, and our fellow citizens.
- we slavishly suspend reason and accept the irrational guidance of ignorant, unprincipled parents, teachers, peers, and leaders, and we hold them up as role models.

If these don't constitute mental illness, I don't know what does. Once-sensible societal and personal norms have been distorted beyond reason, and we have to reestablish our standards through integrity and reason. These will all be major considerations in future volumes of *Notes from a Dying Planet*.

Like any good tragedy, our story has had its heroes and villains. Let's bring them back on stage to take a bow.

Here are the villains (hiss):

- George W. Bush, erstwhile president, and his corporate, government, and religious cronies, who have squandered a precious opportunity to lead the world toward sustainability, squandered the wealth we needed to do it, and quite possibly squandered our future. Here are a few names from this motley crew, alphabetized for your reference:
    - Senators Alexander, Allard, Allen, Baker, Bartlett, Bennett, Bond, Breaux, Brownback, Bunning, Burns, Campbell, Chambliss, Cochran, Coleman, Cornyn, Craig, Crapo, DeWine, Dole, Domenici, John Edwards, Ensign, Enzi, Fitzgerald, Frist, Graham, Grassley, Hagel, Hatch, Hutchison, Inhofe, Kerry, Kyl, Landrieu, Lieberman, Lott, Lugar, McConnell, Zell Miller, Murkowski, Ben Nelson, Roberts, Santorum, Sessions, Shelby, Specter, Stevens, Sununu, Talent, Thomas, Voinovich, and Warner.
    - Representatives Aderholt, Alden, R. Alexander, Akin, Bachus, Ballinger, Barrett, Barton, Beauxprez, Bereuter, Biggert, Bilirakis, Boehner, R. Bishop, S. Bishop, Blackburn, Blunt, Bonilla, Bonner, Bono, Boozman, Boyd, K. Brady, H. Brown, Brown-Waite, Burgess, Burns, Burr, Burton, Buyer, Cabot, Calvert, Camp, Cannon, Cantor, Capito, Carter, Chocola, Coble, M. Cole, Collins, Cox, Cramer, Crane, Crenshaw, Cubin, Culberson, Cunningham, Jo Ann Davis, T. Davis, Deal, L. DeLay, DeMint, L. Diaz-Balart, M. Diaz-Balart, Doolittle, Dreier, Duncan, Dunn, C. Edwards, Emerson, English, Everett, Feeney, Ferguson, Flake, Foley, Forbes, Franks,

Gallegly, Garrett, Gephart, Gibbons, Gilmor, Gingrey, Goode, Goodlatte, Goss, Granger, Graves, M. Green, Gutknecht, R. Hall, Harris, Hart, D. Hastings, Hayes, Hayworth, Hefly, Hensarling, Herger, Hobson, Hoekstra, Houghton, Hostettler, Hulshof, Hunter, Hyde, Isakson, Issa, Istook, Jenkins, John, Sam Johnson, W. Jones, Keller, M. Kennedy, P. King, S. King, Kingston, Kline, Knollenberg, Kolbe, LaHood, LaTourette, Latham, Jerry Lewis, R. Lewis, Linder, F. Lucas, Manzullo, McCotter, McCrery, McHugh, McInnis, McKeon, Mica, Gary Miller, C. Miller, J. Miller, Mollohan, Jerry Moran, Murphy, Musgrave, Myrick, Nethercutt, Neugebauer, Ney, Northup, Norwood, Nunes, Nussle, Ortiz, Osborne, Ose, Otter, Oxley, Paul, Pearce, Pence, C. Peterson, J. Peterson, Pickering, Pitts, Platts, Pombo, Porter, Portman, D. Pryce, Putnam, Quinn, Radanovich, Regula, Rehberg, Renzi, Reynolds, H. Rogers, Michael Rogers, Rohrabacher, Ros-Lehtinen, P. Ryan, J. Ryun, Sandlin, Schrocke, Sensenbrenner, P. Sessions, Shadegg, Shaw, Sherwood, Shimkus, Bill Shuster, Simpson, N. Smith, Stearns, Stenholm, Sullivan, Sweeney, Tancredo, Tauzin, C. Taylor, Terry, Thomas, Thornberry, Tiberi, Tiyahrt, Toomey, J. Turner, M. Turner, Upton, Vitter, Walden, Walsh, Wamp, C. Weldon, Weller, Whitfield, Wicker, H. Wilson, J. Wilson, and B. Young.

- Other national leaders, who have done far too little to educate our citizens and bowed instead to corporate and religious pressure.
- Many religions and their leaders, most notably the Catholic Church and the Pope, for denying overpopulation and opposing population reduction though family planning including abstinence, contraception, sterilization and abortion. Unholy preachers include Jerry Falwell, Oral Roberts, Pat Robertson, and all the ministers and priests who have used their pulpits to deny the onslaught of the three horsemen and to oppose those who would fight them.
- Corporations, especially the fossil fuel industry, especially ExxonMobil, who have spread misinformation and manipulated governments for their own greedy purposes.
- Professional deniers: Richard Balling, Michael Crichton, John Christy, Sherwood and Keith Idso, Richard Lindzen, Pat Michaels, William O'Keefe, Donald Pearlman, and S. Fred Singer.

Here are the heroes (applaud):

- The scientists and other scholars who have met the three horsemen face to face, documented their progress, and advised us how to fight them:
  - The biologists, climate scientists, and other front-line researchers.
  - The historians, anthropologists, sociologists, and students of religion and the arts who provide us with key insights into human behavior.

- The writers whose books I've recommended.
- The journalists whose investigative reporting has uncovered the dirty deals and the conspiracies.
- Environmentalists, our most unsung heroes.
- Civic leaders, especially Al Gore, who are sounding the alarm and showing us how to get from here to sustainability.
- A few religions, most notably the Anglican Church, and their leaders, for sounding the alarm and calling for action.
- Enlightened businesses, including some of the major oil companies, for recognizing the three horsemen and starting to go green.
- Ordinary people like you and me who are taking individual and collective action. Stand firm.

REFERENCES

Abernethy, Virginia, *Population Politics*. New Brunswick, NJ: Transaction Publishers, 2000.

Arendt, Hannah, *On Revolution*. Westport, CT: Greenwood Press, 1982.

Armstrong, Karen, *A Short History of Myth*. Edinburgh: Canongate, 2005.

Barney, Gerald O., et al., *The Global 2000 Report to the President of the United States. Entering the 21st Century: A Report*. New York: Pergamon Press, 1980.

Bent, Robert, et al., *Energy: Science, Policy, and the Pursuit of Sustainability*. Washington: Island Press, 2002.

Blumrosen, Alfred W., and Blumrosen, Ruth G., *Slave Nation: How Slavery United the Colonies and Sparked the American Revolution*. Naperville, IL: Sourcebooks, 2005.

Board on Sustainable Development, *Our Common Journey: A Transition toward Sustainability*. Washington, DC: National Academy Press, 1999.

Bogucki, Peter, *The Origins of Human Society*. Malden, MA: Blackwell Publishers, 1999.

Boulter, Michael Charles, *Extinction: Evolution and the End of Man*. New York: Columbia University Press, 2002.

Brewer, Richard, *Principles of Ecology*. Philadelphia: Saunders, 1979.

Calvin, William H., and Ojemann, George A., *Conversations with Neil's Brain: The Neural Nature of Thought and Language*. Reading, MA: Addison-Wesley Pub. Co., 1994.

Campbell, Joseph, et al., *The Power of Myth*. New York: Doubleday, 1988.

Cavalli-Sforza, Luigi Luca, *Genes, Peoples, and Languages*. New York: North Point Press, 2000.

Chance, Michael R. A., and Jolly, Clifford J., *Social Groups of Monkeys, Apes and Men*. New York: Dutton, 1970.

Clark, Peter, *Zoroastrianism: An Introduction to an Ancient Faith*. Brighton: Sussex Academic Press, 1998.

Daly, Herman E., and Townsend, Kenneth N., *Valuing the Earth: Economics, Ecology, Ethics*. Cambridge, MA: MIT Press, 1993.

Damasio, Antonio R., *Descartes' Error: Emotion, Reason and the Human Brain*. New York: Vintage, 1995.

Dawkins, Richard, *The Blind Watchmaker: Why the Evidence of Evolution Reveals a Universe Without Design*. New York: Norton, 1986.

Dawkins, Richard, *The Selfish Gene*. Oxford: Oxford University Press, 1989.

de Rivero, B. Oswaldo, *The Myth of Development: The Non-Viable Economies of the 21st Century*. London: Zed Books, 2001.

Department of Economic and Social Affairs, Population Division, *Review and Appraisal of the Progress Made in Achieving the Goals and Objectives of the Programme of Action of the International Conference on Population and Development: The 2004 Report*. New York: United Nations, 2004.

Diamond, Jared M., *The Third Chimpanzee: the Evolution and Future of the Human Animal*. New York: Harper Collins, 1992.

Diamond, Jared M., *Guns, Germs, and Steel: The Fates of Human Societies*. New York: W. W. Norton & Co., 1997.

Diamond, Jared M., *Collapse: How Societies Choose to Fail or Succeed*. New York: Viking, 2005.

Dissanayake, Ellen, *Homo Aestheticus: Where Art Comes from and Why*. New York: Maxwell Macmillan International, 1992.

Dorsen, Norman, "Rights in Theory, Rights in Practice," in Berlowitz, Leslie (ed.), *America in Theory*. New York: Oxford University Press, 1988.

Durham, William H., *Coevolution: Genes, Culture, and Human Diversity*. Stanford, CA: Stanford University Press, 1991.

Earle, Timothy K., *How Chiefs Come to Power*. Stanford, CA: Stanford University Press, 1997.

Edey, Maitland A., and Johanson, Donald C., *Blueprints: Solving the Mystery of Evolution*. Boston: Little, Brown, 1989.

Editors of E/The Environmental Magazine, *Green Living: The E Magazine Handbook for Living Lightly on the Earth*. New York: Plume, 2005.

Ehrlich, Paul R., *The Population Bomb*. New York: Ballantine Books, 1968.

Ehrlich, Paul R., *Human Natures: Genes, Cultures, and the Human Prospect*. Washington, DC: Island Press, 2002.

Ehrlich, Paul, and Ehrlich, Anne, *Extinction: the Causes and Consequences of the Disappearance of Species*. New York: Random House, 1981.

Ehrlich, Paul, and Ehrlich, Anne, *One with Nineveh: Politics, Consumption, and the Human Future*. Washington: Island Press, 2004.

Eldredge, Niles, *The Miner's Canary: Unraveling the Mysteries of Extinction*. Princeton: Princeton University Press, 1991.

Eliade, Mircea, *Patterns in Comparative Religion*. Lincoln: University of Nebraska Press, 1958.

Fagan, Brian M., *The Long Summer: How Climate Changed Civilization*. London: Granta Books, 2004.

Feldt, Gloria, *The War on Choice: The Right-Wing Attack on Women's Rights and How to Fight Back*. New York: Bantam Books, 2004.

Field, John G., et al. (eds.), *Oceans 2020: Science, Trends, and the Challenge of Sustainability.* Washington: Island Press, 2002.

Finkelstein, Israel, and Silberman, Neil Asher, *The Bible Unearthed: Archaeology's New Vision of Ancient Israel and the Origin of its Sacred Texts.* New York: Free Press, 2001.

Flanders, Laura, (ed.), *The W effect: Bush's War on Women.* New York: Feminist Press at City University of New York, 2004.

Flannery, Tim F., *The Weather Makers: How Man is Changing the Climate and What it Means for Life on Earth.* New York: Atlantic Monthly Press, 2005.

Foreman, Dave, *Rewilding North America: A Vision for Conservation in the 21st Century.* Washington: Island Press, 2004.

Frank, Thomas, *What's the Matter with Kansas? How Conservatives Won the Heart of America.* New York: Metropolitan Books, 2004.

Gardner, Howard, *Art, Mind and Brain: a Cognitive Approach to Creativity.* New York: Basic Books, 1982.

Gelbspan, Ross, *The Heat is on: The Climate Crisis, the Cover-up, the Prescription.* Reading, MA: Perseus Books, 1998.

Gelbspan, Ross, *Boiling Point: How Politicians, Big Oil and Coal, Journalists, and Activists are Fuelling the Climate Crisis—and What We Can Do to Avert Disaster.* New York: Basic Books, 2004.

Geller, Howard S., *Energy Revolution: Policies for a Sustainable Future.* Washington: Island Press, 2003.

Goodland, Robert (ed.), *Race to Save the Tropics: Ecology and Economics for a Sustainable Future.* Washington, DC: Island Press, 1990.

Graedel, Thomas E., and Crutzen, Paul J., *Atmosphere, Climate and Change.* New York: W. H. Freeman, 1997.

Grant, Lindsey, *Juggernaut: Growth on a Finite Planet.* Santa Ana, CA: Seven Locks Press, 1996.

Greider, William, *The Soul of Capitalism: Opening Paths to a Moral Economy.* New York: Simon and Schuster, 2003.

Harkavy, Oscar, *Curbing Population Growth: An Insider's Perspective on the Population Movement.* New York: Plenum Press, 1995.

Harman, Gilbert, *Change in View: Principles of Reasoning.* Cambridge, MA: MIT Press, 1986.

Hartmann, Thom, et al., *The Last Hours of Ancient Sunlight: The Fate of the World and What We Can Do Before It's Too Late.* New York: Three Rivers Press, 2004.

Harvey, Danny, *Global Warming: The Hard Science.* Upper Saddle River, NJ: Prentice Hall, 2000.

Hawking, Stephen W., *A Brief History of Time.* New York: Bantam Books, 1998.

Heintzman, Andrew, and Solomon, Evan, *Fueling the Future: How the Battle over Energy is Changing Everything*. Toronto: House of Anansi Press, 2003.

Hollingsworth, William G., *Ending the Explosion: Population Policies and Ethics for a Humane Future*. Santa Ana, CA: Seven Locks Press, 1996.

Johnson, Chalmers, *The Sorrows of Empire: Militarism, Secrecy, and the End of the Republic*. New York: Henry Holt, 2004.

Kagin, Edwin, "The Gathering Storm," in Blaker, Kimberley, et al., *The Fundamentals of Extremism: The Christian Right in America*. New Boston, MI: New Boston Books, 2003.

Katz, Leonard D., *Evolutionary Origins of Morality: Cross-Disciplinary Perspectives*. Bowling Green, OH: Imprint Academic, 2000.

Kennedy, Robert Francis, *Crimes against Nature: How George W. Bush and his Corporate Pals are Plundering the Country and High-Jacking our Democracy*. New York: Harper Collins, 2004.

Kirschner, Mark W., and Gerhart, John C., *The Plausibility of Life: Great Leaps of Evolution*. New Haven, CT: Yale University Press, 2005.

Klare, Michael T., *Resource Wars: The New Landscape of Global Conflict*. New York: Henry Holt, 2001.

Klein, Richard G., and Edgar, Blake, *The Dawn of Human Culture*. New York: Wiley, 2002.

Kleveman, Lutz, *The New Great Game: Blood and Oil in Central Asia*. (2003).

Knight, Chris, et al. (eds.), *The Evolutionary Emergence of Language: Social Function and the Origins of Linguistic Form*. Cambridge: Cambridge University Press, 2000.

Leakey, Richard, and Lewin, Roger, *The Sixth Extinction: Patterns of Life and the Future of Humankind*. New York: Doubleday, 1995.

Leisinger, Klaus M., et al., *Six Billion and Counting: Population Growth and Food Security in the 21st Century*. Washington, DC: International Food Policy Research Institute, 2002.

Lewin, Roger, *The Origin of Modern Humans*. New York: W. H. Freeman, 1993.

Lewis-Williams, J. David, *The Mind in the Cave: Consciousness and the Origins of Art*. New York: Thames and Hudson, 2002.

Lieberman, Philip, *Uniquely Human: The Evolution of Speech, Thought, and Selfless Behavior*. Cambridge, MA: Harvard University Press, 1991.

Lovejoy, Arthur O., *The Great Chain of Being: A Study of the History of an Idea. The William James Lectures Delivered at Harvard University, 1933*. Cambridge, MA: Harvard University Press, 1936.

Luker, Kristin, *Abortion and the Politics of Motherhood*. Berkeley: University of California Press, 1984.

Manning, Richard, *Against the Grain: How Agriculture has Hijacked Civilization*. New York: North Point Press, 2004.

Markowitz, Gerald, and Rosner, David, *Deceit and Denial: The Deadly Politics of Industrial Pollution*. Berkeley, CA: University of California Press, 2002.

Mayr, Ernst, *What Evolution Is*. New York: Basic Books, 2001.

Mayr, Ernst, and Provine, William B. (eds.), *The Evolutionary Synthesis: Perspectives on the Unification of Biology*. Cambridge, MA: Harvard University Press, 1980.

Mazur, Laurie Ann (ed.), *Beyond the Numbers: a Reader on Population, Consumption, and the Environment*. Washington, DC: Island Press, 1994.

McKie, Robin, *Dawn of Man: The story of Human Evolution*. New York: Dorling Kindersley Pub., 2000.

McQuaig, Linda, *It's the Crude, Dude: War, Big Oil and the Fight for the Planet*. Toronto: Doubleday Canada, 2004.

Micklethwait, John, and Wooldridge, Adrian, *The Right Nation: Conservative Power in America*. New York: Penguin Press, 2004.

Mooney, Chris, *The Republican War on Science*. New York: Basic Books, 2005.

Patterson, Colin, *Evolution*. Ithaca, N.Y.: Comstock Pub. Associates, 1999.

Pfeiffer, John E., *The Creative Explosion: An Inquiry into the Origins of Art and Religion*. Ithaca, NY: Cornell University Press, 1985.

Phillips, Kevin, *Wealth and Democracy: A Political History of the American Rich*. New York: Broadway Books, 2002.

Pieper, Adi, *The Easy Guide to Solar Electric, Part II: Installation Manual*. Santa Fe, NM: ADI Solar, 1999.

Pinker, Steven, *The Language Instinct*. New York: Harper Perennial, 1995.

Reid, Stephen J., *Ozone and Climate Change: A Beginner's Guide*. Amsterdam: Gordon and Breach Science Publishers, 2000.

Rifkin, Jeremy, *The Hydrogen Economy: The Creation of the Worldwide Energy Web and the Redistribution of Power on Earth*. New York: J. P. Tarcher/Putnam, 2002.

Roberts, Paul, *The End of Oil: On the Edge of a Perilous New World*. Boston: Houghton Mifflin, 2004.

Rogers, John J. W., and Feiss, P. Geoffrey, *People and the Earth: Basic Issues in the Sustainability of Resources and Environment*. Cambridge: Cambridge Univ. Press, 1998.

Rogers, John J. W., and Santosh, M., *Continents and Supercontinents*. New York: Oxford University Press, 2004.

Rudgley, Richard, *The Lost Civilizations of the Stone Age*. New York: Free Press, 1999.

Sagan, Carl, *The Demon-Haunted World: Science as a Candle in the Dark*. New York: Random House, 1996.

Sargant, William Walters, *Battle for the Mind: A Physiology of Conversion and Brain-Washing.* Cambridge, MA: Malor Books, 1997.

Schor, Juliet B., and Taylor, Betsy (eds.), *Sustainable Planet: Solutions for the Twenty-First Century.* Boston: Beacon Press, 2002.

Schwartz, Lewis M., *Arguing about Abortion.* Belmont, CA: Wadsworth Pub. Co., 1993.

Settegast, Mary, *Plato Prehistorian: 10,000 to 5000 B.C. in Myth and Archaeology.* Cambridge, MA: Rotenberg Press, 1987.

Shuman, Michael H., *Going Local: Creating Self-Reliant Communities in a Global Age.* New York: Routledge, 2000.

Smith, Bruce D., *The Emergence of Agriculture.* New York: W. H. Freeman, 1995.

Stearns, Beverley Patterson, and Stearns, Stephen C., *Watching, from the Edge of Extinction.* New Haven, CT: Yale University Press, 1999.

Stiglitz, Joseph E., *Globalization and its Discontents.* New York: W. W. Norton, 2002.

Tainter, Joseph A., *The Collapse of Complex Societies.* Cambridge: Cambridge University Press, 1988.

Tennesen, Michael, *The Complete Idiot's Guide to Global Warming.* Indianapolis: Alpha, 2004.

Tribe, Laurence H., *Abortion: the Clash of Absolutes.* New York: Norton, 1990.

Tudge, Colin, *The Time Before History: 5 Million Years of Human Impact.* New York: Scribner, 1996.

Tudge, Colin, *Neanderthals, Bandits, and Farmers: How Agriculture Really Began.* New Haven: Yale University Press, 1999.

United Nations Human Settlements Programme, *The Challenge of Slums: Global Report on Human Settlements.* Sterling, VA: Earthscan Publications, 2003.

Ward, Peter D., *The End of Evolution: on Mass Extinctions and the Preservation of Biodiversity.* New York: Bantam Books, 1994.

Ward, Diane Raines, *Water Wars: Drought, Flood, Folly, and the Politics of Thirst.* New York: Riverhead Books, 2002.

Weiner, Jonathan, *The Next One Hundred Years: Shaping the Fate of our Living Earth.* New York: Bantam Books, 1990.

Wilson, David B. (ed.), *Did the Devil Make Darwin Do It? Modern Perspectives on the Creation-Evolution Controversy.* Ames: Iowa State University Press, 1983.

Wilson, Edward O., *Consilience: The Unity of Knowledge.* New York: Knopf, 1998.

Wilson, Edward O., *The Diversity of Life.* New York: W. W. Norton, 1999.

Wilson, Edward O., *The Future of Life.* New York: Knopf, 2002.

Wilson Edward O., *On Human Nature.* Cambridge, MA: Harvard University Press, 1978.

ABOUT THE AUTHOR

Paul Brown, a professor of physiology at West Virginia University, was educated at MIT, the University of Chicago, and Cornell University. He has published three books and has had more than sixty peer-reviewed papers in scientific journals.

INDEX

abortion 6, 43, 55, 88, 99, 101, 102, 126, 138, 142, 165, 166, 168, 269, 277, 278, 279, 281, 288, 321, 326, 328

acid 14, 44, 47, 53, 55, 172, 179, 202, 271, 303

aerosol 70, 204

Afghanistan 7, 36, 62, 138, 173, 174, 188, 227

Africa 18, 27, 34, 45, 46, 48, 55, 56, 57, 58, 62, 69, 70, 77, 84, 90, 93, 95, 107, 119, 131, 133, 139, 142, 149, 150, 151, 165, 167, 190, 193, 194, 200, 205, 207, 214, 217, 219, 221, 225, 236, 254, 256, 257, 262, 270, 278

afterlife 82, 85, 239

age 57, 58, 73, 74, 82, 83, 86, 91, 96, 101, 104, 106, 108, 122, 130, 133, 136, 137, 140, 141, 147, 148, 150, 157, 182, 214, 217, 224, 228, 251, 254, 256, 257, 264, 270, 275, 279, 284, 288, 307, 327

agriculture 42, 52, 56, 90, 91, 93, 94, 95, 96, 97, 99, 101, 103, 104, 107, 109, 135, 167, 168, 175, 176, 177, 180, 194, 199, 200, 201, 203, 208, 244, 248, 251, 255, 273, 278, 313, 326

aid 49, 70, 102, 104, 106, 131, 150, 155, 174, 180, 220, 227, 257, 286, 292, 306

AIDS 25, 27, 48, 166, 205, 221, 257

airline 45, 138, 194

albedo 128, 160, 204, 206, 227, 232, 298, 299

allele 47, 60, 61, 122, 123

altruism 115

Amazon 35, 36, 98, 110, 142, 225, 235, 236, 262, 280

amphibian 156, 298

anarchy 25, 27, 145, 275

Andes 220

Antarctic 16, 26, 79, 94, 129, 154, 296

anthropology 79

aquifers 11, 187

archeology 79

Arctic 12, 26, 38, 62, 96, 99, 102, 111, 148, 161, 162, 164, 189, 190, 217, 232, 233, 242, 246, 248

art 58, 69, 77, 80, 81, 82, 83, 84, 85, 87, 88, 96, 106, 108, 109, 111, 254, 324, 325, 326, 327

Asia 16, 27, 62, 65, 70, 71, 90, 93, 95, 99, 131, 133, 139, 147, 150, 151, 159, 165, 167, 182, 190, 194, 200, 201, 205, 207, 224, 225, 236, 256, 292, 326

asthma 14, 20, 27, 262, 302

atmosphere 4, 7, 13, 14, 16, 42, 52, 54, 70, 91, 106, 116, 142, 146, 148, 191, 199, 200, 202, 204, 206, 209, 214, 216, 220, 241, 243, 244, 247, 249, 274, 298, 301, 325

Australia 34, 90, 93, 180, 201, 219, 224, 290, 307

Australopithecine 56

authority 37, 80, 100, 101, 105, 106, 112, 153, 159, 160, 273, 319

automobile 14, 80, 138, 194, 301
avalanche 171, 311

Babylon 93
bacteria 7, 30, 44, 52, 123, 190
balance of trade 137, 148, 279, 280
band 81, 87, 100, 103, 104
behavior i, xvi, 33, 34, 35, 37, 47, 50, 54, 57, 63, 64, 69, 70, 73, 75, 79, 81, 85, 103, 105, 111, 115, 119, 150, 156, 171, 181, 182, 239, 254, 256, 262, 273, 287, 313, 316, 319, 321, 326
Bible 259, 260, 325
Big Bang 41, 42, 69
biodiesel 179, 201, 298
biodiversity xvi, 8, 9, 20, 48, 59, 119, 120, 124, 139, 167, 176, 214, 248, 271, 281
biological i, xvi, 50, 54, 63, 74, 78, 79, 81, 109, 111, 119, 135, 147, 161, 176, 216, 217, 226, 248, 250, 264, 276, 313
biosphere 4, 7, 33, 171, 191, 279, 283
bipedalism 58
birds 20, 50, 54, 57, 72, 98, 108, 118, 119, 139, 200, 201, 224, 229, 235, 252, 285, 292, 296
brain xv, xvi, 7, 38, 57, 58, 61, 69, 70, 72, 76, 77, 84, 86, 165, 323, 325, 328
Bronze Age 100, 101, 108
bushmen 87

California 7, 17, 42, 62, 82, 96, 99, 118, 135, 168, 178, 183, 195, 201, 207, 222, 228, 230, 231, 236, 240, 271, 296, 302, 305, 326, 327
Cambrian 216

cancer 7, 15, 19, 20, 91, 106, 203, 271
carbon 3, 4, 7, 12, 14, 16, 19, 20, 27, 38, 42, 45, 49, 54, 55, 73, 79, 82, 86, 90, 115, 128, 146, 148, 163, 171, 173, 176, 179, 180, 183, 189, 191, 192, 202, 203, 205, 206, 208, 214, 232, 233, 241, 244, 245, 248, 249, 250, 258, 262, 272, 280, 298, 299, 311, 314, 315
carbon dioxide 3, 4, 7, 14, 16, 20, 27, 38, 45, 49, 54, 55, 73, 128, 146, 148, 163, 189, 205, 214, 232, 241, 244, 245, 248, 258, 262, 272
carbon sequestration 250
caribou 252
carrying capacity 5, 8, 121, 123, 127
caste 256, 261
Catholic 99, 133, 227, 288, 321
cave 69, 81, 82, 83, 84, 85, 95, 107, 254, 257, 326
censor 307
certainty xv, 37, 38, 106, 162, 245, 247, 276, 305
chief i, 86, 103, 105, 120, 157, 161, 162, 164, 260, 265
chiefdom 103, 104, 260
China 16, 20, 34, 36, 42, 43, 55, 62, 64, 65, 70, 73, 77, 93, 98, 130, 131, 133, 142, 143, 149, 153, 165, 168, 200, 201, 220, 236, 249, 254, 271, 278, 290
chlorofluorocarbons 15
Christianity 112
cities 5, 8, 9, 20, 27, 48, 97, 107, 109, 119, 184, 185, 187, 188, 193, 203, 220, 227, 235, 275, 305, 306, 315

civilization ii, 95, 99, 109, 200, 225, 275, 313, 324, 326

climate i, xvii, 3, 7, 12, 13, 14, 16, 18, 21, 29, 37, 42, 45, 46, 47, 49, 51, 56, 57, 59, 61, 64, 70, 73, 75, 79, 82, 85, 86, 90, 91, 96, 97, 98, 99, 102, 104, 106, 107, 108, 111, 120, 124, 129, 133, 135, 143, 146, 149, 150, 157, 160, 161, 162, 163, 164, 166, 167, 172, 173, 180, 185, 189, 190, 192, 196, 197, 199, 200, 204, 205, 206, 208, 213, 214, 215, 216, 219, 220, 221, 222, 223, 224, 225, 234, 235, 240, 241, 242, 243, 244, 245, 246, 247, 248, 249, 250, 252, 253, 262, 263, 265, 270, 271, 272, 276, 281, 283, 284, 285, 290, 292, 296, 304, 307, 308, 311, 321, 324, 325, 327

coal 8, 16, 17, 25, 62, 73, 92, 120, 134, 135, 142, 153, 168, 171, 172, 177, 179, 180, 181, 182, 192, 194, 202, 203, 204, 205, 249, 250, 264, 286, 298, 302, 303, 311, 325

coast 96, 99, 104, 187, 207, 215, 222, 231, 280, 296, 305

coerced procreation 288, 316

cognition 78

collapse i, 3, 8, 28, 42, 52, 79, 104, 110, 119, 120, 126, 127, 134, 156, 157, 181, 182, 203, 226, 272, 296, 303, 324

colonialism 89

commons 49, 161, 218

confusion 239

conservation i, 38, 49, 61, 75, 98, 145, 154, 167, 178, 185, 186, 194, 217, 274, 281, 285, 293, 304, 307, 308, 319, 325

consumer 29, 77, 159, 178, 185, 249, 282

continent 14, 108, 217, 219, 221, 227

contraception 18, 43, 127, 130, 150, 152, 165, 166, 225, 269, 277, 278, 279, 281, 288, 321

coral 26, 142, 156, 296

corporations 25, 26, 29, 36, 138, 148, 151, 155, 159, 167, 181, 195, 201, 269, 288, 299, 301, 304, 308, 316, 321

corruption 89, 148, 273, 279, 288

creationism 52, 112, 264, 276

creole 75

critical period 61, 72, 74, 76, 285

Cro-Magnon 95

crops 10, 14, 52, 98, 127, 166, 167, 168, 176, 177, 179, 199, 203, 263, 278

crowding 8, 9, 90, 110, 119, 121, 254

cultural 49, 50, 60, 61, 63, 64, 67, 70, 72, 74, 78, 79, 81, 82, 83, 86, 93, 94, 95, 109, 110, 111, 119, 224, 256, 292, 313

culture 29, 48, 50, 69, 71, 72, 74, 78, 79, 81, 94, 107, 108, 112, 136, 139, 165, 240, 261, 324, 326

dam 12, 180, 253

Dark Ages 144, 177

dead zone 9, 20, 62, 206, 215, 279

defense 55, 81, 88, 100, 103, 108, 117, 155, 161, 174, 195, 227, 306, 317

deforestation 48, 49, 80, 110, 127, 142, 149, 176, 182, 199, 208, 209, 235, 236, 262, 278, 280, 285, 303

democracy 16, 27, 29, 52, 104, 139, 196, 227, 269, 279, 299, 304, 326, 327 demographic transition theory 133, 144, 152

demographic trap 144, 145, 147

denial 33, 49, 179, 205, 270, 276, 286, 302, 327

Denmark 51, 62, 89, 100, 101, 203, 280

desert 9, 51, 87, 88, 127, 135, 172, 177, 199, 206, 220

desertification 46, 99, 149, 199, 285, 314

deterministic 52

discrimination 8, 60, 147, 278, 279, 288

disease 11, 18, 28, 48, 57, 59, 97, 98, 104, 106, 141, 142, 145, 147, 150, 151, 153, 156, 165, 179, 190, 224, 225, 252, 262, 281

diversity 4, 47, 73, 112, 120, 123, 135, 207, 213, 216, 217, 220, 226, 283, 285, 324

DNA 44, 47, 54, 60, 73, 271

dogma 52, 78, 156, 158

domestication 93, 94, 107, 234

Doppler effect 41

doubling time 115, 121, 127

drama 106, 160

droit du Seigneur 261

drought xv, 6, 11, 20, 53, 59, 90, 99, 107, 123, 132, 168, 180, 203, 214, 215, 246, 251, 257, 262, 281, 285

Easter Island 182

ecology i, 9, 30, 46, 48, 49, 54, 90, 115, 118, 121, 124, 135, 201, 217, 323, 325

economics 9, 118, 158, 182, 196, 205, 314, 323, 325

ecosphere 94, 206, 214

ecosystem 4, 8, 29, 48, 112, 115, 117, 119, 120, 126, 154, 156, 157, 203, 216, 225, 226, 227, 234, 242, 245, 276, 285, 286, 299, 303

Ediacaran 216

education 25, 43, 131, 138, 142, 144, 147, 148, 152, 157, 161, 164, 165, 166, 269, 270, 275, 277, 278, 279, 281, 304, 308, 316

emigration 144, 152, 316

empire 29, 101, 107, 130, 152, 196, 199, 257, 299, 326

endangered species 17, 29, 45, 75, 88, 117, 128, 195, 196, 213, 255, 277, 308, 316

energy i, xvi, 4, 6, 7, 9, 12, 13, 16, 17, 19, 29, 34, 36, 37, 38, 41, 48, 52, 53, 54, 62, 64, 82, 86, 89, 91, 92, 93, 108, 115, 120, 128, 129, 133, 135, 137, 149, 150, 151, 153, 157, 162, 168, 169, 171, 172, 173, 174, 176, 177, 178, 179, 180, 181, 182, 183, 184, 185, 186, 187, 189, 190, 191, 192, 193, 194, 195, 196, 200, 201, 203, 206, 208, 209, 214, 215, 222, 223, 232, 233, 240, 241, 242, 243, 244, 245, 247, 249, 250, 258, 265, 271, 273, 274, 277, 279, 280, 281, 284, 285, 286, 287, 289, 290, 291, 292, 293, 299, 301, 302, 303, 304, 305, 306, 308, 310, 313, 314, 315, 316, 323, 325, 326, 327

Enlightenment 144, 261

entropy 53, 79, 149, 171, 191, 192

environment i, 3, 4, 9, 19, 25, 34, 36, 42, 46, 47, 48, 49, 52, 53, 54, 56, 61, 64, 70, 84, 91, 93, 94, 97, 111, 115, 116, 119, 121, 126, 131, 134, 138, 141, 145, 149, 153, 155, 158, 160, 171, 172, 184, 185, 189, 196, 201, 202,

INDEX 335

203, 206, 213, 214, 215, 222, 223, 224, 225, 226, 233, 247, 252, 254, 255, 262, 275, 279, 280, 281, 285, 286, 306, 308, 310, 311, 327

enzyme 44, 122

epidemic 20, 119, 151, 153, 166, 201, 284, 304

erosion 10, 12, 14, 48, 51, 53, 149, 187, 206, 209, 217, 219, 278

ethanol 179, 201, 298

Europe 9, 11, 12, 26, 49, 58, 62, 64, 77, 85, 86, 90, 93, 95, 98, 99, 102, 107, 119, 133, 144, 146, 151, 167, 193, 194, 201, 203, 205, 224, 245, 246, 254, 256, 263, 278, 296, 302

eutrophication 20

evolution 4, 12, 14, 29, 30, 39, 46, 47, 49, 50, 51, 52, 53, 54, 55, 56, 58, 59, 63, 67, 73, 76, 78, 81, 94, 95, 111, 121, 124, 125, 145, 171, 194, 214, 216, 218, 219, 226, 234, 235, 236, 245, 256, 258, 262, 264, 265, 271, 313, 323, 324, 326, 327

exceptionalism 57, 261

extinction xv, xvi, 3, 4, 11, 14, 15, 16, 17, 21, 48, 49, 58, 59, 75, 82, 91, 93, 94, 98, 99, 115, 120, 121, 124, 128, 130, 154, 156, 161, 167, 176, 195, 213, 214, 215, 216, 217, 218, 219, 220, 225, 226, 227, 228, 233, 234, 252, 263, 264, 272, 273, 274, 285, 303, 313, 323, 324, 326

faith 37, 38, 156, 158, 223, 251, 255, 256, 264, 270, 276, 323

family 6, 25, 30, 43, 49, 55, 58, 64, 69, 88, 96, 101, 103, 111, 112, 130, 144, 147, 152, 153, 155, 156, 158, 159, 165, 173, 260, 269, 277, 278, 279, 281, 283, 284, 288, 308, 321

family planning 6, 25, 43, 55, 88, 130, 165, 269, 277, 278, 279, 281, 284, 308, 321

feedback xv, 9, 13, 28, 59, 61, 78, 79, 120, 125, 126, 145, 147, 149, 153, 156, 172, 202, 204, 206, 208, 227, 232, 233, 245, 272, 274, 298

fertility 6, 18, 134, 150, 152, 153, 203, 220, 277, 278, 288, 299

fertilizer 5, 7, 20, 62, 177, 178, 201, 257, 287

fires 20, 45, 48, 59, 86, 97, 99, 168, 179, 182, 203, 204, 214, 217, 220

fish 7, 9, 14, 18, 19, 20, 30, 54, 128, 134, 135, 154, 160, 179, 184, 207, 215, 228, 255, 260, 285, 286, 296, 298, 300

floods 8, 10, 29, 46, 70, 80, 104, 107, 179, 214, 246, 270, 285, 286, 311, 315

food i, xv, 6, 8, 9, 20, 25, 26, 28, 29, 34, 49, 55, 58, 60, 63, 70, 80, 81, 84, 85, 88, 91, 93, 94, 96, 97, 98, 101, 103, 115, 116, 117, 119, 123, 125, 126, 127, 130, 131, 132, 134, 135, 138, 147, 149, 160, 166, 167, 174, 176, 177, 178, 180, 185, 186, 187, 205, 213, 216, 221, 224, 248, 252, 257, 260, 271, 272, 273, 275, 278, 279, 283, 284, 287, 288, 296, 297, 301, 304, 315, 326

food web 94, 116, 119, 296

forests 5, 10, 13, 17, 20, 25, 28, 29, 45, 46, 48, 56, 59, 61, 91, 96, 108, 123, 131, 148, 175, 176, 177, 179, 182, 199, 200, 202, 206, 236, 251, 275, 278, 280, 286, 299 fossil 4, 7, 8, 13, 14, 16, 17, 44, 46, 49, 50, 51, 59, 90, 91, 92, 106, 115, 116, 120, 135, 137, 138, 148, 150, 160, 168, 171, 172, 173, 177, 179, 181, 183, 184, 191, 192, 193, 194, 201, 202, 203, 209, 232, 234, 241, 244, 249, 264, 265, 269, 272, 273, 276, 290, 295, 299, 301, 302, 304, 305, 306, 308, 315, 316, 321

France 10, 69, 129, 177, 188, 193, 203, 214

front organizations 160

fuel 4, 7, 16, 17, 20, 25, 46, 49, 80, 86, 116, 117, 128, 135, 137, 138, 148, 149, 151, 160, 172, 176, 177, 179, 183, 188, 189, 191, 192, 193, 195, 199, 201, 202, 208, 215, 241, 244, 249, 250, 258, 265, 269, 273, 274, 276, 285, 290, 295, 299, 301, 302, 304, 305, 308, 313, 315, 321

fuel cell 17, 20, 86, 128, 135, 151, 191, 192, 193, 250, 301, 302, 313, 315

fuel efficiency 80

fundamentalism 276

funding 17, 42, 88, 99, 132, 160, 161, 192, 235, 306, 316

gag rule 102, 165

Gaia 91

gene 30, 44, 47, 59, 60, 63, 65, 72, 76, 81, 90, 111, 115, 119, 122, 175, 190, 221, 226, 271, 323

generation 10, 47, 123, 145, 172, 177, 183, 184, 196, 247, 258, 280, 291, 318

genocide 8, 133, 152

geology 148

geopolitical 173, 180, 194, 306

Germany 36, 45, 62, 69, 82, 104, 188, 214, 215, 227, 232

glacier 16, 217

globalization 25, 89, 131, 133, 181, 183, 269, 286, 287, 299, 303, 304, 314

global warming xv, xvi, 9, 10, 12, 13, 16, 18, 20, 21, 38, 46, 48, 51, 70, 84, 88, 91, 92, 96, 106, 116, 120, 126, 128, 129, 142, 143, 150, 157, 159, 160, 161, 162, 163, 164, 165, 166, 168, 172, 179, 180, 182, 188, 189, 190, 191, 192, 193, 200, 201, 202, 203, 204, 205, 206, 207, 208, 209, 214, 222, 223, 232, 242, 262, 264, 269, 270, 271, 273, 276, 285, 290, 292, 298, 303, 305, 307, 308, 313, 314, 325

Gondwana 219, 220

grain 97, 99, 107, 123, 131, 132, 142, 177, 326

grammar 72, 74, 147

Great Chain of Being 95, 226, 261, 262, 263, 326

Greece 199

green architecture 315

greenhouse effect 84, 125, 146, 202, 206, 208, 209, 241, 243, 293

greenhouse gas 4, 13, 14, 16, 17, 42, 55, 73, 80, 88, 98, 116, 125, 143, 146, 150, 157, 162, 178, 186, 189, 190, 192, 196, 200, 202, 204, 205, 206, 208, 214, 241, 244, 248, 249, 262, 263, 274, 290, 298, 301, 302, 303, 307, 311

Greenland 157, 189, 285, 296

groundwater 10, 11, 71, 199, 200

growth 4, 5, 6, 25, 29, 47, 48, 49, 52, 54, 77, 90, 91, 101, 109, 110, 121, 122,

130, 133, 135, 143, 144, 145, 147, 149, 150, 152, 155, 156, 157, 158, 165, 172, 176, 177, 182, 204, 205, 220, 232, 233, 242, 245, 247, 249, 273, 274, 275, 277, 280, 284, 286, 298, 299, 306, 313, 325, 326

Gulf Stream 104, 246

habitat i, xv, 20, 45, 49, 55, 73, 93, 98, 115, 124, 135, 149, 161, 182, 185, 186, 195, 208, 217, 225, 226, 234, 235, 252, 255, 272, 273, 300, 314, 316

Haiti 80, 89, 285, 286

health 6, 7, 14, 18, 25, 27, 36, 42, 45, 46, 48, 57, 61, 77, 88, 95, 99, 101, 117, 126, 131, 134, 135, 138, 142, 144, 147, 150, 152, 153, 154, 157, 165, 168, 185, 187, 189, 190, 192, 193, 204, 221, 225, 235, 241, 248, 262, 269, 270, 271, 275, 277, 279, 284, 290, 304, 310, 316, 317, 319

heat 12, 16, 18, 41, 42, 61, 64, 91, 116, 125, 146, 171, 177, 179, 184, 193, 196, 202, 203, 204, 207, 219, 236, 241, 243, 244, 245, 258, 260, 270, 271, 273, 289, 290, 293, 294, 295, 298, 325

herding 96, 97

hierarchy 72, 95, 261

hominids 56, 58, 70, 72, 81, 82, 109, 111, 176

Homo erectus 58

Homo habilis 58

Homo neanderthalensis 58, 82

Homo sapiens 3, 14, 58, 82, 93, 176

horticulture 96

human nature 33, 34, 62, 89, 111, 324

hunting 11, 30, 56, 58, 72, 81, 84, 85, 94, 97, 103, 108, 111, 127, 135, 239

hurricane 73, 80, 227, 305

hybrid vehicle 128, 192, 249, 282

hybrid vigor 60

hydrogen 14, 17, 20, 42, 54, 86, 135, 151, 173, 177, 179, 189, 191, 192, 193, 201, 203, 249, 250, 258, 269, 301, 302, 304, 306, 315, 316, 327

hydrolysis 191, 301

hypotheses xv, 37, 246

ice 12, 14, 16, 26, 38, 46, 50, 55, 58, 59, 73, 75, 79, 81, 94, 95, 96, 99, 104, 111, 143, 146, 157, 161, 167, 182, 189, 200, 202, 204, 206, 214, 217, 241, 242, 244, 246, 248, 252, 274, 296

ice age 14, 55, 58, 59, 73, 81, 94, 95, 96, 104, 157, 182, 204, 214, 217, 300

ice cap 12, 16, 26, 99, 167, 202, 241, 242, 246

ice core 75, 157, 200, 241, 244, 246, 274

idealism 239

ideology 100, 105, 106, 270, 307

Inca 101

India 5, 62, 65, 70, 98, 107, 133, 139, 142, 143, 149, 167, 168, 200, 201, 219, 220, 235, 249, 254, 257, 271, 290, 292, 311

Industrial revolution 4, 55, 130, 144

inequality 109, 111, 148, 254

information xv, xvi, 35, 54, 61, 72, 74, 75, 78, 79, 82, 85, 87, 91, 100, 102, 110, 129, 148, 155, 159, 166, 246, 248, 265, 281, 284, 308

inheritance 47, 60, 101, 152, 226

initiation 82, 83, 87, 105, 239

insecticide 57, 122, 123

insects 96, 102, 115, 117, 118, 119, 122, 123, 127, 252, 260

instability 89

instinct 75, 130, 327

insurance 82, 126, 248, 262, 269, 317

intelligent design 112, 264

Inuit 189

Iran 107, 108, 227, 254

Iraq 7, 62, 65, 93, 104, 138, 168, 174, 188, 302

irrationality 86, 281

irrigation 10, 42, 101, 176, 177, 199, 248

Italy 193, 203, 227

Japan 46, 62, 89, 90, 104, 151, 154, 205, 215, 258, 290

justice 49, 79, 265, 270, 275, 314, 318, 319

krill 94, 253

Kuwait 227, 270

Kyoto 46, 55, 102, 106, 129, 142, 150, 162, 164, 189, 205, 215, 232, 263

lake 120, 168, 228, 229, 285

language i, 48, 56, 57, 58, 72, 73, 74, 75, 76, 77, 78, 81, 82, 83, 85, 95, 103, 109, 111, 139, 164, 313, 323, 326, 327

large mammals 82, 123, 217

law 30, 38, 51, 52, 53, 80, 99, 134, 145, 159, 186, 187, 195, 209, 214, 255, 278, 320

leaders xv, xvi, 26, 34, 99, 100, 104, 112, 123, 138, 155, 159, 182, 185, 190, 200, 222, 242, 250, 256, 264, 281, 282, 286, 307, 310, 313, 320, 321, 322

light 37, 41, 43, 61, 79, 80, 83, 86, 97, 168, 171, 176, 193, 202, 204, 249, 255, 257, 284, 289, 290, 291, 293, 294

local 20, 36, 63, 82, 84, 86, 93, 107, 108, 123, 127, 138, 156, 157, 158, 178, 182, 183, 185, 190, 191, 195, 196, 218, 226, 246, 248, 257, 273, 279, 283, 284, 286, 287, 288, 304, 306, 315, 316, 319

locust 203

logos 106

Magi 254, 257

magma 171, 184, 219

mammal 11, 69, 156

mantle 12, 219, 270

Mars 7

math 272, 274

Maya 182

media 15, 16, 26, 33, 155, 174, 180, 181, 200, 270, 272, 311

meme 60

mercantilism 130, 131, 144

mercury 7, 19, 131, 134, 179, 202, 303

Mesopotamia 101, 199, 259

methane 3, 4, 13, 14, 148, 176, 179, 202, 206, 214

Middle East 90, 95, 107, 149, 194, 200, 215, 254, 260, 302

migration 8, 18, 27, 78, 90, 93, 96, 97, 108, 118, 148, 167, 184, 201, 205, 252, 253, 316

military 20, 26, 38, 80, 100, 101, 103, 105, 130, 131, 137, 153, 173, 174, 175, 183, 194, 195, 227, 257, 259, 270, 288, 302, 306, 317, 319

mind 6, 67, 78, 84, 86, 106, 162, 164, 225, 294, 299, 325, 326, 328

mining 11, 27, 38, 46, 53, 59, 106, 175, 179, 192, 202, 269, 286, 315

Mithra 256, 257

model 45, 64, 94, 118, 152, 153, 242, 245

monsoon 9, 70, 160

morality 38, 61, 85, 159, 239, 256, 319, 326

mortality 6, 11, 18, 82, 85, 106, 134, 147, 251

mosquito 57, 142, 290

mountain 12, 47, 49, 53, 61, 79, 108, 117, 159, 215, 219, 231, 286

mountaintop removal 8, 286

mud 50, 83, 311

music 61, 77, 83, 85, 88, 106

mutation 70

myth 88, 108, 132, 158, 250, 258, 260, 264, 302, 323

natural gas 16, 25, 79, 148, 154, 171, 172, 177, 178, 179, 181, 182, 191, 192, 201, 202, 209, 258, 287, 289, 290, 302, 304

natural law 37

natural selection 47, 52, 53, 72, 78, 90, 93, 115, 226, 227, 264

Neolithic 85, 91, 100, 101, 107, 108, 224, 251, 252, 254, 256

New Zealand 89, 148, 224, 225

niche 47

nitrogen 9, 12, 14, 116, 131, 177, 179, 192, 199

nomadism 95, 313

North America 12, 26, 64, 70, 89, 93, 119, 131, 220, 224, 232, 307, 325

North Sea 62, 65, 260, 285

nuclear 7, 25, 42, 61, 69, 91, 108, 120, 135, 138, 152, 153, 168, 171, 172, 177, 179, 182, 183, 188, 189, 194, 195, 196, 204, 214, 215, 234, 249, 250, 269, 276, 298, 299, 301, 302, 304, 305, 306, 313

nutrition 95, 147, 251, 277, 279, 283

ocean 9, 12, 14, 20, 26, 28, 47, 51, 62, 64, 94, 99, 102, 104, 138, 157, 160, 184, 188, 189, 202, 203, 207, 219, 241, 242, 243, 246, 247, 249, 252, 272, 274, 296, 298, 304

oil i, 6, 13, 16, 17, 19, 25, 28, 36, 38, 49, 62, 64, 65, 73, 79, 80, 84, 89, 92, 130, 142, 145, 150, 151, 160, 162, 166, 172, 173, 174, 175, 177, 178, 179, 180, 181, 182, 188, 194, 195, 200, 201, 202, 203, 205, 209, 227, 262, 264, 265, 269, 270, 273, 280, 283, 290, 298, 301, 303, 304, 305, 306, 311, 322, 325, 326, 327

oligarchy 148

optimism 155, 158, 159

organ 188

organelles 54

overfarming 99, 299

overfishing 28, 30, 38, 207, 285, 299

overgrazing 8, 149

overpopulation i, xv, xvi, 3, 5, 8, 33, 36, 37, 79, 93, 107, 121, 130, 133, 144, 147, 148, 149, 152, 155, 156, 158, 159, 165, 166, 173, 192, 200, 202, 234, 264, 265, 269, 272, 273, 276, 277, 281, 283, 284, 308, 313, 321

oxygen 7, 9, 12, 14, 15, 20, 28, 54, 55, 62, 82, 83, 86, 160, 176, 191, 192, 202, 206, 215, 216, 219, 220, 227, 279, 301

ozone 7, 14, 15, 43, 166, 192, 243, 271, 272, 273, 327

Pacific 64, 127, 129, 131, 154, 157, 190, 192, 193, 207, 220, 248, 252, 270, 296

Paleolithic 81, 83, 84, 85, 86, 107, 108, 224, 239, 240, 251, 254

Pangea 220

Papua New Guinea 131

park 16, 26, 49, 105, 135, 180, 186, 218, 285

pastoralism 259

patent 167, 221, 318

peak 65, 118, 129, 181, 185, 186, 298, 303

peat 171, 176, 214

peer review xv, 222, 308, 317

permafrost 50, 148, 217, 248

Permian 217

pest 119, 178, 260

photosynthesis 14, 28, 54, 176, 206, 216

photovoltaic 201, 287, 291

pidgin 72, 74, 75

plankton 7, 26, 98, 157, 206, 253, 285, 296, 297

planning 6, 25, 43, 55, 58, 70, 88, 101, 130, 165, 172, 185, 186, 187, 247, 248, 250, 269, 277, 278, 279, 281, 284, 298, 308, 321

plant 28, 46, 48, 73, 75, 82, 90, 93, 96, 98, 116, 134, 167, 172, 176, 179, 180, 186, 188, 189, 192, 195, 215, 233, 241, 242, 247, 257, 289, 292

Pleistocene 82, 217, 224

plume 290, 324

pollution xvi, 6, 7, 9, 14, 19, 27, 29, 38, 42, 45, 48, 51, 62, 79, 92, 98, 117, 120, 124, 134, 135, 137, 141, 148, 149, 167, 171, 179, 187, 190, 192, 199, 200, 201, 202, 203, 232, 234, 258, 269, 271, 272, 273, 278, 283, 289, 290, 298, 299, 301, 302, 314, 315, 327

poor 26, 27, 28, 33, 36, 43, 89, 99, 104, 106, 123, 134, 145, 147, 148, 149, 152, 153, 156, 204, 205, 220, 235, 251, 263, 269, 270, 271, 275, 280, 292, 319

population i, xvi, 3, 4, 5, 6, 8, 9, 11, 16, 18, 25, 26, 27, 28, 29, 33, 41, 42, 43, 47, 48, 55, 59, 60, 62, 64, 71, 73, 77, 79, 81, 85, 88, 90, 91, 93, 94, 97, 101, 103, 105, 108, 109, 110, 117, 119, 121, 122, 125, 127, 128, 129, 130, 131, 133, 134, 135, 136, 137, 140, 141, 143, 144, 146, 147, 148, 149, 150, 152, 154, 155, 156, 158, 159, 165, 176, 177, 179, 182, 183, 184, 187, 189, 191, 193, 199, 200, 201, 204, 205, 206, 208, 213, 220, 225, 242, 245, 247, 249, 251, 269, 272, 273, 274, 275, 276, 277, 278, 279, 280, 281, 283, 284, 287, 288,

290, 298, 299, 300, 303, 305, 306, 308, 313, 314, 316, 321, 323, 324, 325, 326, 327

Portugal 59, 151, 172, 203, 215, 304

poverty 89, 104, 144, 145, 150, 153, 157, 235

power 10, 13, 16, 17, 20, 25, 26, 30, 43, 54, 61, 62, 64, 73, 86, 100, 101, 103, 104, 105, 106, 108, 110, 115, 120, 123, 134, 138, 151, 153, 168, 172, 173, 176, 177, 179, 183, 184, 188, 189, 191, 192, 193, 194, 195, 196, 200, 201, 204, 205, 207, 214, 215, 239, 248, 249, 250, 258, 260, 264, 269, 276, 277, 284, 286, 291, 292, 298, 299, 301, 302, 304, 306, 315, 316, 318, 320, 323, 324, 327

predators 55, 56, 81, 85, 111, 119, 125, 207, 213, 224, 239

prehominids 56, 58, 70, 111, 176

prey 48, 56, 58, 81, 111, 119, 125, 135, 176, 200, 224, 239, 252

primates 56, 72, 74

prisons 27, 137, 270, 306, 317

privatization 15, 20, 25

profiteers 17, 174, 281, 306

Project for the New American Century 17, 130

proof 253

protolanguage 58, 72

psychology 152, 240

publication 164

race 25, 49, 60, 61, 118, 119, 128, 153, 235, 313, 325

radioactivity 188, 189

radioisotope 50, 90, 219

rain 10, 13, 14, 29, 42, 46, 64, 73, 98, 142, 148, 176, 179, 193, 199, 202, 235, 236, 303

random 38, 44, 52, 171, 215, 324, 327

reason 3, 16, 33, 38, 45, 52, 60, 73, 77, 103, 106, 115, 123, 125, 130, 133, 171, 180, 183, 214, 224, 240, 257, 264, 273, 285, 320, 323

rebirth 108, 239

recycling 4, 29, 54, 96, 178, 186, 281, 283, 301, 304, 306, 315, 316, 318, 319

red shift 41

religion 58, 69, 80, 81, 82, 83, 84, 85, 86, 108, 111, 164, 237, 239, 253, 254, 257, 264, 265, 277, 278, 313, 318, 321, 324, 327

Renaissance 144

representation 72, 240, 265, 280

reproduction 47, 52, 54, 55, 63, 123, 145, 176, 224

reproductive health 6, 18, 126, 277, 288

resistance 122, 123, 142, 193, 215, 280

resource 29, 79, 100, 103, 124, 133, 136, 150, 171, 173, 181, 182, 183, 184, 270, 272, 273, 281, 318, 326

respiration 52, 54

rights 18, 26, 38, 43, 101, 138, 144, 148, 152, 160, 168, 189, 195, 270, 277, 278, 279, 288, 308, 314, 316, 319, 324

ritual 69, 103, 251, 252

river 10, 53, 104, 120, 123, 138, 168, 187, 207, 214, 221, 228, 248, 253, 257, 262, 325

RNA 44, 47

Rome 99, 180, 193, 199, 257

runoff 7, 10, 141, 199, 207

sacrifice 33, 115, 152, 308, 314
Sahel 70, 214
salinization 199
Saudi 36, 65, 173, 174, 178
savanna 56, 58, 81, 109, 110
science xvii, 7, 16, 17, 25, 30, 35, 36, 37, 38, 52, 53, 70, 72, 88, 92, 106, 109, 112, 129, 134, 136, 146, 150, 159, 160, 161, 162, 163, 164, 167, 172, 189, 190, 192, 196, 200, 205, 207, 215, 222, 223, 234, 240, 242, 245, 246, 247, 248, 250, 252, 264, 265, 274, 276, 286, 300, 307, 317, 318, 319, 323, 325, 327
scientific method 37, 38
scientist i, xv, 18, 35, 37, 86, 91, 129, 157, 160, 161, 163, 214, 223, 246, 252, 271, 292, 307, 308
sea level 51, 91, 96, 104, 107, 142, 146, 187, 205, 219, 221, 222, 227, 241, 242, 246, 248, 252, 274, 296, 305
sedentism 95, 96, 97, 103, 104, 105, 109, 259, 313
sediment 53, 90, 141, 157, 219, 222, 274
sex education 165, 166, 308, 316
Siberia 214, 220, 224
silt 10, 199, 221, 222
skeptic 106, 160, 223
slave 98, 178, 323
slum 150
smog 14, 20, 82, 179, 186, 302
smoke 45, 204, 302
snow 10, 42, 74, 159, 160, 168, 171, 232, 233, 248

social Darwinism 33, 153
soil 10, 14, 19, 49, 51, 53, 70, 86, 96, 99, 116, 127, 199, 201, 232, 233, 251, 257, 278, 280, 317
solar heating 287, 293, 294
South America 70, 93, 193, 194, 205, 207, 219, 224, 253
Spain 107, 193, 203, 207, 215, 227
species i, 3, 4, 6, 7, 8, 11, 13, 14, 17, 18, 19, 20, 29, 30, 38, 45, 46, 47, 48, 49, 50, 51, 52, 53, 54, 55, 56, 57, 58, 59, 63, 69, 72, 73, 75, 77, 79, 81, 85, 88, 90, 93, 94, 96, 98, 99, 102, 108, 109, 111, 115, 116, 117, 119, 120, 121, 123, 124, 125, 126, 127, 128, 129, 130, 135, 148, 154, 156, 161, 165, 167, 176, 182, 186, 195, 196, 201, 202, 207, 213, 214, 215, 216, 217, 220, 224, 225, 226, 227, 228, 234, 235, 241, 242, 252, 255, 260, 263, 264, 272, 273, 277, 283, 285, 292, 298, 299, 308, 313, 314, 316, 324
stability 4, 12, 13, 125, 137, 261, 272
star 20, 41, 42, 183, 289
starvation 25, 27, 28, 127, 165, 257
sterilization 43, 165, 269, 277, 279, 281, 321
subduction 219
sulphur 98
supercontinent 219, 220, 227
superstition 159
sustainability xv, xvii, 4, 29, 34, 64, 87, 131, 136, 137, 152, 153, 172, 176, 185, 272, 275, 279, 298, 300, 318, 320, 322, 323, 325, 327
syntax 72, 74

tar 50, 178

tax 25, 36, 49, 88, 128, 131, 144, 154, 159, 183, 189, 194, 195, 209, 282, 287, 292, 298, 302, 305, 317

tectonic plates 12, 55, 94, 95, 184, 219

teleology 261, 262

temperature xv, 3, 13, 28, 41, 42, 48, 64, 75, 125, 126, 129, 146, 148, 149, 157, 160, 161, 162, 163, 164, 190, 193, 196, 200, 202, 203, 204, 205, 206, 207, 208, 214, 241, 242, 243, 244, 245, 250, 270, 274, 289, 290, 294, 296

territory 63, 90, 103, 217

theory 37, 38, 41, 50, 133, 134, 144, 152, 168, 242, 245, 246, 255, 258, 264, 305, 318, 324

thermodynamics 52, 53

throwing 37, 56, 57, 58

timber 49, 128, 175

tools 33, 56, 57, 58, 69, 81, 82, 83, 88, 90, 96, 111, 177, 239, 247, 284

tragedy 313, 320

transegalitarian 109

transpiration 199

transportation 25, 64, 110, 135, 137, 138, 151, 177, 180, 181, 185, 186, 191, 241, 244, 246, 248, 249, 262, 287, 304, 306, 314, 315, 318

treaties 17, 137, 270

trees 8, 56, 61, 102, 118, 127, 142, 148, 199, 232, 263, 280, 285, 289, 293

tribe 85, 87, 104, 156, 158, 168

tropics 98, 118, 146, 236, 325

tsunami 139, 226

tundra 95, 202, 232, 248, 252, 299

Turkey 107, 108

U.K. 34, 46, 62, 70, 86, 200, 253

U.N. 9, 42, 43, 70, 71, 124, 130, 131, 132, 138, 139, 145, 173, 189, 317

U.S. i, 7, 10, 16, 17, 20, 25, 26, 27, 29, 34, 36, 42, 43, 46, 48, 70, 96, 101, 102, 104, 108, 123, 124, 128, 130, 134, 135, 136, 138, 148, 151, 157, 162, 165, 173, 174, 175, 180, 188, 189, 190, 193, 194, 200, 201, 217, 227, 228, 229, 230, 236, 240, 247, 249, 250, 262, 277, 278, 282, 298, 300, 305, 306, 317

Urban Environmental Accords 184, 185

Vatican 227, 257, 288

Venezuela 195

Venus 7

volcano 214

war 17, 25, 59, 64, 84, 104, 107, 108, 131, 132, 138, 166, 173, 174, 180, 188, 194, 195, 203, 205, 209, 275, 278, 288, 302, 324, 325, 327

water i, xv, 5, 8, 9, 10, 11, 12, 13, 14, 15, 19, 20, 21, 25, 29, 42, 49, 51, 54, 62, 70, 71, 73, 84, 87, 88, 94, 95, 96, 98, 101, 104, 117, 120, 125, 135, 142, 143, 146, 149, 157, 160, 167, 168, 171, 174, 175, 176, 177, 178, 179, 180, 182, 184, 185, 186, 187, 188, 190, 191, 192, 196, 199, 201, 202, 203, 206, 215, 219, 220, 222, 235, 236, 239, 242, 244, 245, 246, 248, 252, 253, 255, 257, 258, 262, 263, 272, 273, 275, 278, 280, 283, 285, 286, 288, 289, 296, 297, 298, 301, 304, 311, 315, 318 wealth xvi, 25, 100, 101, 109, 130, 137, 144, 153, 158, 159, 194, 196, 206, 248, 264, 269, 270, 273, 299, 307, 320, 327

weather 12, 28, 160, 193, 202, 203, 204, 221, 222, 236, 241, 242, 245, 246, 247, 248, 252, 260, 262, 263, 289, 304, 305, 325

wetland 248

whale 20, 154, 231

wilderness 8, 29, 30, 300

wood 7, 20, 25, 131, 171, 176, 177, 179, 199, 201, 225, 288, 293, 294

writing xvi, 72, 85, 107, 108, 109, 110, 217, 234, 256, 313

x rays 171, 188

Zarathustra 107, 254, 255, 256

zero waste 185, 186, 315, 319

978-0-595-40094
0-595-40094-9

Printed in the United States
64592LVS00003B/99